Introduction to Geometric Computing

Sherif Ghali

Introduction to Geometric Computing

Springer

Sherif Ghali, PhD
University of Alberta
Edmonton
Canada

ISBN: 978-1-84800-114-5 e-ISBN: 978-1-84800-115-2
DOI: 10.1007/978-1-84800-115-2

British Library Cataloguing in Publication Data
A catalogue record for this book is available from the British Library

Library of Congress Control Number: 2008926727

Printed on acid-free paper

9 8 7 6 5 4 3 2 1

Springer Science+Business Media
springer.com

Preface

If geometric computing is neither computer graphics nor computational geometry, then what is it? Briefly, it is the mathematics and the engineering that underlie both.

This text

- discusses how to design libraries for Euclidean, spherical, projective, and oriented projective geometries.
- derives the necessary results in each geometry by appealing to elementary linear algebra.
- gives a concrete implementation in C++ for classes in each library.
- suggests that the time may have come for practitioners in computer graphics and in computational geometry to design and implement a sequel to LEDA [67] and to CGAL [23] based on the four geometries. The main aim would be standardization.
- shows examples for the failure of geometric systems when the types **float** or **double** are taken as a substitute for the set $\mathbb{R}$.
- presents the graphics pipeline passing through oriented projective geometry, which makes it possible to talk sensibly about clipping.
- discusses the notion of coordinate-free geometry, a very simple idea that appears not to have yet entered the mainstream.
- presents the classical raster graphics algorithms that are traditionally introduced in an undergraduate computer graphics course. Doing so can be done at a rapid pace after a base of geometric computing has been introduced.
- briefly connects with the established and vibrant discipline of graph drawing [32].
- discusses elements of the now-classical methods of geometric and solid modeling [24, 65, 51]. This also is done with considerable brevity after the appropriate geometric software layers have been introduced.
- shows how geometric algorithms can be designed and implemented such that combinatorial and geometric parts are separated. The geometry would act as a plug-in for a combinatorial algorithm to yield a concrete implementation.
- presents algorithms for Boolean operations on regular sets using binary space partitioning. An arbitrary geometry satisfying separability can be plugged into this combinatorial structure to perform computation in

the corresponding space. Examples are shown for using Euclidean and spherical geometries as a plug-in with separability as the only requirement for the plug-in.

- introduces methods for computing visibility by projecting from Euclidean n-space to spherical $n-1$-space, for $n = 2, 3$.
- seeks to define a common ground from which systems for computer graphics, visualization, computational geometry, computer-aided design, robotics, geographic information system, and computer vision can be designed and implemented.

The text also fills what appears to be a serious gap in the training of undergraduate students in computer science, in mathematics, in physics, and in mechanical engineering. Many software systems built in these disciplines are geometric, yet no body of knowledge has been collected that discusses the issues involved in designing and implementing a geometric system. The appropriate design of an undergraduate curriculum in computer science should include a course on geometric computing in the second or third year. Such a course would be a required prerequisite to (traditionally third- or fourth-year) computer graphics courses and an optional prerequisite to (traditionally late undergraduate or early graduate) computational geometry courses.

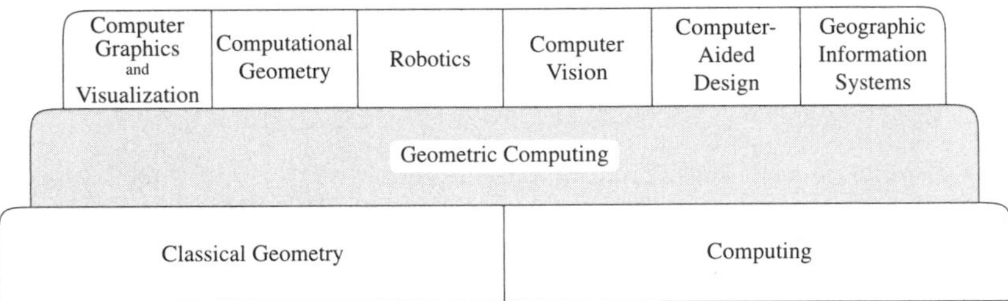

The notion that metric and topological information are isolated conceptually and should remain isolated in a geometric system has long been understood [85]. The genericity of C++ makes this isolation feasible while imposing neither efficiency nor type safety penalties. New questions emerge. Can one, for instance, build CSG trees out of spherical geometry or oriented projective geometry primitives? Knowing the answer as simply an intellectual exercise would be interesting, but it would be at least as interesting to find that doing so is possible and that one can do something with the result.

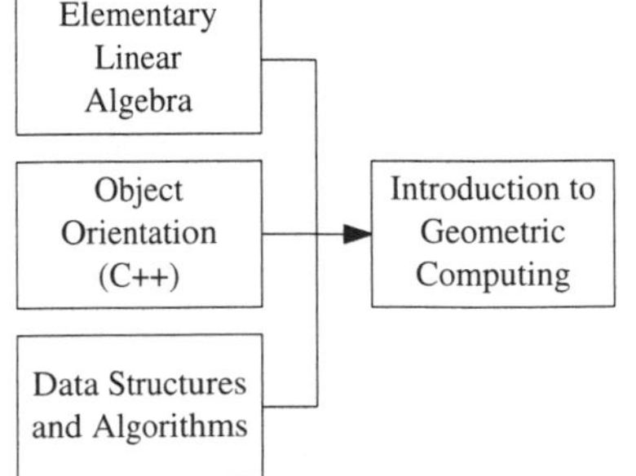

Prerequisites

The background expected is minimal and includes elementary linear algebra, a course on object orientation and C++, as well as familiarity with basic data structures and algorithms. That said, Chapters 1 to 4 take a slow ramp intended as a recapitulation of many of the C++ constructs and of the linear algebra needed.

Geometric Class Naming Convention

The text includes sample C++ code for most topics discussed. The code can be downloaded from http://www.springer.com/978-1-84800-114-5. In the discussion of the basic layers for the various geometries, a concrete sketch is given in C++ of what may arguably be some form of a canonical implementation. Once the beginning layers are established, C++ is used in lieu of pseudo-code. One can indeed argue that an objective of good software design is to abolish the need for pseudo-code. The full implementation contains all the details and should be consulted if needed, but the text contains sufficient excerpts for it to be understandable without referring to additional code. In a sense, the elided C++ code included still acts as a far more precise form of pseudo-code than one using prose.

The exercises frequently refer to start-up code that will make it easier to experiment with an implementation without spending an excessive amount of time on issues unrelated to the topic considered. All start-up code samples compile and run before adding any extra functionality. By releasing start-up code, instructors can expect that a student will be capable of writing a nontrivial geometric program in two to three hours. The chapter granularity is chosen such that each chapter can be discussed in one hour of instruction.

The geometric classes use C++'s genericity and are parameterized by a number type, although the first chapters discuss direct implementations for Euclidean geometry to isolate the issues and to introduce them gradually. Each generic geometric class has a two-letter suffix that identifies the geometry and the dimension. The following examples illustrate the nomenclature.

- Point_E3 is a class for a point in the 3D Euclidean geometry E^3.
- Segment_S2 is a class for a segment in the 2D spherical geometry S^2.
- Line_P2 is a class for a line in the 2D projective geometry P^2.
- Plane_T3 is a class for a plane in the 3D oriented projective geometry T^3.

When instantiating concrete classes, an additional suffix signals the number type used. If, for example, the generic class Point_E2 is instantiated using the **double** built-in type, then the concrete class is called Point_E2d. This is done by the following statement.

```
typedef Point_E2<double> Point_E2d;
```

This convention avoids naming clashes and makes it easy to develop a system involving multiple types, dimensions, and precisions.

Each geometry has its own interpolation operations, intersection functions, predicates, and methods for computing transformations. The file names in the accompanying code also follow the naming convention just mentioned. The files for the Euclidean line, for instance, are shown in the following table.

It will occasionally be clear that an obvious optimization is omitted, but "premature optimization is the root of all evil" and one should first get a design right and an implementation correct, then worry about nonasymptotic optimizations later. It is not necessary to convince those who have tales to tell

File names for the geometry of the Euclidean line
interpolation_e1.h
intersection_e1.h
predicates_e1.h
transformation_e1.h

about a "clever optimization" that actually introduces a performance penalty of the importance of first aiming for an elegant and clean system design. Code optimization should be performed only after code profiling has suggested that it is necessary to do so. In the best of worlds, significant further improvements should be done on the code before it has been optimized since a serious side effect of code optimization is that data lose their type; optimization frequently results in code obfuscation.

Many will consider that the frequent lack of mention of asymptotic complexity is a serious omission. Such discussions are not included to focus on what lies at the intersection of geometry and computing in addition to time and space complexity, which are in any case treated extensively elsewhere. Often complexity is also a contentious issue that separates, and nearly defines, fields. Avoiding complexity also dodges (perhaps too conveniently) debates about whether worst-case complexity is overly pessimistic or whether it is reasonable to assume that information about average time complexity can be measured by a few empirical charts. It is best to argue, also conveniently, that geometric computing is defined such that it remains neutral in this discussion.

One violation of C++ programming style, and one for which I ask the forgiveness of the reader, is the formatting of function arguments and parameters. I hope that the increased readability afforded by the use of proportional fonts to typeset the code more than offsets the use of formatting that will seem unorthodox to many.

I am grateful to Sylvain Pion, Stefan Schirra, and Michiel Smid for having kindly commented on sections of this text. Needless to say, errors or inaccuracies are my responsibility alone. This text has been designed using Peter Wilson's memoir, which may well represent the next generation of LaTeXpackages—the package of compatible packages. Positioning figures in the margins is inspired by *Computational Geometry: Algorithms and Applications* and I am grateful to the authors for sharing their book design. I would also like to acknowledge the funding of the Natural Sciences and Engineering Research Council of Canada.

I owe much gratitude to Otfried Cheong for having developed the ipe drawing package, which is used for all hand-drawn illustrations in this text. That line concurrency and point colinearity in the figures are indeed accurate is entirely due to this superb drawing package. Readers will no doubt wish to write C++ code to try some of the ideas discussed in the pages that follow, but as much as this text tries to show just how enjoyable it can be to write a geometric program, simply redrawing some of the illustrations using ipe will be enlightening. The graphical user interface scene surrounding C++ remains sorely lacking in standardization. This text relies on Daniel Azuma's well-crafted GLOW user interface library. A nice course project would be to use GLOW to write an initial ipe3d—for extra credit, the output would remain purely vectorial (by relying on Chapter 32, itself depending on nearly all preceding chapters).

A risk in technical writing is "telling your readers what you're going to tell them, telling them, and then telling them what you've told them" [119]. This brief introduction is necessary, but since I'll spare the reader a conclusion, it is worthwhile to conclude by asking several questions that will put the "introduction" part of the title in perspective.

- As discussed in § 14.4 and § 18.6, the possibility of a library that is more concise than the one described here, yet one that, crucially, sacrifices neither type safety nor efficiency is intriguing.
- How can an extension that captures software design for what can be tentatively called kinetic geometric computing [44] and for computer animation [77] be designed?
- How can a library for hyperbolic geometry be designed and implemented such that, for instance, tree and graph drawing algorithms can move gracefully between Euclidean, elliptic, and hyperbolic spaces?
- How can geometric algebra [33] be brought under the umbrella of the framework presented here (or the other way around)?

My hope is to have succeeded in not committing one of the cardinal sins of technical writing, boring you, dear reader. I await your comments and hope you will get as much enjoyment reading this text as I had writing it.

Sherif Ghali
February 2008

Contents

Appendices 297

Part I

Euclidean Geometry

1 Computational Euclidean Geometry in Two Dimensions

Consider that we are about to embark on the design and implementation of a software system that has a geometric component. We decide to take a long-term look and to craft a set of geometric classes with enough care so that there would be a high likelihood that we reuse the classes in the next project we tackle. This chapter discusses the decisions that need to be made when designing a set of classes for points, lines, and other objects in planar Euclidean geometry.

Since the geometry tackled is familiar, there is little need to formalize the properties of the Euclidean plane and space. Although the details for designing classes for Euclidean geometry in the plane are quite simple, it is instructive to go through these details, partly because it will then be easier to appreciate the need for more complex structures, but also because doing so has the parallel benefit of acting as a review of C++. It is clear, for example, that we need classes for a point, a line, and a segment and that Cartesian coordinates will be used.

1.1 Points and Segments

The first decision we need to make is whether a point in the plane should be represented using **float**s, **double**s, or **long double**s. We look at a good deal of geometric software written in C++ and find that double is the type most often chosen—possibly because it offers for many applications the right amount of balance between precision and compactness. We would thus write

```
class Point_E2d {
private:
    double _x, _y;
public:
    Point_E2d( double x, double y ) : _x(x), _y(y) {}
    ...
};
```

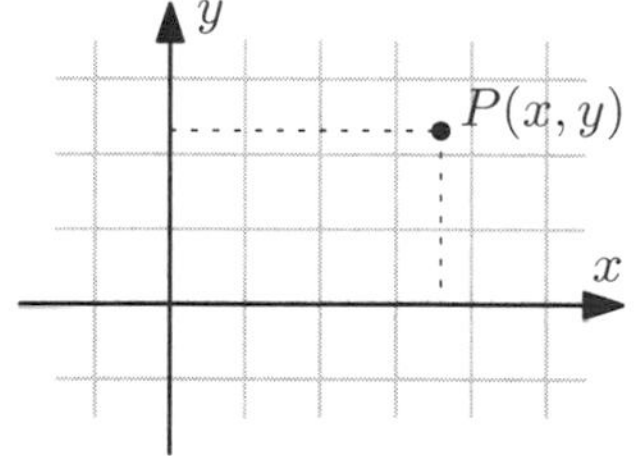

This is good enough for a start, and will in fact be sufficient as the basis for a great many applications. A detail about the naming convention is needed: Point_E2d will refer to a point in two dimensions using double precision. If we need to implement another class that uses less precision—say one that consumes four bytes per coordinate—we could also implement a class Point_E2f and thus ensure that the two names would not clash.

The next decision is whether to provide a default constructor (one taking no arguments) and, if so, which default point should be chosen.

Since defining a C-array of n objects requires n calls to a default constructor, the compiler would in any case have generated a default constructor, but our defining one or more constructors halts that default generation, so we need to supply one.

Should the default constructor initialize the point to the origin? Since built-in types are in theory not initialized [105], we spare the clients of our class the risk of reading uninitialized memory, which leads us to the following revised declaration:

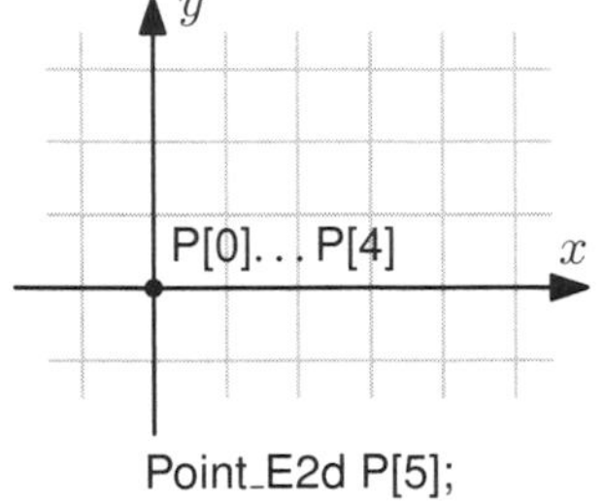

Point_E2d P[5];

```
class Point_E2d {
private:
  double _x, _y;
public:
  Point_E2d() : _x(0), _y(0) {}
  Point_E2d(double x, double y) : _x(x), _y(y) {}
  ...
};
```

Observe that a separate constructor needs to be declared and that it is incorrect to use two default initializations.

```
class Point_E2d { private:
  double _x, _y;
public:
  // incorrect:
  Point_E2d( double x = 0, double y = 0 ) : _x(x), _y(y) {}
  ...
};
```

Using two initializations would make it legal to create a point with only one coordinate Point_E2d A(1.0);. Such code would initialize the y-coordinate to 0. This may not appear to be so terribly bad, but it is. The danger lies in the implicit promotion rules of C++, which would make it possible to initialize another object that can be constructed using a Point_E2d parameter with a single **double** even when such object initialization makes no sense. Using the first of the two options above can thus be seen as a way to push a potential logical error to become a compilation error instead.

To determine whether two point objects are equal, we first attempt to short-circuit the geometric test and determine whether the same object is being tested against itself (either directly or via a distinct pointer/reference). If the two objects are distinct, the two coordinates are compared to report equality.

```
class Point_E2d {
  ...
  bool operator==(const Point_E2d& p) const {
    return (this == &p) ||
      (_x == p._x) && (_y == p._y);
  }
  ...
};
```

To represent a segment in the plane, we need to decide whether our segments will be *directed*. Another way of asking this question is the following: If A and B are two points and we define the two segments $[A, B]$ and $[B, A]$ and subsequently check their equality, would we want the answer to be true or false? At this point it will seem that choosing one decision or the other is arbitrary—not unlike choosing the axioms (Euclidean or otherwise) for a geometry—and just as with the selection of axioms, many options are possible, but some turn out later to be more "elegant" or "useful" than others. Let us assume that our intuition tells us that segments should be directed (which will turn out in fact to be the wiser of the two decisions).

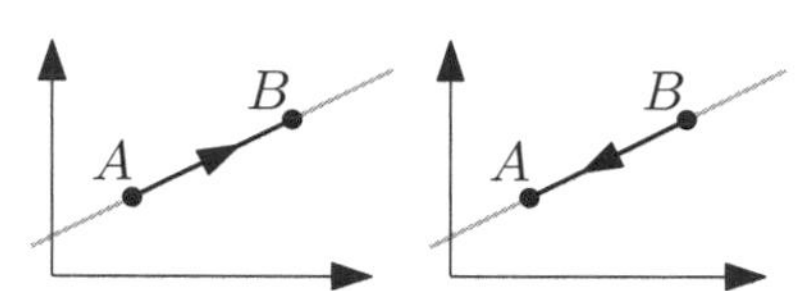

If we refer to the two segment endpoints as *source* and *target*, we have

```
class Segment_E2d {
private:
    Point_E2d _source, _target;
public:
    Segment_E2d() : _source(), _target() {}
    Segment_E2d( const Point_E2d& source, const Point_E2d& target )
        : _source(source), _target(target) {}
    Segment_E2d( const Segment_E2d& seg )
        : _source(seg._source), _target(seg._target) {}
    ...
};
```

Notice that the list of constructors provided in the code above could be increased. Would we rather pamper the clients of this API and provide others? The next obvious choice would be to add a constructor that takes four parameters for the x- and y-coordinates of the two endpoints. When making this decision it is frequently useful to consider whether additional functions are necessary or, conversely, whether the ones already provided are sufficient. And so we decide to keep the interface of Segment_E2d light by not defining more constructors, even at the risk of being accused by some programmers using our API that the list of constructors is incomplete. We return to this question in Chapter 17.

Input/Output Functions

The clients of any C++ library expect that the input and the output operators are overloaded for the classes in the library so that it is possible to write code such as the following:

```
Point_E2d ptA, ptB;
cin >> ptA >> ptB;
...
cout << "Point␣A:␣" << ptA << "␣␣Point␣B:␣" << ptB << endl;
...
```

C++ makes it possible to overload the input (">>") and the output ("<<") binary operators, such that the left operand is an input stream or an output stream, respectively, and such that the right operand is an object. These operators, which are by necessity nonmember functions, are best overloaded as friend functions, giving direct access to member variables. After declaring friendship

```
class Point_E2d {
    ...
    friend std::istream&
    operator>>(std::istream& is, Point_E2d& p);
    friend std::ostream&
    operator<<(std::ostream& os, const Point_E2d& p);
    ...
};
```

an implementation for the class Point_E2d would be as follows:

```
std::istream& operator>> (std::istream& is, Point_E2d& p) {
    is >> p._x >> p._y;
    return is;
}

std::ostream& operator<< (std::ostream& os, const Point_E2d& p) {
    os << p._x << " " << p._y;
    return os;
}
```

1.2 A Separate Type for Vectors

We next argue that a class is needed for the abstraction of a vector in the plane. Object-oriented programmers who have seen too many systems that are far too bloated with unnecessary classes may quickly protest that there is no need to introduce such an abstraction. A point and a vector are the same. Many programmers of geometric systems do in fact take that route and keep the number of classes minimal, but that would be a bad idea. One argument is that an additional vector type would prevent the programmer using our library from writing absurd programming snippets such as

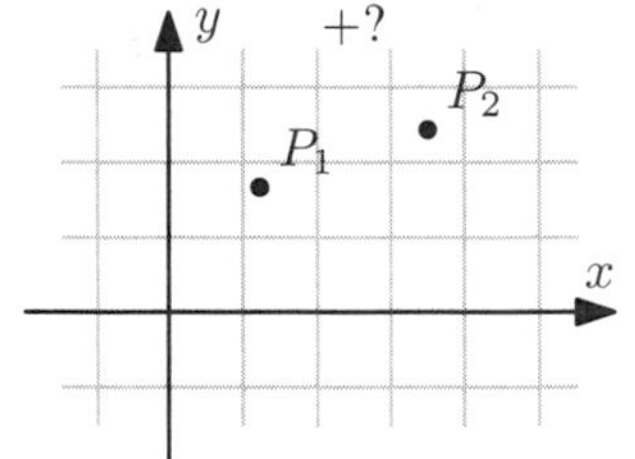


```
Point_E2d p1, p2;
...
Point_E2d p3 = p1 + p2;
```

Unfortunately, the person protesting the loudest that vectors and points should share the same abstraction is also likely to protest that the above code is perfectly legitimate and that it should be allowed. The two types are in fact similar, but we would like to introduce just enough heterogeneity to make statements such as

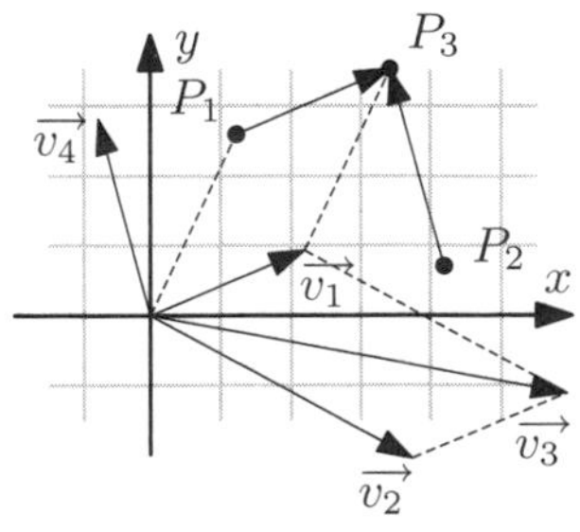


```
Point_E2d p1, p2;
Vector_E2d v1, v2;
...
Point_E2d p3 = p1 + v1;
Vector_E2d v3 = v1 + v2;
Vector_E2d v4 = p3 − p2;
```

legal, while making statements such as

```
Point_E2d p1, p2;
Vector_E2d v1, v2;
...
Point_E2d p5 = p1 + p2;
```

illegal. C++ operator overloading can be used to make some statements legal and keep others illegal. The vector abstraction can best be thought of as a *translation* in the plane. A point may be translated, which is why p1 + v1 is legal; two translations may be combined into a third, which is why v1 + v2 is legal; and the difference between two points returns the *vector* difference, which is why p3 − p2 is legal and its return type is Vector_E2d. It is crucial for a geometric library to define a distinct type for a vector, but in case the preceding argument is unconvincing, we will have a chance to discuss a more compelling one in § 4.2.

One way to think of classes (or types—the two words are synonymous) is that they act somewhat like units in physics equations. If we attempt to substitute the quantities in the equation $F = ma$ (force = mass × acceleration), we would want to maintain a sanity check by confirming that the unit for acceleration is indeed distance over time squared (and that the units otherwise match). Type checking acts as the sanity check in an object-oriented system. The library designer has the benefit of being able to enforce a set of reasonable constraints. Once such constraints are imposed, the code written by application programmers will have this check done for free.

Thinking about vectors as modeling translation is not the only option. We can also think of vectors as modeling a force. Indeed, it would be quite suitable in a physical system to use a vector object to model both the direction and the magnitude of a force.

The declaration of a class Vector_E2d is quite similar to that of a Point_E2d. We even continue to use the zero vector for the default constructor.

```
class Vector_E2d {
private:
    double _x, _y;
public:
    Vector_E2d( ) : _x(0), _y(0) {}
    Vector_E2d( double x, double y ) : _x(x), _y(y) {}

    ...
};
```

The classes Point_E2d and Vector_E2d are so similar that one may consider defining an abstract class Tuple_2d that captures most of the commonality and then derive the two classes Point_E2d and Vector_E2d from Tuple_2d.

Programmers trained in schools of programming preceding object orientation are particularly prone to making this choice. When writing in Pascal or in the C language, it is enough for a few lines of code to look identical for the programmer to extract them into a procedure or a function. Carrying this approach over to object orientation could lead one to extract the common code into a base class. Such an extraction would be pragmatically sound but would be flawed, if only philosophically. Object-oriented extraction of commonality by inheritance from a base class should be used only when there is type commonality that justifies the inheritance.

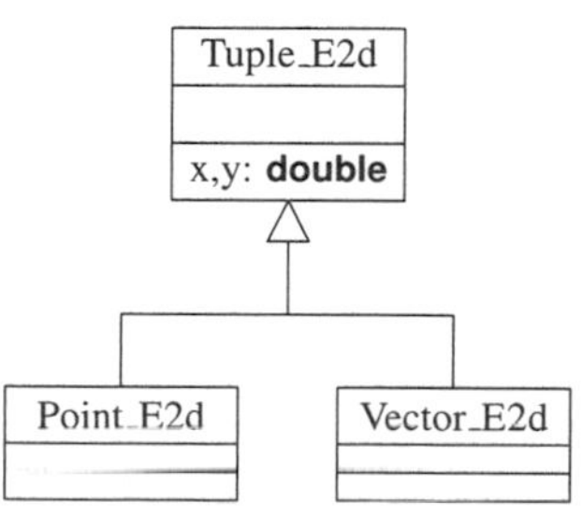

flawed design

But leaving philosophical soundness aside, there is another important reason why using inheritance from a common base would be a poor choice. Using inheritance means incurring a virtual function table, which in turn means that the compiler may not be able to determine which function will be invoked. Since points and vectors constitute the lowest layer in a geometric system, such a minor cost could have significant effect percolating throughout the system. We will need to make a related decision, with similar lines of thought, in § 3.1.

With these definitions for the classes Point_E2d and Vector_E2d, we can now write

```
Point_E2d operator+(const Point_E2d& p, const Vector_E2d& v)
{
  return Point_E2d(p.x()+v.x(), p.y()+v.y());
}

Vector_E2d operator−(const Point_E2d& p1, const Point_E2d& p2)
{
  return Vector_E2d(p1.x()−p2.x(), p1.y()−p2.y());
}


Vector_E2d operator+(const Vector_E2d& v1, const Vector_E2d& v2)
{
  return Vector_E2d(v1.x() + v2.x(), v1.y() + v2.y());
}
```

In the same vein, operators for scalar multiplication and division need to be defined for Vector_E2d, but not for Point_E2d:

```
Vector_E2d operator*(double d, const Vector_E2d& v)
{
  return Vector_E2d(d * v.x(), d * v.y());
}

Vector_E2d operator/(const Vector_E2d& v, double d)
{
  return Vector_E2d(v.x() / d, v.y() / d);
}
```

Vectors have historically taken a long time to be accepted, and so readers who remain unconvinced that the notion of a vector needs to be separate from that of a point can take comfort that even in 1866, Julius Plücker needed to use this many words [81] (although the notion he had in mind included moments):

> *We usually represent a force geometrically by a limited line, i.e. by means of two points (x', y', z') and (x, y, z), one of which (x', y', z') is the point acted upon by the force, while the right line passing through both points indicates its direction, and the distance between the points its intensity.*

1.3 Vector Normalization and Directions

The vector abstraction defined so far is for a vector of an arbitrary magnitude, but a *normalized* vector—a vector of unit magnitude—is often needed to indicate, for example, a direction in the plane for the purpose of lighting (Chapter 21). It would be possible to define a function that normalizes a vector:

```
void Vector_E2d::normalize()
{
    double d = std::sqrt( _x*_x + _y*_y );
    x /= d; y /= d;
}
```

Such a function at the outset does not seem like such a bad choice. The trouble is that if such a function is provided, the client programmer will need to remember whether each instance of a vector is normalized. Recalling which objects have already been normalized (of those that should) is a nuisance. More importantly, if the programmer thinks a vector is normalized when it is not, there is an error in the code. If the programmer thinks a vector is not normalized when it is, run-time performance will be lost.

Implementing a class Direction_E2d that stores a vector of unit magnitude avoids this problem altogether. We are content for now to assume that normalization is performed at construction, but we will see starting in § 8.1 that it may be preferable in many applications to perform the normalization on demand. The designer will have to decide whether the storage penalty needed to cache the normalized vector in addition to saving the unnormalized vector can be afforded. If it is known that each direction will be used for shading computation (Chapter 21) and if no predicates (Chapter 2) will depend on the direction, normalizing at construction will indeed be the approach needed. In this, geometric computing deviates from the classical treatment. Classical geometry texts always assume that normalization is innocuous and use *direction cosines* (our "direction") as a normalized object. Indeed, it would be awkward to keep the term "cosines" without normalization being implicit.

```
class Direction_E2d {
private:
    double _x, _y;
public:
    Direction_E2d(const Vector_E2d& v)
    {
        double d = std::sqrt( v.x()*v.x() + v.y()*v.y() );
        _x = v.x() / d;
        _y = v.y() / d;
    }
    ...
};
```

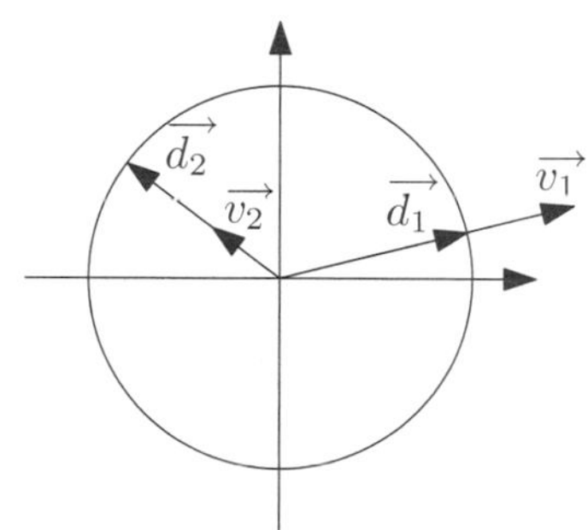

We will be content at this stage to assume that the square root function, **std::sqrt**, is indeed going to be invoked, although, as we will see in Chapter 7, number types other than **double** may frequently need to be used. If the square root function is not available for the number type, then an alternative test is needed. This approach is described in § 8.1 for points in spherical geometry.

The two spaces captured by the classes Vector_E2d and Direction_E2d are distinct. It would be quite all right to talk about a *zero vector*, or a *null vector*—one in which both the x and y elements are zero. If we are using vectors to capture forces or translations in the plane, it still makes sense to talk about applying a zero force or of applying a zero translation on a point. A direction, on the other hand, cannot be null. The set of representable directions can be captured by the unit circle. The vector used to construct a direction can be of arbitrarily small magnitude; normalizing it will still result in a vector of unit magnitude, but its magnitude cannot vanish. Perhaps most crucially if one wishes to be only pragmatic, the same transformation is applied differently on vectors and on directions: Vectors transform as a point, whereas directions transform as a (hyper-) plane—a line in E^2. We continue this discussion in § 4.1 and § 12.6.

1.4 Affine Combinations

Consider animating a particle R moving from a point P to a point Q and using a scalar α to generate the motion. R is at P when $\alpha = 0$ and at Q when $\alpha = 1$. The expression

$$R = (1-\alpha)P + \alpha\, Q$$

would determine R as a function of α, but it also evaluates the product of a scalar and a point, an operation earlier deemed illegal. To satisfy our type system, the expression could be recast as a function of the origin O into the more baroque

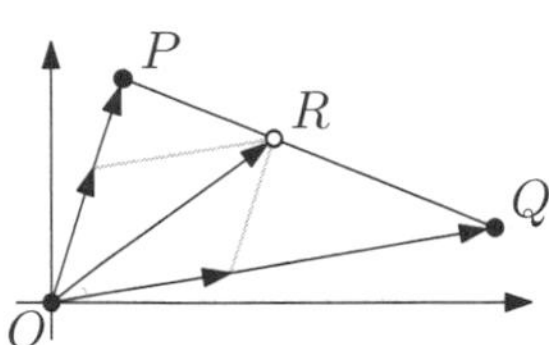

$$R = O + (1-\alpha)(P-O) + \alpha(Q-O).$$

Since in this expression the differences between two points $(P-O)$ and $(Q-O)$ are vectors, they can be scaled and then added to the origin O to construct the point R. It would be unreasonable if constraints from our self-imposed type system would result in a penalty in efficiency (compare the number of floating point operations needed). To avoid the efficiency hit, we declare the first, more direct, expression for the *affine combination* of two points to be legal so long as the scaling factors sum to unity [40]. Even though ultimately either operation must resolve to computations on individual coordinates, there is another important yet subtle difference between the two expressions: As will be discussed in Chapter 17, the second expression has the advantage that it can be implemented while abiding by coordinate freedom, while the first must have access to individual coordinates.

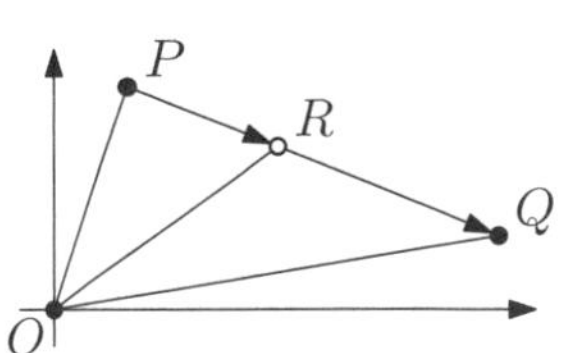

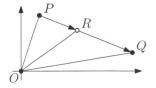

The example serves to illustrate the kind of mundane decisions that need to be made, but ones that influence the overall feel of a geometric library. In this case it is purely illustrative since a third expression

$$R = P + (Q-P)\alpha$$

has the simultaneous advantages ofbeing coordinate free and using only two multiplications for points in the plane implemented using Cartesian coordinates.

1.5 Lines

The next fundamental class to look at is that of a line in the plane. Since we decided that a segment in the plane is directed, it makes sense that a line in the plane should also be directed.

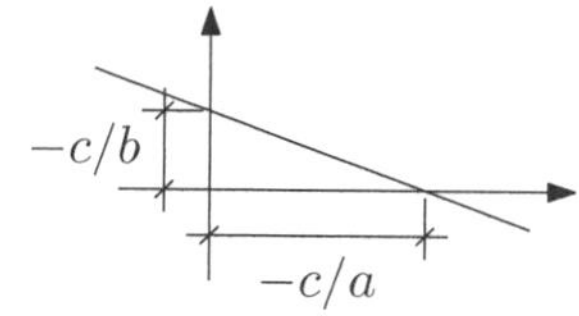

Taking our cue from classical analytic geometry, it is tempting to use $ax + by + c = 0$ as the fundamental equation for a line in the plane. (Let's leave the orientation out for a second.) We ponder whether to adopt this representation for a line by considering an obvious optimization.

If we are eager to minimize the storage for a line object, we may consider normalizing the equation above to $Ax + By + 1 = 0$ (where $A = a/c$ and $B = b/c$). This normalization is rather troublesome: What if $c = 0$? According to the IEEE 754 standard [54], we can indeed represent $\pm\infty$ in a **double** (more in § 7.5), but we would rather avoid relying on such a representation and conclude that normalizing is not an option.

But what is the set of lines that we would be unable to represent if we excluded $c = 0$? These are of course the lines passing by the origin—a rather important subset that we cannot do without. What about other special cases such as $a = 0$ or $b = 0$? At first sight these cases (lines parallel to the x- and the y-axes, respectively) are representable with no trouble. We would, however, run into trouble later on: If we work through what would need to be computed to determine, say, the intersection of two lines in the plane, we will find that our code will be unnecessarily complicated since these special cases will need to be explicitly handled (which is inelegant, error-prone, and time-consuming—both for us and for systems built using our library).

The reader who has gone through a comparison between classical analytic geometry and analytic geometry through vector computation will be quite familiar with the preceding discussion. These preceding arguments can all be seen when a (classical) analytic geometer decides to adopt vector calculus. We may then rush to conclude that the canonical representation we are seeking is none other than representing a line by a point in addition to its direction, or angle with a fixed axis. Using a point and a direction as the member data for a line in the plane may indeed be sensible, but how will a line in the plane be typically constructed?, one may ask. Since a line object is usually constructed from two distinct points, we will simply store these two points as the member variables for a line. Let's confirm that this decision will satisfy all the constraints we have looked at:

- A line remains directed.
- The intersection of two lines is easily computed and the intersection code will contain no special cases.
- Determining whether two lines are equal can be performed reliably without incurring the reduced exactness or the reduced efficiency of trigonometric functions.

We observe that storing the two endpoints requires four floating point numbers, whereas we saw that we could get by with just half that much. Ignoring for a while the difference, we are content to write

```
class Line_E2d {
private:
  Point_E2d source, target;
public:
  Line_E2d() : source(), target() {}
  Line_E2d( const Point_E2d& _source, const Point_E2d _target )
    : source(_source), target(_target) {}
  Line_E2d( const Line_E2d& line )
    : source(line.source), target(line.target) {}
  Line_E2d( const Segment_E2d& seg )
    : source(seg.source()), target(seg.target()) {}
  ...
};
```

Our line class ended up looking so similar to our segment class that it is tempting to consider whether the two abstractions actually need to be separated, but the answer is clear when we think about the intersection of a pair of lines or of segments—the result of the intersection may depend on whether a line or a segment is involved.

Line Construction

Representing a line using two points is a good first design, but there are two reasons why we should not be content with such a choice. One reason is that the line may be constructed using other than two points. The line may be defined as passing by a point while parallel or perpendicular to another line, for instance. It would be inelegant to construct an artificial second point in this case to accommodate the class definition. Another reason is that one line may be used multiple times in intersection computations. Since each would require finding the coefficients of the line equation, it would be advantageous to *cache* the coefficients or to have them simply replace the pair of points, thus avoiding finding a second point if none is given during construction.

Storing the coefficients of the parametric form of the line $ax + by + c = 0$ is superior to the single-valued form $y = mx + b$ since lines parallel to the y-axis would be captured easily by setting $a = 0$.

The $ax + by + c = 0$ could be normalized to $Ax + By + 1 = 0$, but, as discussed, doing so would make it impossible to capture lines passing by the origin, and so we choose the following representation for a line:

```
class Line_E2d {
private:
  // The source and target of the line are not stored
  double a, b, c;
public:
  Line_E2d(const Point_E2d& source, const Point_E2d& target);
  ...
};
```

To find the coefficients given two points defining the line, we solve for A and B in $Ax + By + 1 = 0$. If S and T are the two points, we solve the two equations

$$AS_x + BS_y + 1 = 0,$$
$$AT_x + BT_y + 1 = 0$$

as

$$A = +\frac{\begin{vmatrix} S_y & 1 \\ T_y & 1 \end{vmatrix}}{\begin{vmatrix} S_x & S_y \\ T_x & T_y \end{vmatrix}}, \qquad B = -\frac{\begin{vmatrix} S_x & 1 \\ T_x & 1 \end{vmatrix}}{\begin{vmatrix} S_x & S_y \\ T_x & T_y \end{vmatrix}}.$$

We then avoid the division by storing instead the coefficients of $ax + by + c = 0$, where

$$a = +\begin{vmatrix} S_y & 1 \\ T_y & 1 \end{vmatrix}, \qquad b = -\begin{vmatrix} S_x & 1 \\ T_x & 1 \end{vmatrix}, \qquad c = \begin{vmatrix} S_x & S_y \\ T_x & T_y \end{vmatrix}, \qquad S \neq T.$$

We would define a line by saying that the two points used for its construction are distinct. The word *distinct* appears as a constraint to the solution: The determinant of the matrix defining c above vanishes exactly when the two points S and T coincide. It may be wise to guard against that case by using an assert statement or by throwing an exception (§ 12.2), while ensuring that no run-time penalty is paid by, for example, turning assertions off for production code.

Line Default Constructor

The default constructors used for points (the origin), for vectors (the zero vector), and for segments (two coincident points at the origin) declared somewhat sensible objects. When choosing a default constructor for a line we have a chance to define a line that makes no sense. If the three coefficients are set to zero, the equation $0x + 0y + 0 = 0$ no longer defines a line. One option is to choose some line as the default one (perhaps coinciding with the x-axis). The alternative, not defining a default line, puts application programmers at the risk of reading uninitialized variables.

1.6 Vector Orthogonality and Linear Dependence

We are frequently interested in determining whether two vectors are perpendicular, or *orthogonal*, and whether one is a multiple of the other—whether they are *linearly dependent*. See Figure 1.1.

Figure 1.1
Angle between two vectors

Unit Vectors

The length, or *magnitude*, of a vector can be determined by Pythagoras' theorem. The magnitude of a vector $\overrightarrow{v}$ is $|\overrightarrow{v}| = \sqrt{v_x^2 + v_y^2}$. A vector is said to be of unit length, to be of unit magnitude, or is simply referred to as a *unit vector* if $|\overrightarrow{v}| = 1$, which in turn means that $v_x^2 + v_y^2 = 1$.

Orthogonality

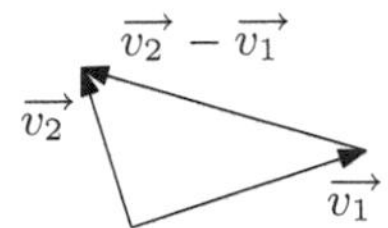

If two vectors $\overrightarrow{v_1}$ and $\overrightarrow{v_2}$ are orthogonal, then their magnitude must be related to that of the hypotenuse $\overrightarrow{v_2} - \overrightarrow{v_1}$ by Pythagoras' theorem. The hypotenuse is also related to the two vectors since, evidently, $\overrightarrow{v_1} + (\overrightarrow{v_2} - \overrightarrow{v_1}) = \overrightarrow{v_2}$. The magnitude of each of the three vectors can itself be found also by Pythagoras' theorem. We can write

$$\begin{aligned}
(\sqrt{x_1^2 + y_1^2})^2 + (\sqrt{x_2^2 + y_2^2})^2 &= (\sqrt{(x_2 - x_1)^2 + (y_2 - y_1)^2})^2, \\
(x_1^2 + y_1^2) + (x_2^2 + y_2^2) &= ((x_2 - x_1)^2 + (y_2 - y_1)^2), \\
(x_1^2 + y_1^2) + (x_2^2 + y_2^2) &= (x_2^2 - 2x_2x_1 + x_1) + (y_2^2 - 2y_2y_1 + y_1^2), \\
0 &= x_1x_2 + y_1y_2.
\end{aligned}$$

The last expression is of course the *inner product*, also called the *dot product* because it is frequently written as $\overrightarrow{v_1} \cdot \overrightarrow{v_2}$. The dot notation is convenient because it makes it possible to write the product as one directly between two column vectors. Writing it as one between two matrices, one would write $\overrightarrow{v_1}^T\overrightarrow{v_2}$. Notice that it would be incorrect to write simply $\overrightarrow{v_1}\overrightarrow{v_2}$. Because we look at vectors as matrices of order 2×1, the inner product is obtained from the product of the transpose of the first with the second.

Recall that the dot product is a scalar-valued function that measures the magnitude of the projection of one vector on the other. If the angle between two vectors $\overrightarrow{v_1}$ and $\overrightarrow{v_2}$ is θ, then

$$\overrightarrow{v_1} \cdot \overrightarrow{v_2} = |\overrightarrow{v_1}||\overrightarrow{v_2}| \cos\theta.$$

Linear Independence

We say that two vectors are linearly independent if there does not exist a scalar k such that $\overrightarrow{v_1} = k\overrightarrow{v_2}$. If such a k exists, the vectors are termed linearly dependent.

Since

$$\begin{bmatrix} x_1 \\ y_1 \end{bmatrix} = k \begin{bmatrix} x_2 \\ y_2 \end{bmatrix} \implies \frac{x_1}{x_2} = \frac{y_1}{y_2}$$

we can conclude the following familiar condition for vector orthogonality:

$$x_1y_2 = x_2y_1 \implies x_1y_2 - x_2y_1 = 0.$$

If we construct a 2×2 matrix using the two column vectors $\overrightarrow{v_1}$ and $\overrightarrow{v_2}$, the last expression is, of course, the determinant of the matrix.

Basis

Any pair of linearly independent (and nonzero) vectors is suitable for use as a *basis*. One point in the plane is declared the origin, its coordinates are set at $O(0,0)$, and the two vectors are used as a frame of reference. The coordinates of the point $O + \overrightarrow{v_1}$ will be defined as $(1,0)$ and those of the point $O + \overrightarrow{v_2}$ as $(0,1)$. This frame of reference makes it possible to uniquely determine a pair of coordinates for all points in the plane. It is clear though that not all bases are equally useful. The closer the determinant of the matrix $M = [\overrightarrow{v_1}\,\overrightarrow{v_2}]$ is to zero, the less suitable the pair of vectors is for use as a frame of reference.

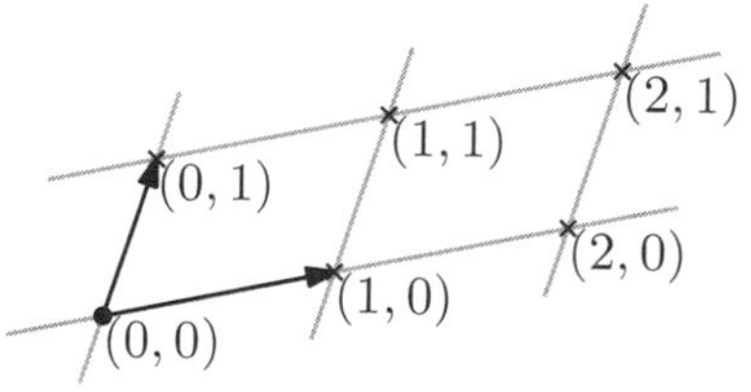

If the pair of vectors we use has the properties:

1. each vector is of unit length and

2. the two vectors are orthogonal,

we say that they form an *orthonormal basis*. It is easy to confirm that if only the first condition is satisfied, the determinant is in the range $[-1 \ldots 1]$ and that if both are satisfied—see Figure 1.2—the determinant is either 1 or -1. If the determinant is 1, we say that the matrix is *proper orthogonal* and if it is -1, the two vectors are still orthogonal, but we refer to the matrix as an *improper orthogonal* matrix (§ 4.6).

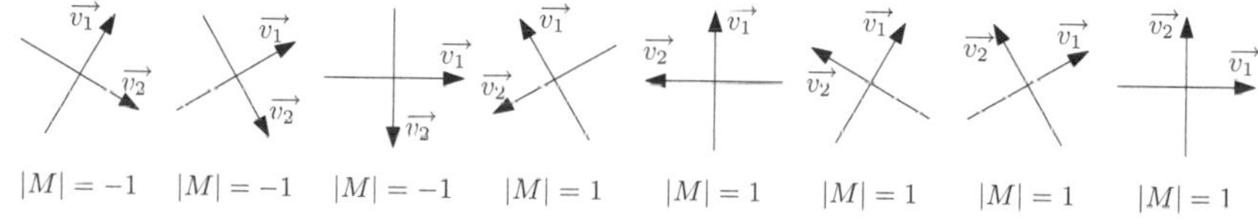

Figure 1.2
Orthonormal basis vectors

2 Geometric Predicates

Now-ancient books on computing frequently use *flow charts*, which conveniently introduce predicates. At the time when FORTRAN in particular, and imperative programming in general, were at the forefront of computing, the use of flow charts was widespread (see, for example, those for polygon clipping [107]). A flow chart illustrates rather pointedly the path that control may take during computation. This path is sketched using straight lines that connect rectangles and diamonds. Assignment statements appear inside rectangles and if-statements appear inside diamonds. Other elements also exist, but we concentrate here on the parts where the linear path of program control is broken, or *branches*. The functions that are evaluated and that decide the path taken at such branches are called *predicates*. Flow charts have since been replaced by pseudo-code, where changing the linear program control appears in the form of an indentation.

System design has gone back to schematics with the advance of techniques for object-oriented design. One such popular visual language and accompanying methodology, the Unified Modeling Language, promotes that system design should be tackled at a higher granularity. Objects and the messages they pass to each other are identified, but the advance of UML did not supplant—it merely enlarged—pseudo-code and the algorithm design that it captures.

The objective of this chapter is to argue that crafting good geometric predicates and using them properly is at the center of geometric computing.

2.1 Predicate Return Type

We generally think of predicates as functions with a Boolean return type. The Boolean type might identify whether a counter has reached some bound, whether a predetermined tolerance has been satisfied, or whether the end of a list has been reached. Such predicates arise in geometric computing, but an additional type of test is frequently needed. Because this geometric test has *three* possible outcomes, we refer to it as a ternary branching test. Yet most often, we are interested in forming a binary predicate from the three possible outcomes.

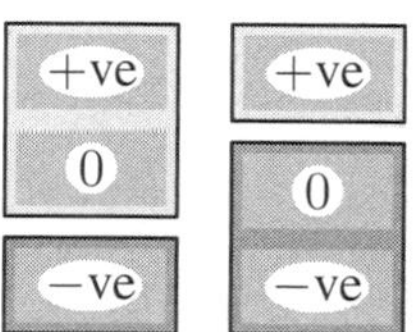

The need for three branches in a test can be seen when we consider an oriented line splitting the plane. The plane is split into the points that lie on the positive halfplane, the points that lie on the negative halfplane, as well as those that lie on the line itself. A geometric library will offer such ternary outcomes to clients, and the application programmer will decide how the predicate

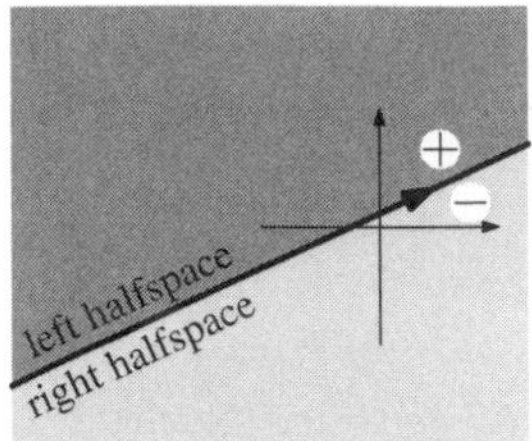

should be formed. An application might quite suitably need to capture only two cases, the set of points lying on the positive halfplane or the line and the set of points lying in the negative halfplane, for example. But the geometric tests should be offered in such a way that if the application programmer wishes to provide different handling for each of the three cases, it is possible to do so.

Just as we refer to an interval being *open* if it does not include its extremities and refer to it as *closed* if it does, we can also talk about either open or closed halfspaces. A left open halfspace consists of the points lying to the left of the line, not including the points on the line itself. A left closed halfspace does include the points on the line. Whether open or closed, we define the *boundary* of the halfspace as the points on the line. Thus, a closed halfspace includes its boundary and an open halfspace does not. The *interior* of an interval is the corresponding open interval. A set is termed *regular* if it is equal to the closure of its interior—an interval is regular if it is closed. By thinking of the predicate as a ternary rather than as a binary predicate we simplify the design of a predicate and leave the decision of choosing among the different representable sets to the client.

But before discussing the turn predicate, it is worthwhile to discuss an even simpler predicate, testing whether two points in the plane coincide. Saying that the two points coincide is less ambiguous than saying that the two points are equal, since the latter may be interpreted to mean that the object representing the points is identical. Implementing a two-point coincidence predicate is simple enough: It suffices to test whether the x-coordinates are equal and the y-coordinates are equal. The computation needed for a three-point and a four-point predicate are described next. These 2-, 3-, and 4-point predicates together capture the vast majority of the geometric tests that arise in a geometric system in the plane.

2.2 The Turn Predicate

Determining the orientation of a point with respect to the line defined by two other points is easily defined by appealing to a function that will take us momentarily to a third dimension.

The Cross Product

There is more than one way to define the *cross product* of two vectors $\overrightarrow{v_1}$ and $\overrightarrow{v_2}$. In this text we take the classical—in computer graphics—view that the cross product $\overrightarrow{v} = \overrightarrow{v_1} \times \overrightarrow{v_2}$ is a vector that is simultaneously orthogonal to $\overrightarrow{v_1}$ and $\overrightarrow{v_2}$, that obeys the right-hand rule with respect to the two vectors, and whose magnitude is related to that of the two vectors by

$$|\overrightarrow{v}| = |\overrightarrow{v_1}||\overrightarrow{v_2}| \sin\theta,$$

where θ is the angle between the two vectors. Defining and using the cross product is in general awkward [33]. Its awkwardness in our context is that we would have liked to be able to consider geometric structures as standalone objects worthy of study even if they are not embedded in a higher dimension (which, at least linguistically, affects the choice of prepositions—see § 11.1).

Defining the cross product as a vector lying in a third dimension breaks that uniformity, but we can live with that breach. To develop an intuition about cross products one needs only to consider how it varies when one of the two vectors, say $\overrightarrow{v_2}$, moves. Consider positioning $\overrightarrow{v_1}$ such that it coincides with the positive x-axis. If $\overrightarrow{v_2}$ also coincides with the x-axis, the cross product will be the zero vector. This is natural: The two vectors do not define a plane or, alternatively, the parallelogram they define has zero area.

Now consider that $\overrightarrow{v_2}$ rotates toward the y-axis. The magnitude of $\overrightarrow{v_1} \times \overrightarrow{v_2}$ increases until it reaches a maximum when $\overrightarrow{v_1}$ and $\overrightarrow{v_2}$ are orthogonal. As $\overrightarrow{v_2}$ rotates beyond the y-axis, the magnitude of $\overrightarrow{v}$ retracts. It reaches zero when $\overrightarrow{v_2} = -\overrightarrow{v_1}$. Once $\overrightarrow{v_2}$ goes past the $-x$-axis, the direction of $\overrightarrow{v}$ is aligned with the $-z$-axis.

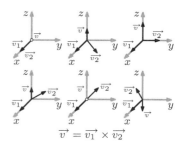

Design of the Turn Predicate

One could argue that a predicate that reports whether three points are colinear would be needed. But rather than implement such a predicate by itself, it is more convenient to implement a more general one that will also determine colinearity. Such a *turn predicate* might have the signature

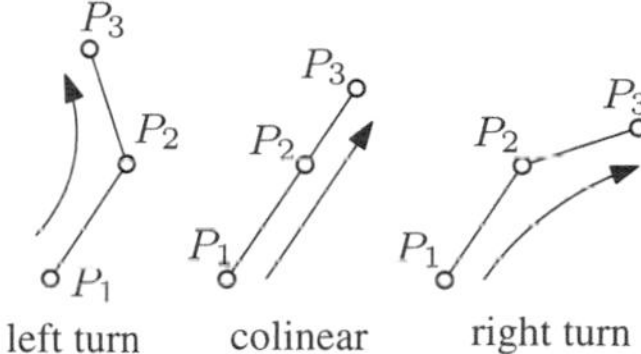


```
Oriented_side oriented_side(
        const Point_E2d& p1,
        const Point_E2d& p2,
        const Point_E2d& p3);
```

where the return type is defined as

```
enum Oriented_side {
    ON_NEGATIVE_SIDE = −1,
    ON_ORIENTED_BOUNDARY,
    ON_POSITIVE_SIDE
};
```

These names are those defined by CGAL [23]. Whenever possible when choosing names for Euclidean objects we choose ones that match the terminology established by CGAL, which simplifies the task of moving from one geometry module to another.

If necessary, the implementation of various convenience predicates is now easy. The following binary predicates delegate the requests they receive to the oriented_side function:

```
bool is_left_turn(
            const Point_E2d& p1,
            const Point_E2d& p2,
            const Point_E2d& p3) {
    return oriented_side(p1, p2, p3) == ON_POSITIVE_SIDE;
}
bool are_colinear(
            const Point_E2d& p1,
```

```
            const Point_E2d& p2,
            const Point_E2d& p3) {
    return oriented_side(p1, p2, p3) == ON_ORIENTED_BOUNDARY;
}
bool is_right_turn(
            const Point_E2d& p1,
            const Point_E2d& p2,
            const Point_E2d& p3) {
    return oriented_side(p1, p2, p3) == ON_NEGATIVE_SIDE;
}
```

Matrix Form of the Turn Predicate

As the oriented line $\overrightarrow{P_2P_3}$ divides the plane into the points lying on, to the left, or to the right of the line, the sign of the expression

$$\overrightarrow{P_1P_2} \times \overrightarrow{P_2P_3}$$

identifies the location of the point P_3. If the sign is positive, P_3 is to the left; if it is zero, P_3 is on the line; and if it is negative, P_3 is to the right of the line. The vector product above evaluates to the determinant

$$\begin{vmatrix} x_2 - x_1 & x_3 - x_2 \\ y_2 - y_1 & y_3 - y_2 \end{vmatrix},$$

which can in turn be expanded into the 3×3 determinant

$$\begin{vmatrix} x_1 & x_2 - x_1 & x_3 - x_2 \\ y_1 & y_2 - y_1 & y_3 - y_2 \\ 1 & 0 & 0 \end{vmatrix},$$

where the two values x_1 and y_1 could be chosen arbitrarily. Adding the first column to the second and the resulting second to the third, we obtain the equivalent homogeneous form

$$\begin{vmatrix} x_1 & x_2 & x_3 \\ y_1 & y_2 & y_3 \\ 1 & 1 & 1 \end{vmatrix}.$$

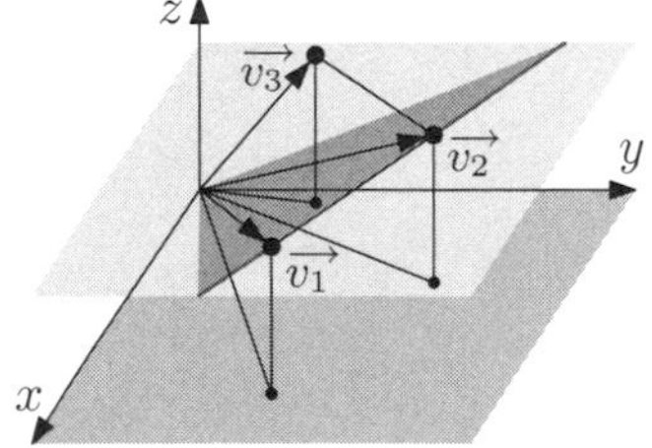

We will encounter the homogeneous forms in Part II, but we can stop briefly at this time and interpret this expression as three vectors in 3D rather than as three points in 2D.

The particular test we will use will depend on whether we are testing for the inclusion of the query point in an open or a closed halfplane. If we wish to determine, for instance, whether a point lies in the open left halfplane, we simply test oriented_side(...) == ON_POSITIVE_SIDE. If we wish to determine whether a point lies in the closed left halfplane, we test instead oriented_side(...) != ON_NEGATIVE_SIDE.

2.3 Side of Circle Predicate

Design of the Side of Circle Predicate

Just as two points naturally define a line that splits the plane into two regions in addition to the line separating them, three points in the plane P_1, P_2, and P_3 define a circle that splits the plane into two regions in addition to the circle itself. The following inside_circle predicate and accompanying constants can therefore be defined:

```
Orientation_to_circle
inside_circle(
        const Point_E2d& p0,
        const Point_E2d& p1,
        const Point_E2d& p2,
        const Point_E2d& p3);

enum Orientation_to_circle
{ OUTSIDE_CIRCLE = −1, COCIRCULAR, INSIDE_CIRCLE };
```

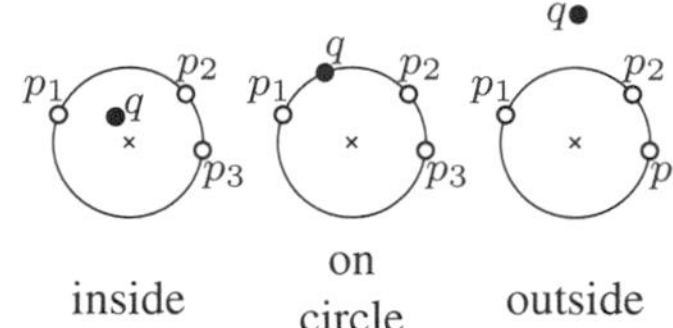

Matrix Form of the Side of Circle Predicate

A circle with center (x_c, y_c) and radius r in the plane has the equation

$$(x - x_c)^2 + (y - y_c)^2 = r^2,$$

which expands to

$$(x^2 + y^2) - 2(xx_c + yy_c) + (x_c^2 + y_c^2 - r^2) = 0.$$

More generally,

$$A(x^2 + y^2) + Bx + Cy + D = 0$$

is the equation of a circle in the plane provided that $A \neq 0$.

The equation above can be written as the determinant

$$\begin{vmatrix} x^2 + y^2 & x & y & 1 \\ x_1^2 + y_1^2 & x_1 & y_1 & 1 \\ x_2^2 + y_2^2 & x_2 & y_2 & 1 \\ x_3^2 + y_3^2 & x_3 & y_3 & 1 \end{vmatrix} = 0, \tag{2.1}$$

where

$$A = \begin{vmatrix} x_1 & y_1 & 1 \\ x_2 & y_2 & 1 \\ x_3 & y_3 & 1 \end{vmatrix}, \qquad B = \begin{vmatrix} x_1^2 + y_1^2 & y_1 & 1 \\ x_2^2 + y_2^2 & y_2 & 1 \\ x_3^2 + y_3^2 & y_3 & 1 \end{vmatrix},$$

$$C = \begin{vmatrix} x_1^2 + y_1^2 & x_1 & 1 \\ x_2^2 + y_2^2 & x_2 & 1 \\ x_3^2 + y_3^2 & x_3 & 1 \end{vmatrix}, \qquad D = \begin{vmatrix} x_1^2 + y_1^2 & x_1 & y_1 \\ x_2^2 + y_2^2 & x_2 & y_2 \\ x_3^2 + y_3^2 & x_3 & y_3 \end{vmatrix}.$$

It is clear that the determinant in Eq. (2.1) vanishes if the point $P(x, y)$ coincides with any of the three points $P_1(x_1, y_1)$, $P_2(x_2, y_2)$, or $P_3(x_3, y_3)$. Moreover, we know from § 2.2 that $A \neq 0$ if and only if the three given points are not colinear.

It is also clear that all the points lying either inside or outside the circle generate a positive determinant and that the points lying on the other side generate a negative determinant. Since exchanging any two rows in Eq. (2.1) would flip the sign of the determinant, the order of the three given points does matter. Clients of this predicate would likely rather not be careful in selecting a particular order for the three points and so it would be appropriate to take a small efficiency hit and compute the 3×3 determinant for the orientation of the three points in addition to computing the 4×4 determinant in Eq. (2.1). And so a point $P(x, y)$ can be classified with respect to a circle defined by three points by evaluating the following equation:

$$\begin{vmatrix} x^2 + y^2 & x & y & 1 \\ x_1^2 + y_1^2 & x_1 & y_1 & 1 \\ x_2^2 + y_2^2 & x_2 & y_2 & 1 \\ x_3^2 + y_3^2 & x_3 & y_3 & 1 \end{vmatrix} \times \begin{vmatrix} x_1 & y_1 & 1 \\ x_2 & y_2 & 1 \\ x_3 & y_3 & 1 \end{vmatrix}$$

$$= \text{side\_of\_circle}(P,P_1,P_2,P_3) \begin{cases} < 0 & \text{inside,} \\ = 0 & \text{on the circle boundary,} \\ > 0 & \text{outside.} \end{cases}$$

2.4 Order Predicate

Consider the following problem in the plane. Given an oriented line L and a set S of lines, we wish to sort the intersections of S with L along the orientation of L.

Our sorting implementation will need an *order predicate*. Given a line L defined by two points, the predicate will report either that one of the two points is encountered first along the line, or that the two points coincide.

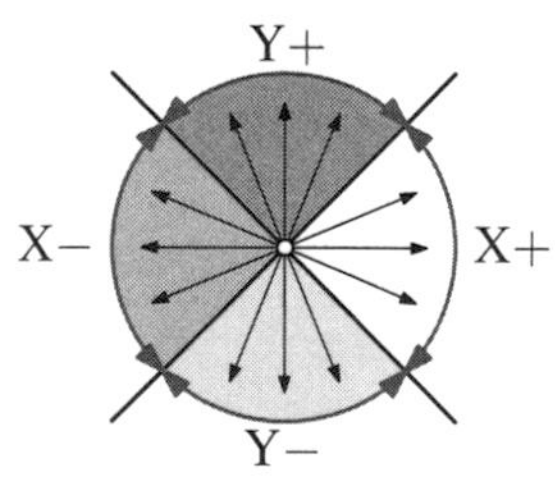

If the line has the orientation illustrated in the adjacent figure, then comparing the x-coordinate of the two points will produce the result needed. If the slope of the line with the x-axis is large while the x values still increase along the line, it would be more reliable to use the y-axis instead. For a general line in the plane, we divide the plane into four quadrants. When the quadrant in which a vector along the line is determined, we will know the line's *dominant* direction. We thus introduce a new data type Dominant_E2 that stores this direction.

```
enum Dominant_E2 {
  E2_POSX, E2_NEGX, E2_POSY, E2_NEGY
};
```

The following function will determine the dominant direction of L:

```
Dominant_E2 dominant(const Segment_E2d& segment)
{
```

```
    double dx = segment.target().x() − segment.source().x();
    double dy = segment.target().y() − segment.source().y();

    double dxabs = dx >= 0 ? dx : (−dx);
    double dyabs = dy >= 0 ? dy : (−dy);

    if(dxabs >= dyabs)
        return ((dx > 0.0) ? E2_POSX : E2_NEGX);
    else
        return ((dy > 0.0) ? E2_POSY : E2_NEGY);
}
```

We return to this function in § 17.4 and move now to the implementation of a function that reports the order of two points P_1 and P_2 along a line L in the plane.

```
enum Comparison {LessThan, Equal, GreaterThan};

Comparison
find_order(const Point_E2d& p1, const Point_E2d& p2, const Dominant& D)
{
    switch(D) {
    case POSX:
        return ( p1.x() < p2.x()
                ? LessThan : (p1.x() == p2.x()
                ? Equal : GreaterThan); break;
    case NEGX:
        return ( p1.x() > p2.x()
                ? LessThan : (p1.x() == p2.x()
                ? Equal : GreaterThan); break;

    case POSY:
        return ( p1.y() < p2.y()
                ? LessThan : (p1.y() == p2.y()
                ? Equal : GreaterThan); break;
    case NEGY:
        return ( p1.y() > p2.y()
                ? LessThan : (p1.y() == p2.y()
                ? Equal : GreaterThan);
    }
}
```

2.5 The Geometry of the Euclidean Line E^1

Simple as they are, objects and predicates for the geometry of the Euclidean line E^1 are frequently needed. A point object is merely a wrapper for a (real) number. But before contemplating the geometry of E^1 it is instructive to consider a question that will appear to border on pedantry. Should we say that a point is *in* E^1 or *on* E^1? The difference is not just a matter of language. If we refer to a point that lies *on* E^1, we have in mind a Euclidean line E^1 that is itself embedded, or lying, somewhere in a higher-dimensional space, as a line in E^2 or in E^3. But E^1 is a perfectly fine structure by itself; there is no need for it to lie inside another. If we say that a point is in E^1, we have in mind a setting in which some small creature (the point) is only endowed with the ability of "looking" in one of two directions, and of moving and distinguishing distances in only these two directions. The issue of prepositions is revisited in the context of projective geometry in § 11.1.

One useful operation for points in the Euclidean line is the $\leq$ operation. It satisfies some elementary properties on the set of points. In particular, the relationship is transitive; given three points A, B, and C,

$$A \leq B \wedge B \leq C \implies A \leq C.$$

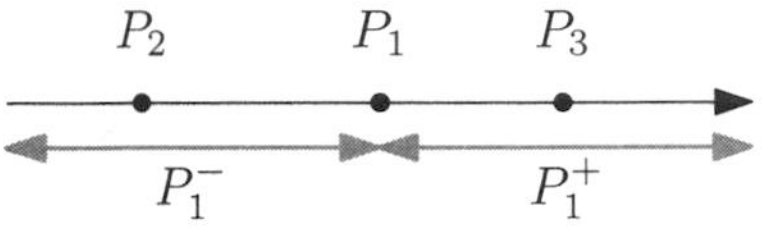

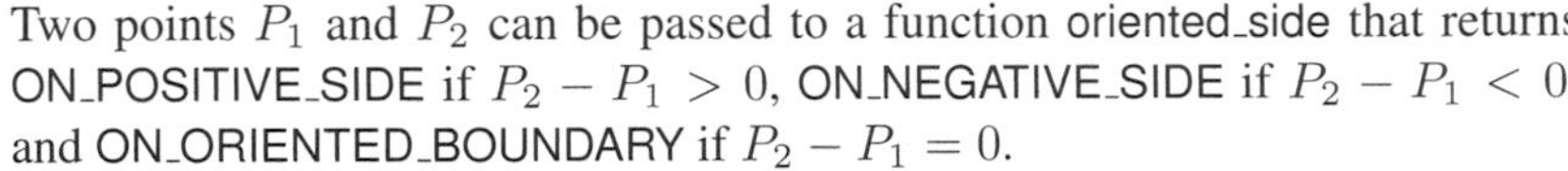

Two points P_1 and P_2 can be passed to a function oriented_side that returns ON_POSITIVE_SIDE if $P_2 - P_1 > 0$, ON_NEGATIVE_SIDE if $P_2 - P_1 < 0$, and ON_ORIENTED_BOUNDARY if $P_2 - P_1 = 0$.

Likewise a segment in E^1 is a wrapper for two real numbers. We define a segment using an ordered tuple of a *source* and a *target* with the constraint that *source* < *target*.

A point P_1 splits the Euclidean line into two parts. A point P_2 lies on the *positive side* of P_1 if $P_1 - P_2 > 0$. Many applications benefit from abstracting the Euclidean line. We will get a chance in § 28.2 to discuss some operations that can be implemented based on the abstraction of Point_E1 and Segment_E1 for objects in the Euclidean line.

2.6 Immutability of Geometric Objects

We conclude the chapter by arguing whether geometric objects should be *immutable*. An object is immutable if it is not possible to change its state—or in the case of geometric objects, its coordinates—after construction. Consider the following code:

```
Point_E2d P1 = Point_E2d(5,7);
Point_E2d P2 = P1;
Point_E2d & P3 = P1;
```

Were we to implement a member function Point_E2d::set_x(...), it would be possible to modify the coordinate of P1 after the execution of the code above. Afterwards, a programming bug will continue to lurk, waiting to arise [67, LEDA Rule 4]. The trouble is that the client programmer must subsequently make the conscious effort to recall that P2 is merely a copy of P1 and is unaffected by modifications to P1, whereas P3 is a reference and would mirror changes to P1's data members. The safest way to avoid this error is to declare geometric objects immutable and to provide no set functions. Client programmers who need a new geometric object need to create a distinct object for it. Disallowing the modification of a single coordinate can also be seen as an instance of a larger set of suggestions for a geometric library, coordinate freedom, the topic of Chapter 17.

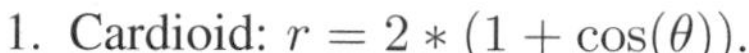

2.7 Exercises

2.1 Write a program that creates suitably named encapsulated postscript files (with an extension of "eps") showing drawings of the following polar or parametric equations of a curve. First read § 4.5 as well as Appendix A.

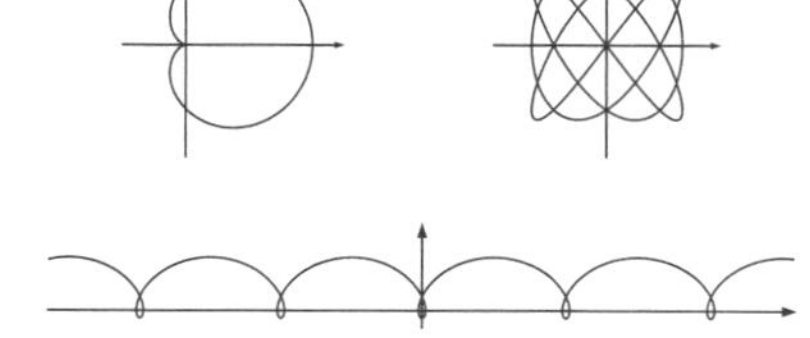

1. Cardioid: $r = 2 * (1 + \cos(\theta))$.
2. Lissajous: $x = \sin(a\theta + c), y = \sin(b\theta)$ for constant reals a, b, and c.
3. Trochoid: $x = a\theta - h\sin(\theta), y = a - h\cos(\theta)$ for constant reals a and h.

2.2 The four Boolean variables s2s, s2t, s1s, and s1t determine the turn from one segment to the endpoints of another. Write an expression on the four variables that evaluates to true if and only if the two segments intersect in their interior.

2.3 Write an expression that evaluates to true when the segments intersect in any configuration.

2.4 Suppose we wish to partition a set of points into those lying to the right, on, and to the left of the line formed by a pair of points in the plane. The most natural approach is to start by constructing an instance of Line_E2d and then to iterate over the points. It now helps to know whether the line is stored internally as $ax + by + c = 0$ or as simply the given pair of points. Suggest an efficient solution for this problem and count the number of floating point multiplications that your answer would incur to partition a set of n points.

2.5 Consider the following design option for the function calculating the intersection between two segments. The function returns an array—perhaps implemented as **std**::**vector**<Point_E2d>—and the size of the vector would be zero, one, or two depending on whether the result of the intersection is no object, a point, or a segment. Evaluate this approach in two scenarios of your choice.

3 Computational Euclidean Geometry in Three Dimensions

This chapter discusses the design of classes for geometric objects in three dimensions and the predicates and intersection functions that arise in space. Since many of the ideas encountered while designing classes for 2D in Chapter 1 are similar to those in 3D, the pace is somewhat faster.

3.1 Points in Euclidean Space

When asked to implement a class for a point in three dimensions after implementing a class such as Point_E2d, it is tempting to write

```
// poor design
class Point_E3d {
    Point_E2d pnt;
    double z;
    ...
};
```

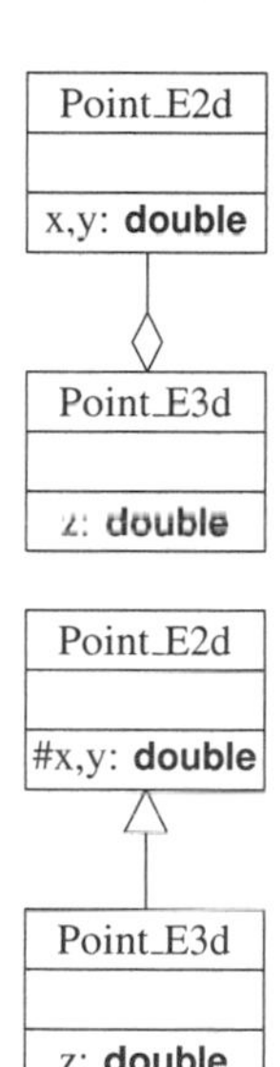

or perhaps to modify the access specifier for the x- and y-coordinates in Point_E2d to be **protected** and write

```
// poor design
class Point_E3d : public Point_E2d {
    double z;
    ...
};
```

Both designs are flawed. The first consists of adopting an *aggregation* relationship from Point_E3d to Point_E2d. Saying that an instance of a Point_E3d is an aggregate of a Point_E2d in addition to some third dimension is hard to justify geometrically. Likewise, connecting Point_E3d to Point_E2d through inheritance is also hard to justify.

The proper approach is then to keep the two classes separate and write the following:

```
class Point_E3d {
private:
    double _x, _y, _z;
public:
    Point_E3d( double x, double y, double z ) : _x(x), _y(y), _z(z) {}
    ...
};
```

3.2 Vectors and Directions

As for Vector_E2d, an implementation for a vector in three dimensions would start as follows:

```
class Vector_E3d {
private:
  double _x, _y, _z;
public:
  ...
};
```

An implementation for a class Direction_E3d would capture the notion of a *normalized* vector, or one of unit magnitude. The three components may be divided by the length of the vector during construction:

```
class Direction_E3d {
private:
  Vector_E3d v;
public:
  Direction_E3d(double x, double y, double z) {
    double L = x*x + y*y + z*z;
    v = Vector_E3d(x/L, y/L, z/L);
  }
};
```

But as we will see shortly, it is not necessary to perform the normalization step. An operation such as testing for the equality of two directions can be reduced to a variation of testing for linear dependence.

3.3 Vector Orthogonality and Linear Dependence

Vector Magnitude

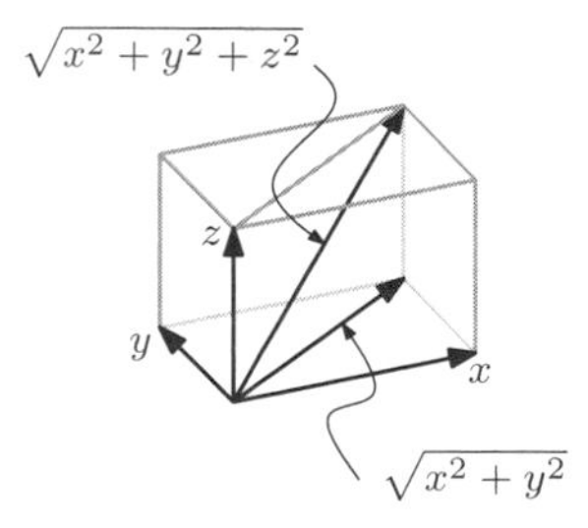

The magnitude of a vector $\overrightarrow{v}(x, y, z)$ in three dimensions can be determined by applying Pythagoras's theorem twice. The projection of $\overrightarrow{v}$ on the xy-plane is $\sqrt{x^2 + y^2}$ and the magnitude is

$$\sqrt{\left(\sqrt{x^2 + y^2}\right)^2 + z^2} = \sqrt{x^2 + y^2 + z^2}.$$

If $\sqrt{x^2 + y^2 + z^2} = 1$, then $|\overrightarrow{v}| = 1$ and $\overrightarrow{v}$ is termed a unit vector.

Orthogonality

As in § 1.6, the condition for two vectors $\overrightarrow{v_1}$ and $\overrightarrow{v_2}$ in three dimensions to be orthogonal can be derived by relating their magnitude to that of their difference. As

$$|\overrightarrow{v_1}|^2 + |\overrightarrow{v_2}|^2 = |\overrightarrow{v_2} - \overrightarrow{v_1}|^2,$$

we expand the left side to

$$\left(\sqrt{x_1^2+y_1^2+z_1^2}\right)^2+\left(\sqrt{x_2^2+y_2^2+z_2^2}\right)^2$$

and the right side to

$$\left(\sqrt{(x_2-x_1)^2+(y_2-y_1)^2}+(z_2-z_1)^2\right)^2,$$

which yields

$$x_1x_2+y_1y_2+z_1z_2=0.$$

Since this expression is that for computing the inner product, we conclude that $\overrightarrow{v_1}\cdot\overrightarrow{v_2}=0$ if and only if the two vectors are orthogonal.

Linear Independence

Two vectors $\overrightarrow{v_1}$ and $\overrightarrow{v_2}$ in three dimensions are linearly dependent if there exists a k such that $\overrightarrow{v_1}=k\overrightarrow{v_2} \implies x_1=kx_2$, $y_1=ky_2$, and $z_1=kz_2$, which yields

$$\frac{x_1}{x_2}=\frac{y_1}{y_2}=\frac{z_1}{z_2}$$

and

$$x_1y_2=x_2y_1,\qquad y_1z_2=y_2z_1. \tag{3.1}$$

Two vectors are linearly dependent if $\exists k\in\mathbb{R}, k>0, \overrightarrow{v_1}=k\overrightarrow{v_2}$. Eqs. 3.1 tell us that two vectors are linearly dependent, but do not identify whether $k>0$ can be satisfied.

To determine whether $k>0$ can be satisfied, we can proceed in one of two methods:

1. We confirm that the dot product of the two vectors is positive.
2. We determine whether the coordinates with the largest absolute magnitude have the same sign. Determining which of x, y, or z is largest in absolute terms is needed often enough. We call it the "dominant" direction in § 2.4 and § 3.7.

Orthonormal Basis

Any three linearly independent unit vectors are suitable as a basis in three dimensions. If the three vectors are mutually orthogonal, they form an orthonormal basis.

3.4 Planes in Space

As with lines in 2D, we consider *oriented* planes in 3D. This amounts to making a distinction between the two sides of space split by a plane. One side is termed the positive halfspace and the other the negative halfspace. A plane can be constructed using three noncolinear points lying on it. The order of the three points matters. If any two of them are swapped, a plane coincident with the first, but of opposite orientation, is defined.

The vector normal to the plane is $\overrightarrow{N} = (P2 - P1) \times (P3 - P2)$ (see § 2.2). Since any vector lying in the plane is orthogonal to the normal $\overrightarrow{N}$, a point P on the plane must satisfy $(P - P_1) \cdot \overrightarrow{N} = 0$ (where P_2 or P_3 could equally well be used). The general form of this equation, in which a plane is used to partition space, is discussed in § 3.6.

Whenever possible, we prefer to use vectorial computation over ones using scalar quantities (more in Chapter 17), but it is easy in this case to avoid paying the time and space penalty by extracting the constant in the plane equation. The normal $\overrightarrow{N}$ captures the coefficients a, b, and c in the general equation of first degree in three variables $ax + by + cz + d = 0$ representing a plane. The coefficient d is obtained by substituting with the coordinates of the origin O, which leads to $d = (O - P_1) \cdot \overrightarrow{N}$.

```
class Plane_E3d {
private:
    Vector_E3d N;
    double d;
public:
    Plane_E3d(const Point_E3d& p1, const Point_E3d& p2, const Point_E3d& p3)
    {
        Vector_E3d v1 = p2 − p1;
        Vector_E3d v2 = p3 − p2;
        N = cross_product(v1, v2);
        d = dot_product((O − p1), N);
    }
    ...
};
```

3.5 Lines in Space

An implementation for Segment_E3d will parallel that for Segment_E2d and need not be discussed, but implementing a class for a line in space offers a baffling question. A line can be *constructed* as that passing by two points or as the one at the intersection of two planes. Any number of other options are possible. One could ask for the line passing by a point and parallel to another line, and so on.

But what member variables should a line object have? Certainly two points can be stored. If two planes are used to construct the line, one can determine and then store two points at the intersection.

Yet a pair of points are merely two points on the line. They are not intrinsic to the line. Two points are convenient. They make it possible to parameterize the line (for example, to sort intersections—see Chapter 23). The parametric equation of a line passing by two (distinct) points S and T can be written as

$$P = S + \alpha(T - S).$$

The types of the variables in this equation parallel those in an implementation. $T - S$ is a vector, α is a (real) scalar, and the sum of a point and a vector, $S + \alpha(T - S)$, is a point.

We leave the *intrinsic* form for a line in space to the general case in projective geometry (see Plücker coordinates in § 12.4), although the idea can be applied (less elegantly) in Euclidean geometry. We will be content instead right now with the following as a start for an implementation for a line in space:

```
class Line_E3d {
private:
  Point_E3d _source, _target;
public:
  Line_E3d( const Point_E3d& source, const Point_E3d& target )
    : _source(source), _target(target) {}
  ...
};
```

3.6 Sidedness Predicates in 3D

The three-dimensional predicates will resemble their two-dimensional variants described in § 2.1.

Matrix-Form of the Side-of-Plane Predicate

Consider that we are given four points in space

$$P_i(x_i, y_i, z_i), \text{for } i = 1, \ldots, 4,$$

and that we would like to determine whether P_4 lies on the positive or the negative halfspace defined by P_1, P_2, P_3, which we assume not to be colinear. We further define the positive side of the plane passing by these three points such that it lies on the same side as the normal vector

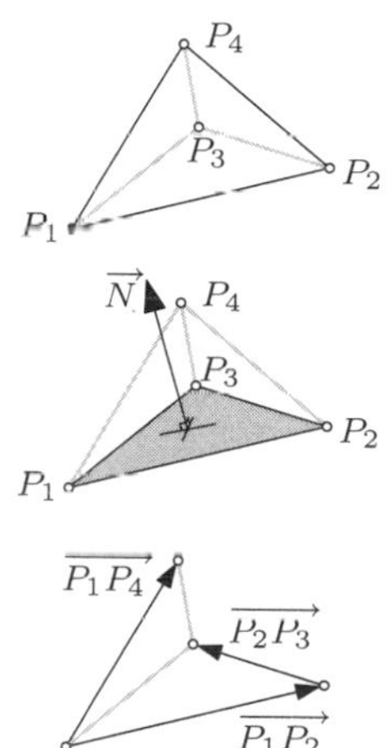

$$\overrightarrow{N} = \overrightarrow{P_1P_2} \times \overrightarrow{P_2P_3}.$$

We say then that a point on the same side as the normal vector lies on the plane's positive halfspace. To determine which side P_4 lies in, we compute the dot product

$$\overrightarrow{N} \cdot \overrightarrow{P_1P_4} = (\overrightarrow{P_1P_2} \times \overrightarrow{P_2P_3}) \cdot \overrightarrow{P_1P_4}$$

and that leads to the determinant:

$$\begin{vmatrix} x_2 - x_1 & x_3 - x_1 & x_4 - x_1 \\ y_2 - y_1 & y_3 - y_1 & y_4 - y_1 \\ z_2 - z_1 & z_3 - z_1 & z_4 - z_1 \end{vmatrix},$$

which is

$$\begin{vmatrix} x_1 & x_2 - x_1 & x_3 - x_1 & x_4 - x_1 \\ y_1 & y_2 - y_1 & y_3 - y_1 & y_4 - y_1 \\ z_1 & z_2 - z_1 & z_3 - z_1 & z_4 - z_1 \\ 1 & 0 & 0 & 0 \end{vmatrix},$$

resulting in the homogeneous form

$$\begin{vmatrix} x_1 & x_2 & x_3 & x_4 \\ y_1 & y_2 & y_3 & y_4 \\ z_1 & z_2 & z_3 & z_4 \\ 1 & 1 & 1 & 1 \end{vmatrix}. \tag{3.2}$$

Matrix Form of the Side-of-Sphere Predicate

A sphere with center (x_c, y_c, z_c) and radius r has the equation

$$(x - x_c)^2 + (y - y_c)^2 + (z - z_c)^2 = r^2,$$

which expands to

$$(x^2 + y^2 + z^2) - 2(xx_c + yy_c + zz_c) + (x_c^2 + y_c^2 + z_c^2 - r^2).$$

More generally,

$$A(x^2 + y^2 + z^2) + Bx + Cy + Dz + E = 0$$

is the equation of a sphere in space provided that $A \neq 0$. This equation can be written as the determinant:

$$\begin{vmatrix} x^2 + y^2 + z^2 & x & y & z & 1 \\ x_1^2 + y_1^2 + z_1^2 & x_1 & y_1 & z_1 & 1 \\ x_2^2 + y_2^2 + z_2^2 & x_2 & y_2 & z_2 & 1 \\ x_3^2 + y_3^2 + z_3^2 & x_3 & y_3 & z_3 & 1 \\ x_4^2 + y_4^2 + z_4^2 & x_4 & y_4 & z_4 & 1 \end{vmatrix} = 0, \tag{3.3}$$

where

$$A = \begin{vmatrix} x_1 & y_1 & z_1 & 1 \\ x_2 & y_2 & z_2 & 1 \\ x_3 & y_3 & z_3 & 1 \\ x_4 & y_4 & z_4 & 1 \end{vmatrix}, \qquad B = \begin{vmatrix} x_1^2 + y_1^2 & y_1 & z_1 & 1 \\ x_2^2 + y_2^2 & y_2 & z_2 & 1 \\ x_3^2 + y_3^2 & y_3 & z_3 & 1 \\ x_4^2 + y_4^2 & y_4 & z_4 & 1 \end{vmatrix}, \ldots.$$

As with a circle in the plane, we conclude that it is best to incur a small efficiency hit and spare clients from having to worry about the order with which they define the four points describing the sphere. That leads to the following equation:

$$\begin{vmatrix} x^2 + y^2 + z^2 & x & y & z & 1 \\ x_1^2 + y_1^2 + z_1^2 & x_1 & y_1 & z_1 & 1 \\ x_2^2 + y_2^2 + z_2^2 & x_2 & y_2 & z_2 & 1 \\ x_3^2 + y_3^2 + z_3^2 & x_3 & y_3 & z_3 & 1 \\ x_4^2 + y_4^2 + z_4^2 & x_4 & y_4 & z_4 & 1 \end{vmatrix} \times \begin{vmatrix} x_1 & y_1 & z_1 & 1 \\ x_2 & y_2 & z_2 & 1 \\ x_3 & y_3 & z_3 & 1 \\ x_4 & y_4 & z_4 & 1 \end{vmatrix}$$

$$= \text{side\_of\_sphere}(P, P_1, P_2, P_3, P_4) \begin{cases} < 0 & \text{inside,} \\ = 0 & \text{on the sphere boundary,} \\ > 0 & \text{outside.} \end{cases} \tag{3.4}$$

3.7 Dominant Axis

As in E^2 (§ 2.4), the *dominant* axis of a vector in E^3 is frequently needed in geometric computing. Determining whether a ray intersects a polygon is one classic example (§ 23.2).

If the vector (or direction) in question is the normal to a plane, then the dominant axis suggests the principal plane on which one could project orthogonally to yield the least distortion. Yet since for some applications we may need to distinguish between positive and negative dominant orientations, six enumerations are declared.

```
enum Dominant_E3 { E3_POSX, E3_NEGX, E3_POSY, E3_NEGY, E3_POSZ, E3_NEGZ };
```

Finding the axis (and orientation) in question requires a few simple comparisons.

```
Dominant_E3 dominant(double dx, double dy, double dz)
{
  const double zero = 0.0;
  double dxabs = dx >= zero ? dx : (−dx);
  double dyabs = dy >= zero ? dy : (−dy);
  double dzabs = dz >= zero ? dz : (−dz);

  if(dxabs >= dyabs && dxabs >= dzabs)
    return ((dx > zero) ? E3_POSX : E3_NEGX);
  else if(dyabs >= dzabs)
    return ((dy > zero) ? E3_POSY : E3_NEGY);
  else
    return ((dz > zero) ? E3_POSZ : E3_NEGZ);
}
```

Finally, a geometric library would provide (nonmember) functions that report the dominant axis.

```
Dominant_E3 dominant(const Vector_E3d& V)
{
  return dominant(V.x(), V.y(), V.z());
}
```

Occasionally, the *least dominant* axis is also needed (§ 9.3).

3.8 Exercises

3.1 Write a constructor for a Line_E3d class that takes two planes as input and determines two points on the line to initialize the member variables. Also write test cases that cover potential errors.

3.2 The C language (and, by extension, the C++ language) makes it possible to define an anonymous, or nameless, type.

```
struct { float x; float y; float z; } myPoint;
```

In the above expression the type remains unnamed—only the object is named. The use of anonymous *types* is generally undesirable, but *objects* can be nameless through the use of pointers. In the following code, A is a named Point_E3d object, whereas B is the name of a pointer. B points to a nameless, or anonymous, object. The use of a reference (&) creates an additional name for an object (regardless whether the object was previously named).

```
Point_E3d A = Point_E3d();
Point_E3d ∗ B = new Point_E3d();
```

Draw a sketch that illustrates the following code while showing the name(s) of each object or pointer.

```
Point_E3d p1 = Point_E3d();
Point_E3d & p2 = p1;
Point_E3d * p3 = &p1;
Point_E3d *p4 = new Point_E3d();
Point_E3d **p5 = &p4;
```

3.3 Implement a class Sphere_E3d that represents a sphere in E^3 and implement the side-of-sphere predicate.

3.4 Implement an intersection function between a Sphere_E3d object and a Ray_E3d object, which represent a sphere and a ray in E^3, respectively.

3.5 A useful, perhaps necessary, member function in a Vector_E3d is the unary negation operation. Yet simply writing

```
class Vector_E3d
{
    ...
    Vector_E3d
    operator-() { return Vector_E3d(-x,-y,-z); }
};
```

is flawed. The function must be defined as a **const** function. Give two (conceptually different) examples of client code that are reasonable, yet that would only work if unary negation is a **const** function.

3.6 This exercise is from Coxeter's *Introduction to Geometry* [27].

Show that the following statement holds.

> *The plane through three given points* $(x_i, y_i, z_i)(i = 1, 2, 3)$ *is*
>
> $$\begin{vmatrix} x_1 & y_1 & z_1 & 1 \\ x_2 & y_2 & z_2 & 1 \\ x_3 & y_3 & z_3 & 1 \\ x & y & z & 1 \end{vmatrix} = 0.$$
>
> *If the requirement of passing through a point is replaced (in one or two cases) by the requirement of being parallel to a line with direction* (X_i, Y_i, Z_i)*, the corresponding row of the determinant is replaced by* $X_i, Y_i, Z_i, 0$.

4 Affine Transformations

What are the properties of applying linear functions on the Cartesian coordinates of a point? The resulting *affine transformations* can be conveniently expressed in matrix form and can be classified according to the type of transformation they produce.

Affine transformations are canonical in visual computing. Viewport mapping and orthogonal view transformations are needed in computer graphics and the reverse problem, finding the mapping given matched sets of points, is needed in computer vision and in computer animation. An important degenerate transformation that is used for generating fake shadows is discussed in Exercise 12.6, where it is cast as an instance of the more general set of projective transformations [13].

4.1 Affine Transformations in 2D

Transforming Points

Consider the transformations defined by linear functions in x and y that can be applied on a given point $P(x, y)$ in the plane. All linear functions T can be represented using the two equations

$$x' = ax + by + e,$$
$$y' = cx + dy + f$$

We say that a point $Q(x', y')$ is the *image* of P under the transformation T and write $Q = T(P)$. For convenience, we can write the two equations in matrix form as

$$\begin{bmatrix} x' \\ y' \end{bmatrix} = \begin{bmatrix} a & b \\ c & d \end{bmatrix} \begin{bmatrix} x \\ y \end{bmatrix} + \begin{bmatrix} e \\ f \end{bmatrix}$$

or as $Q = MP + \overrightarrow{v}$, where

$$M = \begin{bmatrix} a & b \\ c & d \end{bmatrix}, \qquad \overrightarrow{v} = \begin{bmatrix} e \\ f \end{bmatrix}.$$

We first confirm that our notation is sensible. We define the product of the matrix M and the point P, MP, as a point and, as seen in Chapter 1, the addition of a point and a vector, $MP + \overrightarrow{v}$, results in a point that is geometrically the translation of the point by the magnitude and orientation of the vector.

Observe that we have a choice between premultiplying

$$P' = MP$$

and postmultiplying

$$P' = PM$$

a point P to yield the transformed point P'. The second would appear to be more intuitive since concatenating transformations can be performed by appending transformations, yet the former offers the advantage of coinciding with function composition and is the one used in this text.

Types of Affine Transformations

Translation

A translation in the plane can be achieved by setting the transformation matrix M to the identity matrix

$$I = \begin{bmatrix} 1 & 0 \\ 0 & 1 \end{bmatrix}$$

and setting $\overrightarrow{v}$ to the desired translation.

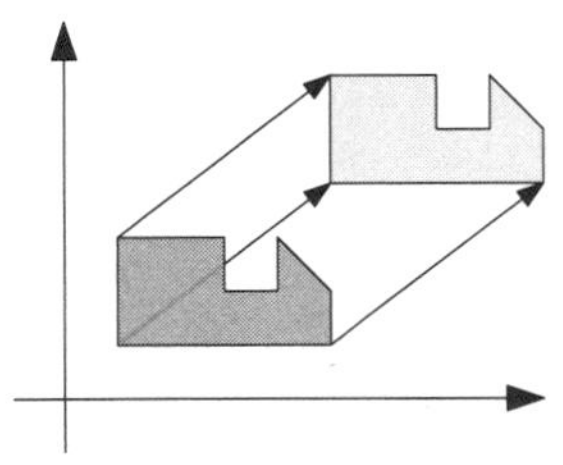

Despite its simplicity, it is useful to talk about several properties of translation. A translation preserves angles between lines and so it also preserves parallelism—two parallel lines remain parallel after translation. And because translations also preserve distances, areas are preserved. We say that translations are a rigid-body transformation (§ 4.2) because an animation of a figure under translation would generate a faithful simulation of the body animated with no deformation.

To animate a figure under translation, it suffices to scale the translation vector $\overrightarrow{v}$ by $\alpha \in [0..1]$. A translation can be inverted using a vector $-\overrightarrow{v}$.

Scale

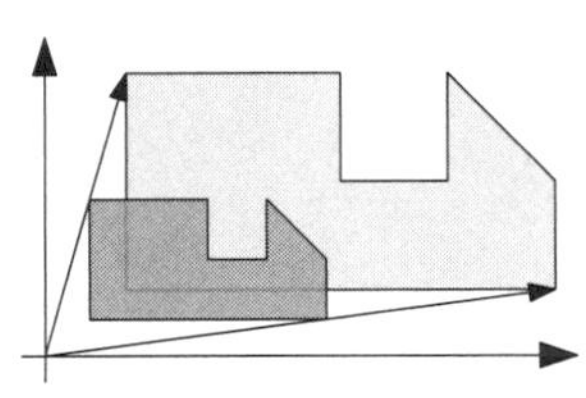

Uniformly scaling a figure can be achieved by applying a factor k to the identity matrix I. Clearly, uniform scaling preserves angles and parallelism, but does not preserve areas. If a matrix

$$\begin{bmatrix} k & 0 \\ 0 & k \end{bmatrix}$$

is used to scale a closed figure, the area of the figure after scale will be k^2 times its area before scaling.

Nonuniformly scaling a figure by k_x in the x-direction and by k_y in the y-direction can be performed by multiplication by the matrix

$$\begin{bmatrix} k_x & 0 \\ 0 & k_y \end{bmatrix}.$$

The ratio of areas after scaling to the areas before scaling is $k_x k_y$.

Rotation

It is simplest to consider first rotation about the origin. To find the coordinates of a point $P(x, y)$ after rotation by an angle ϕ, we express P in polar coordinates $(r\cos\theta, r\sin\theta)$ and write the coordinates of the rotated point $P'(x', y')$ as $(r\cos(\theta+\phi), r\sin(\theta+\phi))$.

Because

$$x' = r\cos(\theta + \phi) = r(\cos\theta\cos\phi - \sin\theta\sin\phi),$$
$$y' = r\sin(\theta + \phi) = r(\cos\theta\sin\phi + \sin\theta\cos\phi),$$

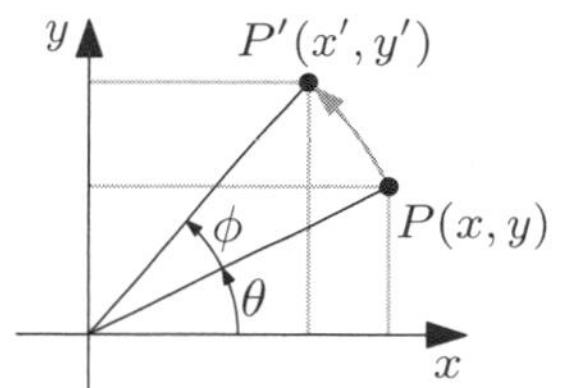

a rotation by an angle ϕ can be expressed in matrix form as

$$\begin{bmatrix} x' \\ y' \end{bmatrix} = \begin{bmatrix} \cos\phi & -\sin\phi \\ \sin\phi & \cos\phi \end{bmatrix} \begin{bmatrix} r\cos\theta \\ r\sin\theta \end{bmatrix} = \begin{bmatrix} \cos\phi & -\sin\phi \\ \sin\phi & \cos\phi \end{bmatrix} \begin{bmatrix} x \\ y \end{bmatrix}.$$

From that we conclude that the determinant of a transformation matrix that represents rotation is 1:

$$|M| = \begin{vmatrix} \cos\phi & -\sin\phi \\ \sin\phi & \cos\phi \end{vmatrix} = 1.$$

The formula for a rotation arises sufficiently often that it is worthwhile recalling it. What is being rotated in this formulation is a point—not the axes. If the latter are to be (also counterclockwise) rotated, the transpose of the matrix should be used instead.

Transforming Vectors

It is reasonable to define rotation and scale transformations on vectors in the same way they are defined for points, but what is the result of applying a translation on a vector? If vectors are used to model forces, accelerations, or even translations themselves, we would not want either the magnitude or the orientation of the vector to be modified by a translation.

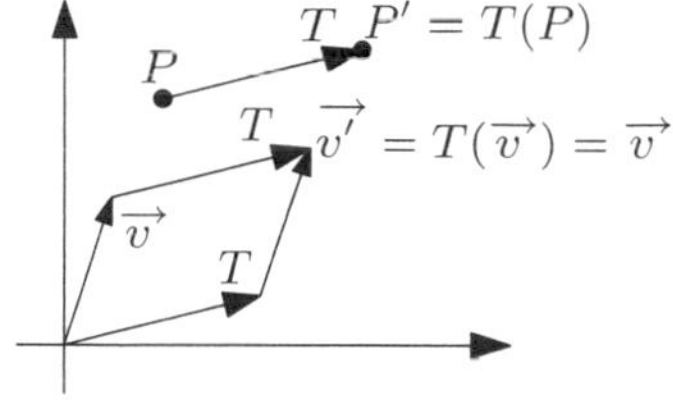

When we observe that the result of applying a linear transformation on a point differs from that of applying it on a vector, another advantage for having separate types for points and vectors (§ 1.2) becomes apparent: The separation ensures that transformations can be applied differently depending on the object transformed.

Applying scale and rotation transformations on vectors and points, and applying translations on points, we write

$$\overrightarrow{v'} = M\overrightarrow{v} \quad = \begin{bmatrix} a & b \\ c & d \end{bmatrix} \begin{bmatrix} v_x \\ v_y \end{bmatrix},$$
$$P' = MP + T = \begin{bmatrix} a & b \\ c & d \end{bmatrix} \begin{bmatrix} x \\ y \end{bmatrix} + \begin{bmatrix} t_x \\ t_y \end{bmatrix}.$$

Transforming Directions and Lines

After we observe in Chapter 16 the many advantages that homogeneous coordinates have for Euclidean geometry, it will become clear that transforming directions and lines in Euclidean geometry is but a special case of transforming lines in projective geometry, so we leave both out at this time and take the topic up again in § 12.6.

Notice that points and vectors transform similarly, but as suggested in § 1.3, directions do not transform as vectors. We view directions in E^2 as placeholders for an oriented line (the direction is orthogonal to the line) and directions in

E^3 as a placeholder for an oriented plane (the direction is the plane's normal vector). Directions should be seen as an incomplete object waiting to be associated with a line or a plane. Occasionally, we are truly only interested in the abstract notion of a direction (such as when used for shading; see Chapter 21). Regardless of whether the direction object will remain "incomplete" and used for shading or is used to create a line, a plane, or a set of lines or planes, directions transform as a line or a plane would have (described on page 133).

4.2 Properties of Affine Transformations

It is interesting to ask which properties of figures are preserved by a linear transformation. From the previous discussion, and as seen by applying various linear transformations to a triangle or a grid as shown in Figure 4.1, it is clear that linear transformations do not preserve lengths, angles, nor, consequently, areas. Illustrating the image of a uniform grid reveals the effect of linear transformations; parallelism is the only property preserved. As we will see in Part II, the distinguishing characteristic of such *affine transformations* is that they preserve parallelism and that they map affine points (those not at infinity) to other affine points. According to Coxeter, Blaschke credits Euler for having coined the term "affine" (German "affin") [27, p. 191].

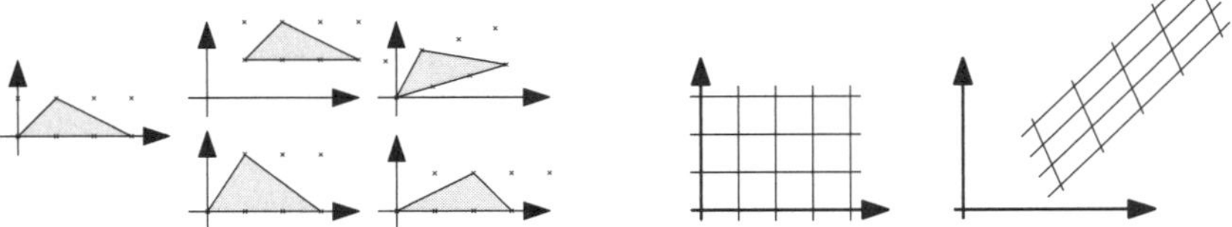

Figure 4.1
Affine transformations only preserve parallelism.

It is frequently useful to move a solid object without deforming it, or without changing the distances between its points or the angles between its lines. The subset of linear transformations that preserves rigidity is the set of *rigid-body transformations*. Only rotations and translations are rigid-body transformations. The characterization needed for a linear transformation to be rigid is for the condition $ad - bc = 1$ to be satisfied, or for $|M| = 1$. This suggests the following types of affine transformations.

Transformation	Rigid-body transformation	Affine transformation
Preserves	parallelism distances and angles	parallelism
Types	rotation translation	rotation translation scale shear

The reader who was not convinced in § 1.2 that points cannot be added, but that they can be subtracted (yielding a vector), may be convinced by a different argument based on transformations. The sanity of any transformation we apply means that applying the transformation before and after a geometric operation ought to yield the same result. But consider what would happen if adding two points is allowed [16]. The expression $T(P_1 + P_2)$ means that the coordinates would be added and that the translation component of the transformation would appear once in the outcome. If we instead write $T(P_1) + T(P_2)$, then both points will be transformed and *two* translations would appear in the outcome.

That also suggests why subtracting points should be legal. The two translations would cancel out, which is exactly the effect we desire for a transformation on a vector (the result of subtracting two points); the translation component does not affect the vector.

4.3 Composition of Affine Transformations

The composition of two affine transformations T_1 and T_2 may be performed in two steps by evaluating $T = T_2(T_1(P))$, but if the combined transformation is to be applied frequently, it is advantageous to capture it in an object $T = T_2 \circ T_1$. The composition can be determined by combining the linear functions

$$\begin{bmatrix} x' \\ y' \end{bmatrix} = \begin{bmatrix} a_1 & b_1 \\ c_1 & d_1 \end{bmatrix} \begin{bmatrix} x \\ y \end{bmatrix} + \begin{bmatrix} e_1 \\ f_1 \end{bmatrix}$$

and

$$\begin{bmatrix} x'' \\ y'' \end{bmatrix} = \begin{bmatrix} a_2 & b_2 \\ c_2 & d_2 \end{bmatrix} \begin{bmatrix} x' \\ y' \end{bmatrix} + \begin{bmatrix} e_2 \\ f_2 \end{bmatrix},$$

leading to

$$\begin{bmatrix} x'' \\ y'' \end{bmatrix} = \begin{bmatrix} a_2 & b_2 \\ c_2 & d_2 \end{bmatrix} \begin{bmatrix} a_1 & b_1 \\ c_1 & d_1 \end{bmatrix} \begin{bmatrix} x \\ y \end{bmatrix} + \begin{bmatrix} a_2 & b_2 \\ c_2 & d_2 \end{bmatrix} \begin{bmatrix} e_1 \\ f_1 \end{bmatrix} + \begin{bmatrix} e_2 \\ f_2 \end{bmatrix},$$

which reveals that the first translation combines with the second 2×2 matrix into a translation component.

Likewise, the composition of two affine transformations in Euclidean space is

$$\begin{bmatrix} x'' \\ y'' \\ z'' \end{bmatrix} = \begin{bmatrix} a_2 & b_2 & c_2 \\ d_2 & e_2 & f_2 \\ g_2 & h_2 & i_2 \end{bmatrix} \begin{bmatrix} a_1 & b_1 & c_1 \\ d_1 & e_1 & f_1 \\ g_1 & h_1 & i_1 \end{bmatrix} \begin{bmatrix} x \\ y \\ z \end{bmatrix} + \begin{bmatrix} a_2 & b_2 & c_2 \\ d_2 & e_2 & f_2 \\ g_2 & h_2 & i_2 \end{bmatrix} \begin{bmatrix} j_1 \\ k_1 \\ l_1 \end{bmatrix} + \begin{bmatrix} j_2 \\ k_2 \\ l_2 \end{bmatrix}.$$

After hand-programming this expression, we would be tempted to use two-dimensional arrays and three nested loops for matrix multiplication. But the tedium in coding the expression by hand will likely pay off. We can confirm that that is the case by comparing the number of cycles needed for a direct memory access with those needed for an indirect access, as would be needed if indexing is used.

4.4 Affine Transformation Objects

In the following possible design for a class for affine transformations in the plane, four matrix elements and a vector are stored individually. Whether storing a vector is a wise choice (as opposed to an additional two **double** variables) depends on the compiler; the designer should ensure that no penalty is caused by the compiler used. Using four variables is almost surely preferable over using an array of four variables **double** m[4] on any compiler, however, unless the designer can confirm that an optimizing compiler will replace expressions such as m[0] with variables that do not use indirection (and therefore require two memory accesses). These issues would not matter if efficiency is not a main concern or if all applications of affine transformations will occur outside performance bottleneck loops.

Since we chose for our point and vector objects to be immutable (§ 2.6), the two transformation functions must return new objects and leave the original objects unchanged.

```
class Affine_transformation_E2d
{
  double m00, m01;
  double m10, m11;
  Vector_E2d translation;
public:
  Affine_transformation_E2d(
    const double& _m00 = 1.0,
    const double& _m01 = 0.0,
    const double& _m10 = 0.0,
    const double& _m11 = 1.0,
    const Vector_E2d& _translation = Vector_E2d(0.0,0.0))
  {
    m00 = _m00; m01 = _m01;
    m10 = _m10; m11 = _m11;
    translation = _translation;
  }

  Point_E2d transform(const Point_E2d& P)
  {
    double x, y;
    x = m00 * P.x() + m01 * P.y() + translation.x();
    y = m10 * P.x() + m11 * P.y() + translation.y();
    return Point_E2d(x,y);
  }
  Vector_E2d transform(const Vector_E2d& V)
  {
    double x, y;
    x = m00 * V.x() + m01 * V.y();
    y = m10 * V.x() + m11 * V.y();
    return Vector_E2d(x,y);
  }
};
```

To spare clients of our code the trouble of hand-constructing the entries, convenience constructors for various types of transformations are needed. We could contemplate determining at run time which transformation is requested and write the following code:

```
enum Affine_transformation_E2d_types {
  IDENTITY, SCALE, TRANSLATION, ROTATION
};
```

```
Affine_transformation_E2d::Affine_transformation_E2d(
                                    Affine_transformation_E2d_types type,
                                    const T& x, const T& y)
{
  if(type == TRANSLATION) {
    m00 = 1.0; m01 = 0.0;
    m10 = 0.0; m11 = 1.0;
    translation = Vector_E2d(x,y);
  }
  else if(type == SCALE) {
    m00 = x;  m01 = 0.0;
    m10 = 0.0; m11 = y;
    translation = Vector_E2d(0.0,0.0);
  }
}
```

But it is a simple matter to move the choice of the initialization code used to compile time. We borrow an idea from the CGAL project [23] (see Chapter 18) and use four global and vacuous objects to determine through overloading the type of the affine transformation intended.

```
class Identity    {};
class Scale       {};
class Rotation    {};
class Translation {};
Identity    IDENTITY;
Scale       SCALE;
Rotation    ROTATION;
Translation TRANSLATION;
class Affine_transformation_E2d
{
  double m00, m01;
  double m10, m11;
  Vector_E2d translation;
public:
  void set_to_identity()
  {
    m00 = 1.0; m01 = 0.0;
    m10 = 0.0; m11 = 1.0;
  }
  Affine_transformation_E2d(
    const Identity& t)
  {
    set_to_identity();
  }
  Affine_transformation_E2d(
    const Scale& s,
    const double& xscale,
    const double& yscale)
  {
    m00 = xscale; m01 = 0.0;
    m10 = 0.0; m11 = yscale;
    translation = Vector_E2d(0,0);
  }
  Affine_transformation_E2d(
    const Translation& t,
    const Vector_E2d& _translation)
  {
    set_to_identity();
    translation = _translation;
  }
  Affine_transformation_E2d(
    const Rotation& r,
    double angle)
  {
    double c = cos(angle);
    double s = sin(angle);
    m00 = c; m01 = -s;
    m10 = s; m11 =  c;
    translation = Vector_E2d(0,0);
  }
  ...
};
```

4.5 Viewport Mapping

The next structure we consider has been termed an *axis-parallel rectangular region*, a *viewport*, a *bounding box*, and an *Axis-Aligned Bounding Box*. All refer to a rectangle with sides parallel to the major axes, an often-needed structure in geometric and graphical applications. The term *viewport* is used when referring to some interesting subset of the plane after an orthogonal projection in 3D (see § 4.6), but the term is used by extension for applications operating exclusively in the plane.

The problem is to map a region in an xy-plane bounded by the coordinates $(x_{\min}, x_{\max}, y_{\min}, y_{\max})$ to another region $(x'_{\min}, x'_{\max}, y'_{\min}, y'_{\max})$. Aside from the difficulty one would have in remembering the order of these parameters to a function ("is it not $(x_{\min}, y_{\min}, x_{\max}, y_{\max})$?"), there is another reason—coordinate freedom, discussed in Chapter 17—why a superior approach is to define a viewport using two points. A possible implementation follows:

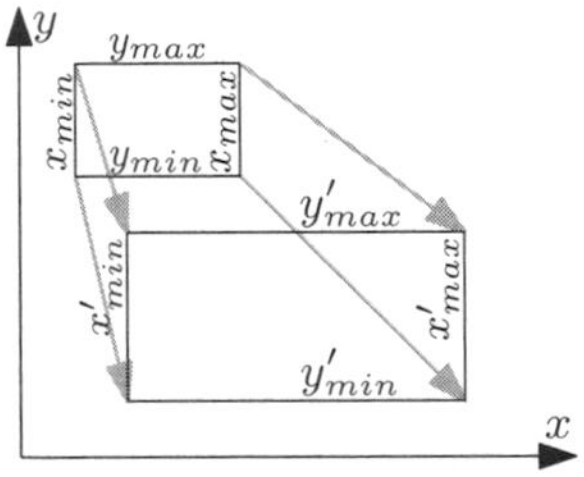

```
class Bbox_E2d
{
  Point_E2d _LL; // lower left corner
  Point_E2d _UR; // upper right corner
public:
  Bbox_E2d() : _LL(), _UR() {}
  Bbox_E2d(const Point_E2d& p) : _LL(p), _UR(p) {}
  Bbox_E2d(const Point_E2d& pLL, const Point_E2d& pUR)
    : _LL(pLL), _UR(pUR) {}
  T get_width()  const { return _UR.x() − _LL.x(); }
  T get_height() const { return _UR.y() − _LL.y(); }
  Point_E2d LL() const { return _LL; }
  Point_E2d UR() const { return _UR; }
  Point_E2d center() const { return Point_E2d(
                              (_LL.x() + _UR.x())/2.0,
                              (_LL.y() + _UR.y())/2.0); }
};
```


Aside from implementation details, the affine transformation in viewport mapping is a combination of scale and translation transformations. The mapping could be either derived directly

$$x' = x'_{\min} + \frac{x - x_{\min}}{x_{\max} - x_{\min}}(x'_{\max} - x'_{\min})$$
$$y' = y'_{\min} + \frac{y - y_{\min}}{y_{\max} - y_{\min}}(y'_{\max} - y'_{\min}),$$

or an affine transformation object could be constructed using a combination of a translation to the origin, a scale, and a translation from the origin.

A class for bounding boxes in Euclidean space could also be similarly defined.

```
class Bbox_E3d
{
  Point_E3d _LL;
  Point_E3d _UR;
public:
  ...
};
```

4.6 Orthogonal Matrices

Consider forming a transformation matrix using two orthogonal unit vectors $\overrightarrow{v_1}$ and $\overrightarrow{v_2}$. We saw that there are several properties to the two vectors.

The inner product of a vector of unit magnitude by itself is unity.

$$x_1x_1 + y_1y_1 = 1 \implies \overrightarrow{v_1} \cdot \overrightarrow{v_1} = 1; \quad |\overrightarrow{v_1}| = 1,$$
$$x_2x_2 + y_2y_2 = 1 \implies \overrightarrow{v_2} \cdot \overrightarrow{v_2} = 1; \quad |\overrightarrow{v_2}| = 1.$$

Also, since $\overrightarrow{v_1}$ and $\overrightarrow{v_2}$ are orthogonal,

$$x_1x_2 + y_1y_2 = 0 \implies \overrightarrow{v_1} \cdot \overrightarrow{v_2} = 0.$$

These equations make it possible to find the product of the transformation matrix M with its transpose M^T:

$$M^TM = \begin{bmatrix} x_1 & y_1 \\ x_2 & y_2 \end{bmatrix} \begin{bmatrix} x_1 & x_2 \\ y_1 & y_2 \end{bmatrix} = \begin{bmatrix} 1 & 0 \\ 0 & 1 \end{bmatrix}. \tag{4.1}$$

Proper and Improper Orthogonality

The previous equations remain valid even if the two vectors are swapped, but the determinant will be of opposite sign. To find the sign (and value) of the determinant, we write the two vectors in polar form:

$$\overrightarrow{v_1} = \begin{bmatrix} \cos\phi \\ \sin\phi \end{bmatrix}, \qquad \overrightarrow{v_2} = \begin{bmatrix} \cos(\phi + \pi/2) \\ \sin(\phi + \pi/2) \end{bmatrix}.$$

Replacing the identities $\cos(\phi + \pi/2) = \ \sin\phi$ and $\sin(\phi + \pi/2) = \cos\phi$, we can rewrite the matrix

$$M = \begin{bmatrix} \cos\phi & -\sin\phi \\ \sin\phi & \cos\phi \end{bmatrix}, \tag{4.2}$$

and it now becomes clear that $|M| = 1$. An orthogonal matrix satisfying the condition $|M| = 1$ is called a *proper orthogonal matrix*. If the two vectors are swapped, or conversely, if the two angles representing the two vectors are ϕ and $\phi - \pi/2$—in that order—the matrix is termed an *improper orthogonal matrix* and its determinant $|M| = -1$.

As is clear from Eq. (4.1), the inverse of an orthogonal matrix is particularly easy to compute:

$$M^{-1} = M^T.$$

Because transposing a matrix does not change its determinant, it is also clear that

$$|M| = 1 \implies |M^T| = 1 \implies |M^{-1}| = 1$$

and

$$|M| = -1 \implies |M^T| = -1 \implies |M^{-1}| = -1.$$

Relating Eq. (4.2) to the discussion in § 4.1, we also conclude that if a proper orthogonal matrix is used as a transformation matrix, the resulting transformation is a rotation about the origin.

Because M satisfies multiple constraints, it is sufficient to know one of the two vectors to determine the second and hence the matrix. If, say, $\overrightarrow{v_1}$ is known, $\overrightarrow{v_2}$ can be determined since its angle in polar coordinates will be $\tan^{-1} y_1/x_1 + \pi/2$. In practice, and as we will see in Chapter 10, there is no need to invoke trigonometric functions.

4.7 Orthogonal Transformations

Orthogonality in Three Dimensions

If three unit orthogonal vectors $\overrightarrow{v_1}$, $\overrightarrow{v_2}$, and $\overrightarrow{v_3}$ are used to define a 3×3 matrix M, then $M^T M = I$. In analogy with the two-dimensional case, that each vector is of unit magnitude results in the unit values along the diagonal and that each pair of vectors are orthogonal results in the zeros outside the diagonal.

Here also $|M|$ remains either ± 1. If $|M| = +1$, M is termed *proper orthogonal* and if $|M| = -1$, it is termed *improper orthogonal*.

If only two of the three basis vectors are known, it is possible to uniquely determine the third. The constraint that the determinant of the matrix be unity makes it possible to deduce the components of the third vector:

$$\begin{bmatrix} x_1 & x_2 & x_3 \\ y_1 & y_2 & y_3 \\ z_1 & z_2 & z_3 \end{bmatrix}.$$

Writing

$$\begin{aligned} x_3 &= y_1 z_2 - z_1 y_2, \\ y_3 &= -x_1 z_2 + z_1 x_2, \\ z_3 &= x_1 y_2 - y_1 x_2 \end{aligned}$$

results in a vector $\overrightarrow{v_3}$ that is orthogonal to both $\overrightarrow{v_1}$ and $\overrightarrow{v_2}$, which can be more concisely written as the cross product $\overrightarrow{v_3} = \overrightarrow{v_1} \times \overrightarrow{v_2}$ (§ 2.2). Note that the other orthogonal vector, $-\overrightarrow{v_3} = \overrightarrow{v_2} \times \overrightarrow{v_1}$, yields an improper orthogonal matrix.

Orthogonal View Transformation

Consider a virtual eye situated at $E(e_x, e_y, e_z)$. The eye models an observer or a virtual camera and we wish to render, or draw, a representation of three-dimensional objects on a two-dimensional surface that would be a faithful rendition of what would be captured by the virtual eye. This problem has long been considered by artists, but here we do not study the general problem, which involves perspective (see Chapter 11), but only *orthographic projection*. Our objective is to define a plane, called the *view plane* or the *image plane*, in 3D and to project points on it. To find the projection of each point, a perpendicular line is erected from the point to the image plane and the point of intersection is the projection of the point.

The orientation of the view plane is defined by the *view direction* $\overrightarrow{d}$, a unit vector orthogonal to the image plane. Our objective is to find a transformation

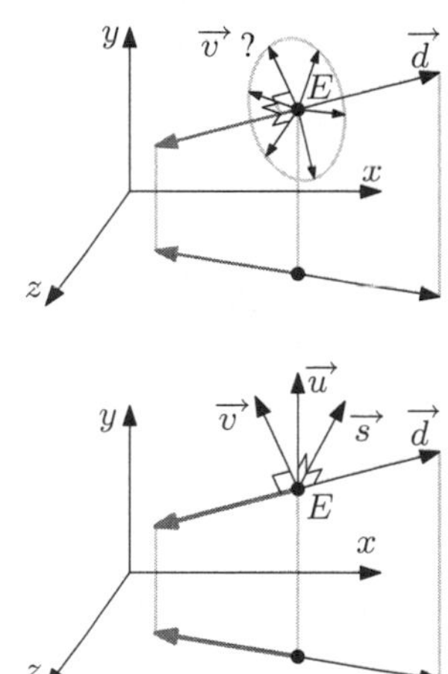

that maps E to the origin and that maps $\vec{d}$ to the *negative* z-axis. The image plane is then parallel to the $z = 0$ plane and the orthographic projection of a point can be found by simply ignoring its z-coordinate after the transformation.

But the two constraints E and $\vec{d}$ do not define a unique rigid-body transformation matrix since the orientation of the virtual camera could be rotated along the view axis without changing either E or $\vec{d}$.

A routine for computing the orthogonal view transformation could make it simpler for a user by not expecting a value for the *side vector* $\vec{s}$, but by using by default a convention that a photographer may use naturally. If we assume that gravity is directed along $(0, -1, 0)$, we can define the *up direction* $\vec{u}(0, 1, 0)$.

The triple of vectors $(\vec{s}, \vec{v}, \vec{d})$ defines an orthogonal transformation matrix. If $\vec{v}$ is constrained to lie in the plane carrying both $\vec{d}$ and $\vec{u}$, $\vec{s}$ can be found by computing $\vec{s} = \vec{u} \times \vec{d}$ and in turn $\vec{v} = \vec{d} \times \vec{s}$.

If the triple of vectors $(\vec{s}, \vec{v}, \vec{d})$ is used as the columns of an orthogonal transformation

$$\begin{bmatrix} s_x & v_x & d_x \\ s_y & v_y & d_y \\ s_z & v_z & d_z \end{bmatrix},$$

the transformation would map the basis axes to the view frame. Since we wish to apply the inverse of this map, we use instead the matrix transpose.

$$V = \begin{bmatrix} s_x & s_y & s_z \\ v_x & v_y & v_z \\ d_x & d_y & d_z \end{bmatrix}.$$

A translation from E to the origin followed by the transformation effected by V yields the desired orthogonal view transformation.

Axonometric View Transformation

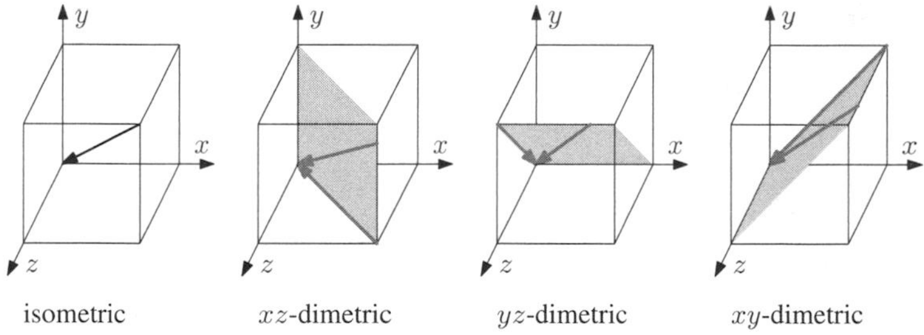

Figure 4.2
Types of axonometric projections

The field of *engineering drawing* has an established graphics tradition that precedes the development of computing. A drafting engineer wishing to communicate to a field engineer can facilitate the task of the latter by using an *orthographic projection*, where the view direction is parallel to one of the main axes, or by using an *axonometric projection*, which is also an orthogonal projection. Axonometric projections are classified according the angle of the projectors as shown in Figure 4.2. If the angles separating the view direction from the three axes are equal, the projection is said to be *isometric*—knowing one scale suffices to recover lengths *parallel to the main axes*. If two scales are needed, the projection is said to be *dimetric* and if three are needed, the projection is said to be *trimetric*.

4.8 Euler Angles and Rotation in Space

One way of applying rotations in 3D is to perform successive rotations around one of the basis axes, referred to as rotation by Euler angles. The matrix needed in each case is akin to performing a rotation in 2D [Eq. (4.2)]—with one of the coordinates remaining invariant, which results in the following matrices:

$$R_x = \begin{bmatrix} 1 & 0 & 0 \\ 0 & \cos\phi_x & -\sin\phi_x \\ 0 & \sin\phi_x & \cos\phi_x \end{bmatrix},$$

$$R_y = \begin{bmatrix} \cos\phi_y & 0 & \sin\phi_y \\ 0 & 1 & 0 \\ -\sin\phi_y & 0 & \cos\phi_y \end{bmatrix},$$

$$R_z = \begin{bmatrix} \cos\phi_z & -\sin\phi_z & 0 \\ \sin\phi_z & \cos\phi_z & 0 \\ 0 & 0 & 1 \end{bmatrix}.$$

Unlike rotations in the plane, rotations in 3D are not commutative. Figure 4.3 shows that the results obtained from reversing the order of two rotations are, in general, different. Hence, three rotations must also be applied in some given order, but there is no canonical order for applying the three rotation matrices. If Euler angles are to be used, a choice must be made for such an order. This in turn makes it difficult to use Euler angles because each rotation is applied on the transformed object, which is unintuitive. Worse, the first two angles of rotation can be chosen (to a positive or a negative quarter turn) such that the third rotation effects in fact a rotation around one of the two previous axes. In the context of gyroscopes losing one degree of freedom, this phenomenon has acquired the name *gimbal lock* [46], a term also used in geometric computing regardless of the application.

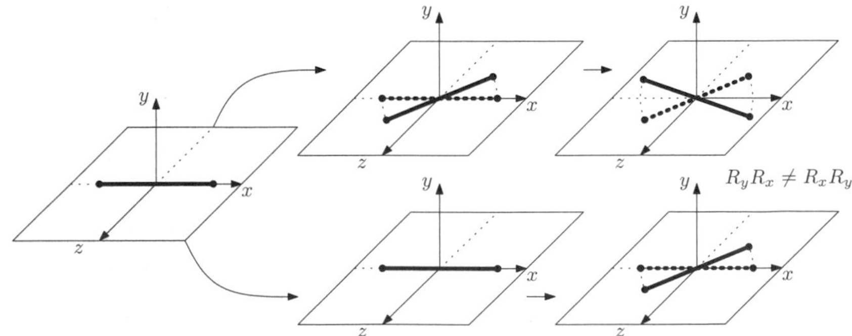

Figure 4.3
Rotation in space is not commutative.

The code labeled "euler-angles" accompanying Exercise 4.6 provides a user interface for appreciating the limitations of using Euler angles in practice. Chapter 10 pursues the natural object for performing rotations in space, the quaternion.

4.9 Rank of a Matrix

Rank of Mappings in the Euclidean Line

If a transformation on the Euclidean line is described by $x' = ax + b$, then the only way for the 1×1 matrix $M = [a]$ to be singular is if $a = 0$. The determinant of M is then zero and the mapping transforms all points on the Euclidean line to the point with coordinate b. It is interesting to consider this trivial case before looking at the plane and in space because there are degrees of singularity of a matrix M, with different resulting transformations.

Rank of Mappings in the Euclidean Plane

A matrix $M = \begin{bmatrix} a & b \\ c & d \end{bmatrix}$ is invertible if and only if the two vectors $\begin{bmatrix} a & c \end{bmatrix}^T$ and $\begin{bmatrix} b & d \end{bmatrix}^T$ are linearly independent. If the two vectors are linearly independent, the rank of M is 2, the determinant is not zero, and M is invertible. If the two vectors are not linearly independent, but are not the zero vector, the rank of M is 1, the determinant is zero, and M is not invertible. In that case the transformation defined by M collapses all points in the plane to a line. Since more than one point, in fact a line or a linear subspace of the plane, maps to a single point, the mapping is no longer one-to-one and it is not possible to find a unique inverse for M.

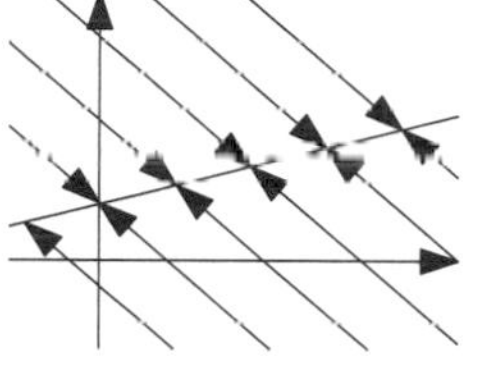

A square matrix M (of arbitrary order) can have rank 0 only if all entries are 0. In that case all points map to a single point, the one determined by the translation part of the affine transformation.

Rank of Mappings in Euclidean Space

The different cases for transformations in 3D should now be clear. An invertible 3×3 matrix M has rank 3. If the rank of M is 2, all points map to one plane in space. An even "more singular" 3×3 matrix M has rank 1. In that case all points map to one line in space.

4.10 Finding the Affine Mapping Given the Points

We have so far considered that we have a set of points along with an affine transformation and looked at how the transformed points can be derived. Suppose that we have instead both the set of points and their image, but wish to find the affine mapping. Since the mapping is linear, the problem reduces to solving a set of linear equations. We consider the cases of finding the mapping in one, two, and three dimensions.

Affine Mappings in the Euclidean Line

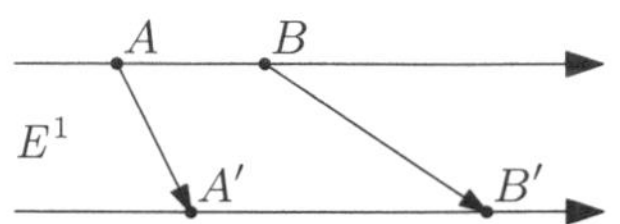

We start by considering the simple case of an affine mapping on the Euclidean line. Since such a mapping is defined by two scalars a and b, $x' = ax + b$, two points and their images are needed to find the mapping. The scalar b acts as a translation and a as a scale. If $a > 0$, then the order of points is preserved after the transformation. If $a = 0$, all points in the domain map to one point in the range and the mapping is not invertible.

If $a \neq 0$, we say that the mapping is one-to-one and onto. Each point in the domain maps to exactly one point and one point in the range also has one pre-image point.

Affine Mappings in the Euclidean Plane

We can reason by analogy in two dimensions. If three distinct points are given in the domain and their images are three distinct points, the affine mapping can be found by solving two sets of three equations in three unknowns:

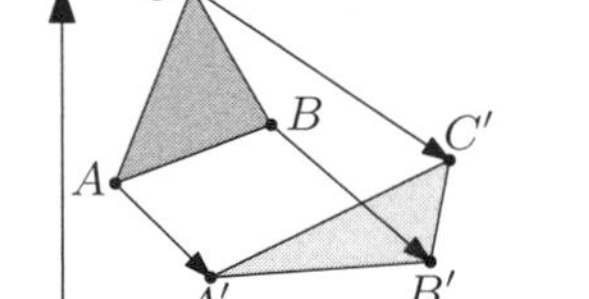

$$x' = ax + by + e,$$
$$y' = cx + dy + f.$$

But in 2D it is no longer sufficient for the three points in either the domain or the range to be distinct; they must also not be colinear. If both are, then not enough information is provided about the points not lying on either line. If one triple of points is colinear, but the other is not, the mapping can be found, but if the mapping (or its inverse) collapses three noncolinear points onto a line, then it must also be the case that it collapses *all* points onto the same line; the transformation matrix is singular.

Affine Mappings in Euclidean Space

Four pairs of points need to be known to find a unique affine mapping in Euclidean space. The four points (in either the source or the image) cannot be coplanar, which also implies that no three of the four points may be colinear and that no two points may be coincident. The numbers of points that need to be known for the affine mapping to be uniquely defined, one more point than the dimension, should be compared with the case in projective geometry, discussed in Chapter 11.

4.11 Exercises

4.1 Identify the transformation that results from each of the following affine transformation matrices.

(a) $M = \begin{bmatrix} 3/5 & -4/5 \\ 4/5 & 3/5 \end{bmatrix}$

(b) $M = \begin{bmatrix} 2 & 0.2 \\ 0 & 1 \end{bmatrix}, \quad T = \begin{bmatrix} -3 \\ 2 \end{bmatrix}$

4.2 The objective of this project is to gain an insight into the significance of the determinant of a 2×2 transformation matrix by manipulating its coefficients. The start-up code labeled "slide-determinant" accompanying this chapter allows you to modify one of the two column vectors of the transformation matrix (while leaving the other column vector constant at $[1, 0]^T$). Modify the source code to have instead four sliders that modify the value of the four matrix coefficients in the range $[0 \cdots 1]$. Also display the value of the determinant at each manipulation of one of the sliders. What conclusions do you draw regarding the connection between the areas of the square before and after transformation and the determinant? Refer to Appendices B and C.

4.3 *"If the determinant of a square matrix M is* 1*, then M is proper orthogonal."*

"If the determinant of a square matrix M is -1*, then M is improper orthogonal."*

For each of these two statements, mention whether the statement is correct and then argue why it is or give a counterexample.

4.4 Even though an order 2 proper orthogonal transformation matrix is defined by *two* vectors, the *first* of the two vectors is sufficient to determine the second. Write a brief code snippet with that objective.

4.5 The objective of this project is to study transformation matrices and the different animations obtained when using different ways of interpolating transformation matrices.

Modify the start-up code labeled "animation-in-2d" to animate the "small house" from one keyframe to another in the plane. Compare the result of animating the house by using the three following approaches:

- Linear interpolation of the house vertices
- Linear interpolation of the entries in the two transformation matrices
- Linear interpolation of the scale, rotate, and translate parameters

Ensure that the interpolation of the rotation is performed on the shorter of the two arcs on a circle. If you see a similarity between two methods, confirm you see why that is the case algebraically also.

4.6 Moving the three sliders in the start-up program labeled "euler-angles" manipulates an object using its Euler angles. Mentally thinking of an orientation in space and then attempting to reach it using the sliders will suggest how difficult it is to use Euler angles to orient objects.

Modify the start-up code so that *two* orientations are defined by two sets of Euler angles. Clicking a push button would thereafter animate the object from one orientation to the other using linear interpolation of the Euler angles. Ensure that for each of the three interpolations, the shortest path from one angle to the other is taken (so that, say, only 20 degrees are traversed if the two angles are 350 and 10).

The objective of implementing this project is to appreciate how unrealistic the motions generated by using Euler angles to animate an object are.

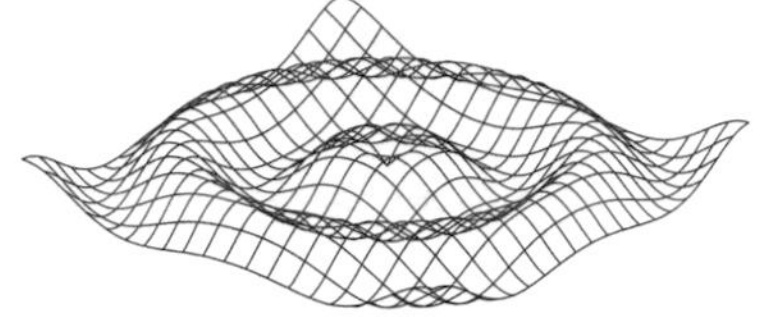

4.7 Write a program that generates an encapsulated postscript file (see Appendix A) consisting of a drawing of the surface $y = \sin(\sqrt{x^2 + z^2})$ by illustrating its intersection with planes perpendicular to the x- and to the z-axes. It should be possible to print the file you generate by directly sending it to a printer (i.e., without necessarily embedding it in another document).

4.8 Two planar polygons are affinely equivalent if one can be obtained from the other by an affine transformation. How would you determine whether

a. a given quadrilateral is affinely equivalent to a square?

b. a given hexagon is affinely equivalent to a regular hexagon? (A hexagon is regular if it has equal sides and equal internal angles—§ 26.3.)

c. a given triangle is affinely equivalent to an equilateral triangle?

5 Affine Intersections

We consider intersection problems between various types of geometric objects in the Euclidean plane. Variations on intersection problems in which one of the two objects is a rectangle have traditionally been called *clipping* problems. In these variations, the rectangle models a viewport and one wishes to discard the set of points lying outside the viewport. Because clipping problems frequently arise in the inner loops of geometric and graphical applications, it is worthwhile to examine minute improvements—even when there is no asymptotic change in run time.

5.1 Nomenclature

The Bbox_E2d and Bbox_E3d objects introduced in Chapter 4 capture only those that are axis-aligned, but a rectangular area of the plane and a parallelepiped in space do not necessarily have to be aligned with the coordinate axes.

Several terms have been used in the literature to refer to these two objects. *Viewport* is used when the use of a Bbox_E2d object as an abstraction for an output device is stressed. *Axis-aligned bounding box* (AABB) is used when its orientation is stressed. AABB is also used in contrast to *Object-oriented bounding box* (OOBB). OOBB is used when the smallest-sized rectangle or parallelepiped is used to replace another object. ("Object" here is used in the physical, not the programming language, sense). By reducing the complexity of the object considered, AABBs and OOBBs are useful in intersection computations: If a ray does not intersect the simpler object and that object bounds more complex objects, the latter do not need to be intersected with the ray. The term *axis-parallel rectangle*, in the obvious sense, is also used.

5.2 Finding Line and Segment Intersections

Line Intersection

In symmetry to the construction of a line from two points (§ 1.5), the intersection of two lines in the plane Line_E2d L1, L2 represented by the equations

$$a_1 x + b_1 y + c_1 = 0,$$
$$a_2 x + b_2 y + c_2 = 0$$

is given by

$$x = +\frac{\begin{vmatrix} b_1 & c_1 \\ b_2 & c_2 \end{vmatrix}}{\begin{vmatrix} a_1 & b_1 \\ a_2 & b_2 \end{vmatrix}}, \qquad y = -\frac{\begin{vmatrix} a_1 & c_1 \\ a_2 & c_2 \end{vmatrix}}{\begin{vmatrix} a_1 & b_1 \\ a_2 & b_2 \end{vmatrix}}.$$

Membership of Intersection Functions

The choice of class membership of the functions for point equality and for the orientation of three points, or the predicates in the Euclidean plane, was not discussed in Chapter 1 and indeed there will be no significant repercussions whether these functions are chosen as member functions in Point_E2d or as nonmember functions. Still, we ponder this issue along with the corresponding one for intersection functions. Should functions for computing intersections be member functions in Segment_E2d? Should they (also) be member functions in Line_E2d?

The repercussions from this decision will little affect the flavor of the library (nor would we regret very much making one choice or another). We still of course would like to choose the best of the following three design options.

1. The functions are member functions.
2. Predicate functions are static functions in some class Predicate and the intersection functions are static functions in another class Intersection.
3. The functions are nonmember functions.

Of these three options, the first one is perhaps the one we would be most prone to regret. To see why, consider that handling the intersection of a segment and a line, say, would lead to the following options.

a. Duplicate the intersection code in both classes.
b. Write the intersection function in only one of the two classes.
c. Make the intersection function a member of one class and write a member function in the other class that delegates to the first.

Option a makes the library a maintenance headache; option b puts the burden on the client to remember the class in which each function is defined, and option c deceives the client into thinking that the two functions are equally efficient. It is sensible then to choose not to define predicates and intersection functions as member functions.

Encapsulating intersection functions as static members would appear to be a cleaner design than simply using nonmember functions, yet the latter option does not lead to any difficulties. Overloading safely takes care of the multitude of intersection functions that will be defined (for both Euclidean and non-Euclidean geometries)—and hence that is the design chosen here.

Return Type of Intersection Functions and Segment Intersections

A more interesting design decision is encountered when writing a routine for the intersection of two segments. Choosing the type of the return object given a signature such as the following will impact both the efficiency as well as the flavor of the library.

```
class Intersection_E2d {
  static returnObject intersect(
                    const Segment_E2d& s1,
                    const Segment_E2d& s2);
  ...
};
```

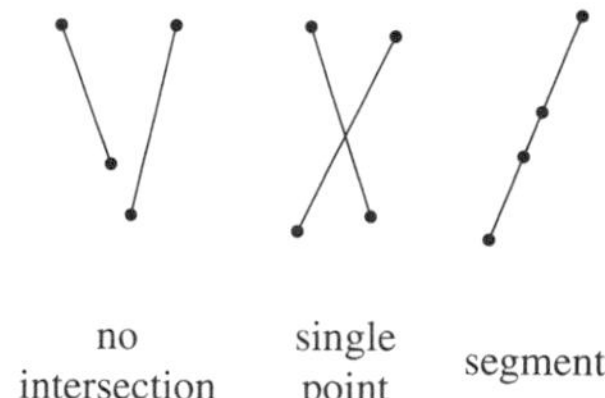

The two line segments s1 and s2 can be in one of three configurations. There may be no intersection; the two segments may intersect at a single point (possibly coinciding with an endpoint of one or both segments); or the two segments may lie on the same carrying line and have an overlap, in which case they would intersect at an infinite number of points: a segment.

Since the type of the object returned may vary, there are a few options for handling the return type:

- Return a flag identifying the object type.
- Introduce an abstract base class (called perhaps Object_E2d for an object in the plane) in the class hierarchy and derive the classes Null_E2d (for no intersection), Point_E2d, and Segment_E2d from Object_E2d. The intersection function would return a pointer to Object_E2d and the object pointed to would determine the result of the intersection.

The segment intersection routine is likely to be called often and so we have the following constraints:

1. The code should handle special cases (e.g., one of the two segments is vertical) as easily as possible.
2. The number of floating point operations should be minimized.

The two constraints are satisfied by determining the intersection using the following four turns:

```
TURN s2s = s1.turn(s2.source());
TURN s2t = s1.turn(s2.target());
TURN s1s = s2.turn(s1.source());
TURN s1t = s2.turn(s1.target());
```

If the four predicates suggest that the two segments intersect in their interior, the problem reduces to finding the point of intersection of two lines.

Intersection of Directions

Another intersection problem is sometimes needed. Consider that we have an observer O located at a vertex of a triangle and that we wish to determine

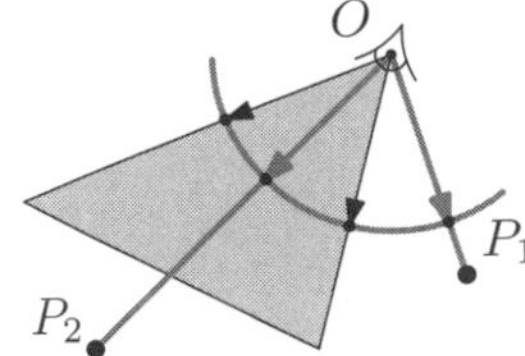

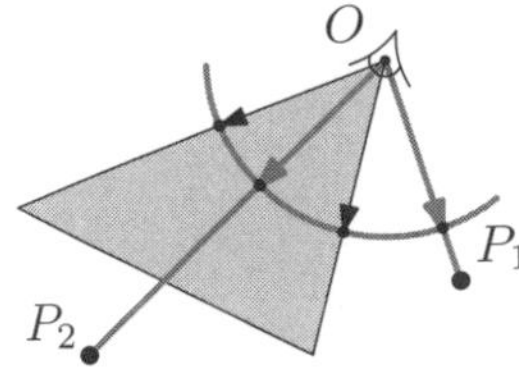

whether a point is visible by the observer. The subset of the plane bounded by the two edges adjacent to the vertex at O is defined as invisible—by imagining perhaps that the triangle models a wall or some such obstacle. To determine that a point P_1 is visible whereas a point P_2 is invisible, one needs to determine whether the directions defined by $\overrightarrow{OP_1}$ and $\overrightarrow{OP_2}$ lie *between* the two directions defined by the edges of the triangle.

Such computations in the space of directions rightly fall under the geometry of the circle—or the one-dimensional spherical geometry S^1 discussed in Chapter 8. The purpose from introducing it at this stage is to suggest that it would indeed be possible to solve such a question adequately in the Euclidean plane, but that a far better approach is to define spherical geometric objects, test that their behavior is as we expect, and then incorporate the predicates defined in that geometry in a system.

5.3 Clipping or Splitting a Segment

splitting line

clipping region

The distinction between splitting and clipping is the following. We say that a geometric object is being *clipped* if one or more portions are eventually discarded, whereas we say that it is being *split* if all fragments are returned by the function. One may, for instance, split by a line or clip by either a halfspace or a viewport. Much of what follows applies directly to 3D.

Halfspace Clipping of Segment

An oriented line partitions the plane into three sets of points: those on its left, those on its right, and the points lying on the line. These cases arise when the sign of a determinant is positive, negative, or zero, respectively (§ 2.2). We use the turn predicate and the enumeration Oriented_side mentioned in § 2.2.

The following function evaluates the side of the splitting line on which each endpoint of a segment lies. The intersection point is determined if the endpoints of the segment lie on the opposite side of the line. The returned segment maintains the same orientation as the input segment.

```
bool
positive_half_space_clip(
      const Line_E2d & splitting_line,
      Segment_E2d & my_segment)
// return the regularized portion of my_segment lying
// in the closed positive (left) halfspace of splitting_line
{
   Oriented_side source_side = oriented_side(
         splitting_line, my_segment.source() );
   Oriented_side target_side = oriented_side(
         splitting_line, my_segment.target() );
   if(
     source_side != ON_NEGATIVE_SIDE &&
     target_side != ON_NEGATIVE_SIDE)
        return true; // no clipping needed: segment is entirely inside
   else if(
     (source_side == ON_POSITIVE_SIDE &&
      target_side == ON_NEGATIVE_SIDE) ||
     (source_side == ON_NEGATIVE_SIDE &&
      target_side == ON_POSITIVE_SIDE))
```

```
    {
        Point_E2d intersectionPoint = intersection(splitting_line, my_segment);
        if(source_side == ON_POSITIVE_SIDE)
            my_segment = Segment_E2d(my_segment.source(), intersectionPoint);
        else if(target_side == ON_POSITIVE_SIDE)
            my_segment = Segment_E2d(intersectionPoint, my_segment.target());
        return true;
    }
    else
        // my_segment is ON_NEGATIVE_SIDE, possibly with (at most)
        // one endpoint ON_ORIENTED_BOUNDARY
        return false;
}
```

All possible positions of the segment endpoints with respect to the splitting line are straightforward, but the two input cases on the right of Figure 5.1 merit an explanation. We assume that we are only interested in a returned *segment*. If the intersection of the input segment with the halfplane results in a point, that point is discarded. If the segment coincides with the boundary of the halfplane, however, the segment is returned intact.

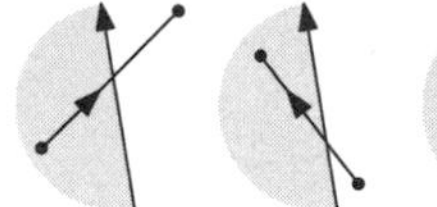

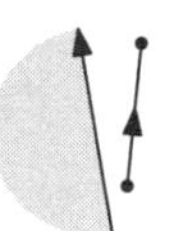
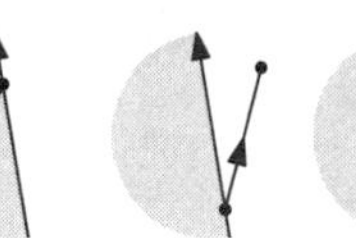

Figure 5.1
Halfspace clipping of a segment

Line Splitting of Segments

Even though segment splitting is a simple exercise, it arises with sufficient frequency to merit inspecting the source code. In the preceding case, the Boolean flag returned by positive_half_space_clip signals whether a fragment remains inside the clipping halfspace. The function split returns instead two flags positive_side_flag and negative_side_flag to give two such signals for either side of the splitting line.

```
void
split(
    const Segment_E2d & my_segment,
    const Line_E2d & splitting_line,
    bool & positive_side_flag, // is there a positive_side?
    Segment_E2d & positive_side,
    bool & negative_side_flag, // is there a negative_side?
    Segment_E2d & negative_side)
{
    Oriented_side source_side = oriented_side( splitting_line, my_segment.source() );
    Oriented_side target_side = oriented_side( splitting_line, my_segment.target() );
    if(
        source_side != ON_NEGATIVE_SIDE &&
        target_side != ON_NEGATIVE_SIDE)
    {
        positive_side_flag = true;
        positive_side = my_segment;
        negative_side_flag = false;
        // negative_side is not modified
        return;
    }
    else if(
            source_side != ON_POSITIVE_SIDE &&
```

```
        target_side != ON_POSITIVE_SIDE)
    {
        positive_side_flag = false;
        // positive_side is not modified
        negative_side_flag = true;
        negative_side = my_segment;
        return;
    }
    else if(
        source_side == ON_POSITIVE_SIDE &&
        target_side == ON_NEGATIVE_SIDE)
    {
        Point_E2d intersectionPoint = intersection(splitting_line, my_segment);
        positive_side_flag = true;
        positive_side = Segment_E2d(my_segment.source(), intersectionPoint);
        negative_side_flag = true;
        negative_side = Segment_E2d(intersectionPoint, my_segment.target());
    }
    else if(
        source_side == ON_NEGATIVE_SIDE &&
        target_side == ON_POSITIVE_SIDE)
    {
        Point_E2d intersectionPoint = intersection(splitting_line, my_segment);
        positive_side_flag = true;
        positive_side = Segment_E2d(intersectionPoint, my_segment.target());
        negative_side_flag = true;
        negative_side = Segment_E2d(my_segment.source(), intersectionPoint);
    }
    else
        assert(false);      // we shouldn't be here
}
```

The final assertion will never evaluate to **false**. It prevents compilers that do not observe that the preceding four conditions are exhaustive within the nine options from issuing a warning.

Viewport Clipping of Segments

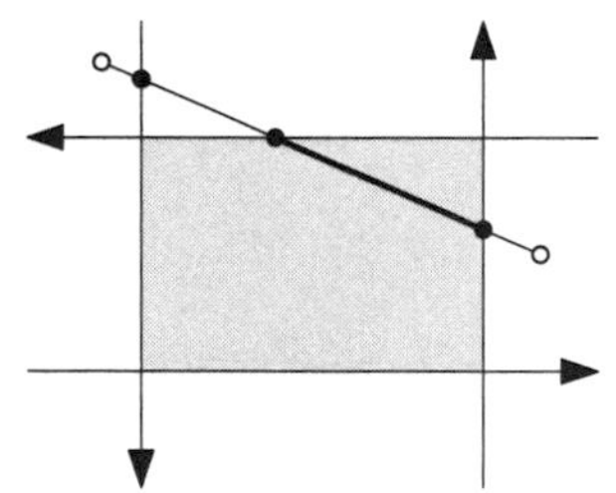

Performing line rasterization, or finding a sampled representation of a line (discussed in Chapter 19), requires that the line be wholly inside the viewport in question. Thus, viewport clipping would precede line rasterization.

Viewport line clipping could be performed easily by repeatedly invoking halfspace line clipping and discarding the portions of the line that lie outside the four halfspaces whose intersection forms the viewport. If this approach is adopted, the best one would hope for is that unnecessary intersections are never computed, but since the problem is symmetrical with respect to the four bounding lines, there is no way in general to guarantee that the optimal number of intersections is found.

Computing the intersection of a segment and an arbitrary line is more costly than determining the side of the line on which the two endpoints of the segment lie. But determining the side of an axis-parallel line on which a point lies is yet simpler; a single coordinate comparison suffices. It is possible to take advantage of this efficiency by identifying the segments that lie wholly on the right of one of the four bounding lines. In that case no fragments remain inside the viewport.

Consider determining a four-bit signature for each point as a concise way to capture the side of each bounding line on which the point lies. A point inside the viewport would have the signature LLLL to identify that it is on the left of each of the four lines. If both endpoints have this signature, the segment is inside the viewport [73].

After computing the signature of the two endpoints of a segment, corresponding bits in the signature are compared. If the two bits in any position are both R, the segment is known to lie on one side of one line and it is discarded.

If the segment is neither wholly inside nor to one side of one bounding line, it is incrementally clipped by the four halfplanes as seen in § 5.3.

Traditionally in computer graphics segment clipping is entangled with operations in a given geometry [14]. One of the advantages of defining non-Euclidean geometries is that the two issues can be kept separate—both conceptually and at the system design level. This isolation makes it simple to define a segment clipping or splitting function that operates without knowing the underlying geometry. Clipping or splitting would proceed in the same way regardless whether the geometry is spherical (Chapter 9), oriented projective (Chapter 15), or Euclidean (this chapter).

Another intersection problem—determining whether a polygon intersects a point (or whether the point is inside the polygon)—arises frequently in computer graphics and is delayed to § 20.2 and 20.3.

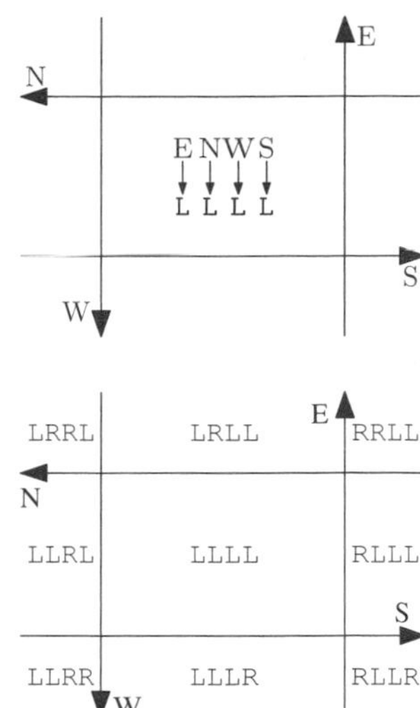

5.4 Clipping or Splitting a Polygon

Because the constraints for designing a polygon class are more varied than those for designing the planar geometric classes discussed in Chapter 1, it is more difficult to argue for a canonical design. A system may even do without a polygon class and simply use an array of points instead. In any case the following clipping and intersection problems will arise in the context of boundary representations for 3D objects discussed in Chapter 26 and the particular representation used will dictate the design constraints.

The discussion that follows assumes a class for a polygon in E^2 such as that sketched in **class** Polygon_E2d. As with segments, intersection problems involving polygons in space are entirely analogous.

```
class Polygon_E2d {
  vector<Point_E2d> points;
public:
  Polygon_E2d(const vector<Point_E2d> _points) : points(_points) {}
  ...
};
```

Halfspace Clipping of Polygons

Clipping can be performed by iterating over the segments bounding the polygon and incrementally generating the fragment inside the halfspace. The image that one should have is that of a pen tracing the boundary of the polygon. The interesting instants of the motion of the pen are those *events* when it changes direction (passing through a polygon vertex) and those when it intersects the

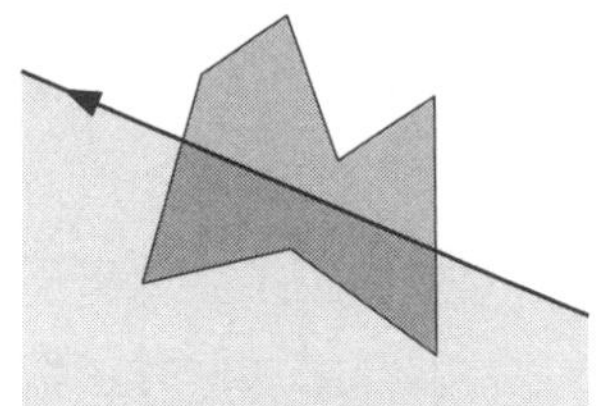

boundary of the halfspace. The events could be handled when the pen is either at the beginning or at the end of each segment bounding the polygon. Adopting the latter is customary, which gives rise to the four cases illustrated in Figure 5.2.

- If both the source and the target of the segment lie inside the halfspace, the target is copied to the clipped polygon's vertices.
- If the source is inside, but the target is outside, the point of intersection with the boundary is copied.
- If both endpoints are outside, no action is taken.
- If the source is outside and the target is inside, both the intersection and the target are copied—in that order.

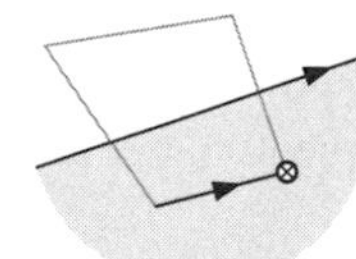
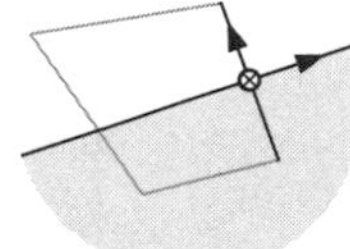
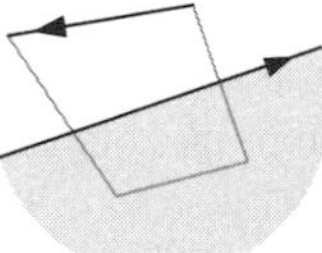
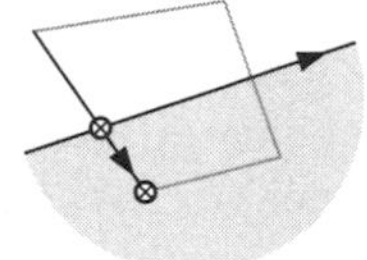

Figure 5.2
Cases for halfspace clipping of a polygon

This procedure is correct if the input polygon is convex. It is also correct if the input polygon is concave, but its boundary intersects the line bounding the halfspace only twice. If there are four or more intersections, the output polygon will no longer be *simple*, but will self-intersect: Some vertices will coincide with edges other than those to which they are adjacent. Whether such an output is adequate depends on how the set of points in the polygon is defined. If it is defined to include the points on its boundary, some points will be incorrectly determined to be in the intersection of the halfspace and the polygon. Still, rasterization algorithms handle the "degeneracy" of such polygons correctly (see Chapter 20). If the polygon is being clipped only as a preparation step prior to rasterization, the potential for degeneracy may be safely ignored. If degeneracy cannot be tolerated, a constraint may be imposed that the input polygon be convex. In the figure, the vertices are offset from the edge for illustration.

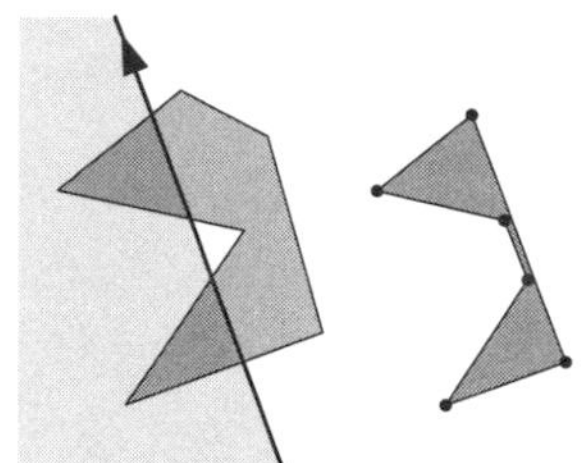

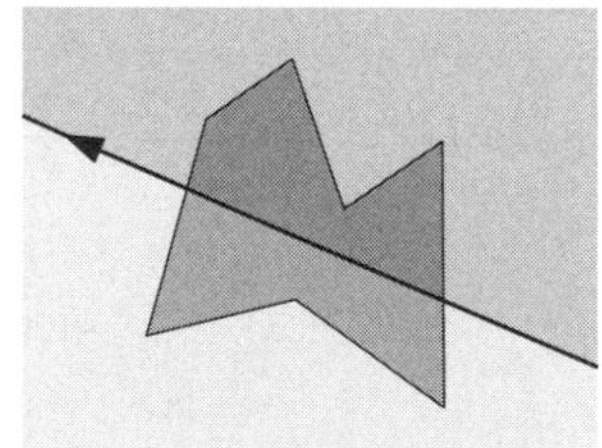

Line Splitting of Polygons

Splitting a polygon by a plane in 3D is often needed in graphical systems (see Chapter 28). The 3D problem can be studied by looking at its 2D version, splitting a 2D polygon by a line. The 2D problem is itself no harder than halfspace polygon clipping. One merely needs to duplicate the events and the output list of points to generate the fragment lying on the right (or negative halfspace) of the splitting line.

If degeneracy is a crucial issue or if the polygon is described by its bounding set of oriented lines, not its vertices, to guarantee that the algebraic degree (§ 6.4) of its vertices remains minimal, then an alternative treatment discussed in § 28.4 is needed.

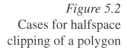

Viewport Clipping of Polygons

The Sutherland–Hodgman viewport polygon clipping algorithm discussed next is also a variation on halfspace polygon clipping [107]. The problem is to compute the fragment of a polygon that lies inside a viewport. As illustrated in Figure 5.3, the input polygon is passed through four processing stages, one for each of the lines bounding the viewport, and the output from one stage is forwarded to the next. The output of the fourth stage is the desired viewport clipped polygon.

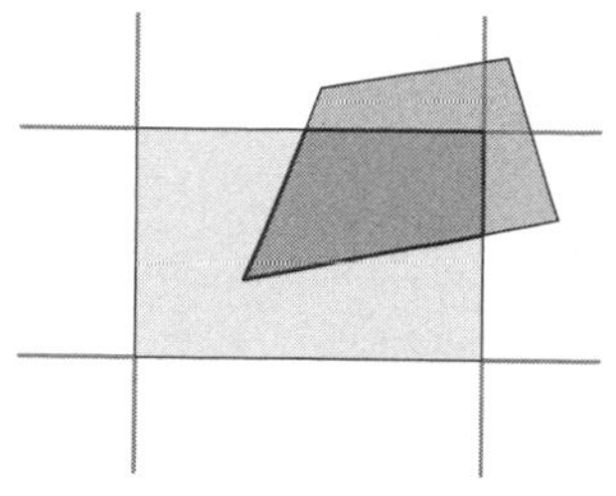

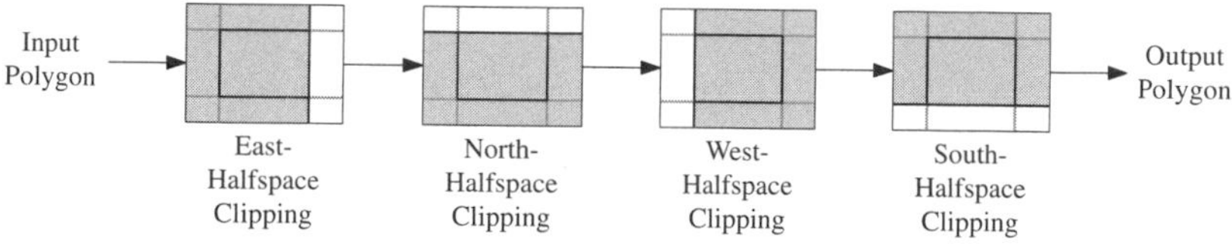

Figure 5.3
Viewport clipping of a polygon

5.5 Exercises

5.1 How many floating point additions, subtractions, multiplications, and divisions are needed to find the intersection point of two nonparallel lines? Assume that the system in question anticipates performing a great deal of such intersection computations and thus attempts to cache as much as possible in advance.

5.2 Were you convinced that the algorithm discussed in § 5.3 is indeed faster than simply calling a viewport line clipping routine four times? Write a program that determines the time ratio gained when both methods are used for a set of random lines lying in the square $[0..1] \times [0..1]$. Does the ratio observed change when the viewport is either larger than the unit square or significantly smaller?

5.3 *Rubber banding* user interfaces make it possible to manipulate a (geometric) figure and interactively observe what the final figure would look like before committing to it. Modify the start-up code labeled "clip-polygon" accompanying this chapter, which makes it possible to draw a polygon using a rubber-banding interface, to implement Sutherland–Hodgman's viewport polygon clipping algorithm.

5.4 Modify the start-up code labeled "clip-polygon" to implement the algorithm for viewport clipping of segments discussed in § 5.3. Modify the rubber banding so that it works for segments, not polygons, and so that segments are not erased after they are clipped.

5.5 Solve Exercise 5.3, paying special attention to properly handle the cases when vertices lie on one of the lines bounding the viewport. Your implementation should make it convenient to enter interactively a polygon with such vertices by *snapping* an input vertex to the nearest grid point and by setting the viewport such that it coincides with grid points.

5.6 Solve Exercise 5.4, paying special attention to the coincidence of a segment with the viewport by implementing snapping as discussed in Exercise 5.5.

6 Genericity in Geometric Computing

By choosing a type such as **double** to represent the coordinates of a point, the Point class and all accompanying classes are forever sealed at that precision. Yet another application may need to use an identical set of classes at a different precision, perhaps **float**, **long double**, or a rational number type from the GMP library such as mpq_class. This chapter discusses how the genericity of C++ can be used to design classes that delay the decision of the number type used for coordinates. Such a delay increases the chance that multiple systems could be developed based on the same set of foundational geometric classes.

It is best to read Chapters 6 and 7 simultaneously, perhaps by reading either of the two chapters before and after the other.

6.1 A Generic Class for a Point

A main tenet of good design is to delay decisions for as long as possible. This must in particular be the case when designing a geometric *library*. The type double was earlier selected as the basis for representing the coordinates of a point in 2D or in 3D, but the C++ language makes it equally easy to delay that selection until the class in question needs to be used. We say in this case that not one, but *two* instantiations take place. Whereas formerly we would instantiate an object from a class, we would now first instantiate a class from a parameterized class and then instantiate an object from the resulting (concrete) class. The code would look as follows.

```
template<typename NT>
class Point_E2 {
private:
  NT _x, _y;
public:
  Point_E2( NT x, NT y )
    : _x(x), _y(y) {}
  ...
};
```

```
...
typedef Point_E2<double> Point_E2d;
typedef Point_E2<float>  Point_E2f;
Point_E2d P;
Point_E2f Q;
...
```

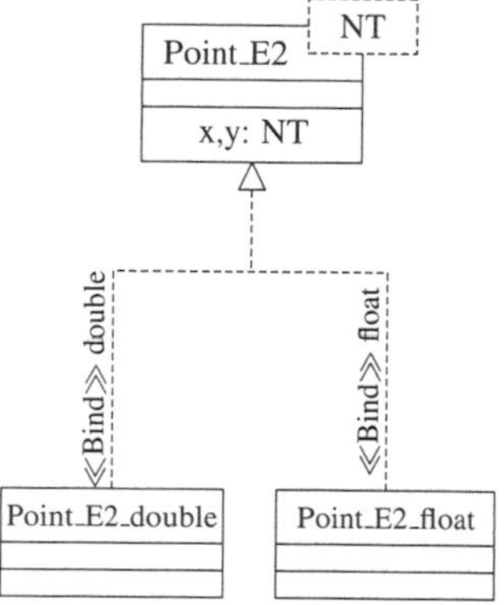

The last lines above instantiate two distinct classes for a point in two dimensions and instantiate two objects from these two classes. The class instantiations can be documented visually using the dashed lines in the figure.

But since C++ supports operator overloading, the number type used to instantiate a class for a point in E^2 is not restricted to primitive number types. One can also design one's own number type and use that type in the instantiation, just so long as the operations used in the library in E^2 are implemented for one's new types. As will be discussed in Chapter 7, one of the useful number types is to use the set of rational numbers $\mathbb{Q}$. One can write the two instantiations below, themselves based on a number type provided by a library (GMP in this case):

```
class mpq_class {
  mpz_class numerator;
  mpz_class denominator;
  ...
};
...
typedef Point_E2<mpq_class> Point_E2q;
Point_E2q R;
...
```

The use of an object of type mpz_class for the components of mpq_class above captures an integer of arbitrary size (subject to available memory). In its most simple incarnation a number for an arbitrarily large integer maintains however many digits are needed in a string. It is capable of computing elementary arithmetic and comparison operators on its internal representation. This flexibility in instantiating classes with various underlying *number types* (float, double, rational numbers, etc.) is convenient when developing a geometric system that needs to perform computations using various precisions.

But how do we know, when starting the design of a system, whether floating point numbers are adequate? The answer is sometimes easy. It is clear that interactive computer games, or systems that need to run in real time in general, cannot afford to use data types not provided by the hardware. This restricts the usable data types to int, long, float, and/or double. It is also clear that systems that perform Boolean operations on polygons or solids, such as the ones discussed in Chapter 28, will need to use an exact number type. In general, however, this is an important decision that needs to be made for each individual system. Genericity is a powerful device at our disposal to attempt to delay the choice of number type as long as possible, but to generate one executable or program, the various compromises have to be weighed and the decision has to be made.

6.2 A Scalar Type

We saw in § 1.2 that an operation such as multiplying a vector by a scalar can be captured by C++'s operator overloading. It suffices then for the client programmer to know that the geometric classes are implemented, say, using the built-in type double. Now that we delay the decision of the number type and let client programmers make the choice, it would be inelegant to expect them to use explicitly the double type as the scalar type; a decision to move from double to float (for space considerations) or to rationals (for precision) should ideally be made in just one line of code. The C++ idiom of using type aliasing via **typedef** is adequate. Another option is to define a wrapper class for the number type. Such a wrapper class [64] will also conveniently make it possible to overload operators symmetrically as member functions.

```
template<typename NT>
class Scalar
{
  NT scalar;

  friend
  Vector_E3<NT> operator*(
    const Vector_E3<NT>& v,
    const Scalar<NT>& s);
  ...
};
```

```
template<typename NT>
class Vector_E3
{
  NT x, y, z;

  friend
  Vector_E3<NT> operator*(
    const Vector_E3<NT>& v,
    const Scalar<NT>& s);
  ...
};
```

```
template<typename NT>
Vector_E3<NT>
operator*(const Vector_E3<NT>& v, const Scalar<NT>& s)
{
  return Vector_E3<NT>(v.x * s, v.y * s, v.z * s);
}

template<typename NT>
Vector_E3<NT>
operator*(const Scalar<NT>& s, const Vector_E3<NT>& v)
{
  return v * s;
}
```

An alternative, less heavy-handed approach than the one above is possible. One may avoid defining a scalar class and rely instead on a suitable typedef.

```
typedef double NT;
```

The disadvantage of this approach is that C++'s type system would not prevent client programmers from declaring scalars using **double** rather than NT, which would make the move to **float** significantly harder since the statements that depend on **double** in a different context would need to be identified and isolated. The developer of foundation geometric classes should emphasize that the adoption of a "heavy-handed approach" does not indicate a lack of confidence in the abilities of the client programmer, but that the systematic transfer of every conceivable error to a compile-time error is a desirable objective even if the two programmers swapped roles.

Yet a third approach, invisible from the client side, is for a vector object to consist of scalar objects.

```
template<typename NT>
class Vector_E3
{
  Scalar<NT> x, y, z;

  ...
};
```

Even though doing so is cleaner in theory, it unnecessarily introduces additional complexity. It could be used after comparing the assembly language code generated to the one generated by the first approach and confirming that no run-time penalty would be incurred.

A class Scalar is seldom needed in practice. The need for it arises only when developers have not yet decided whether the code they are writing will become library code or concrete code. The decision to write a generic geometric system rather than a concrete one is an all-or-nothing decision. Just as introducing **const**-correctness into a C++ system results in repercussions throughout the system, so does introducing genericity into a system result in a requirement for all classes and functions in the system to be made generic as well. Also, because of implicit type conversions, designers and programmers need to be careful about how subtle the resulting computation can be—one must ensure that implicit conversions are either eliminated or carefully maintained.

6.3 Approach to Choosing the Number Type

At this time there is no silver bullet to determine whether to sacrifice efficiency and use an exact number type. A simple rule of thumb is to consider the compromise between speed and accuracy. If the system requirements suggest speed, then we have to sacrifice accuracy, and vice versa. The answer is of course easy if neither is required, but it is more often the case that both are.

It is more often the case that one can determine at the outset which number types would *not* be adequate. For example:

- If the intermediate computations will end up affecting the final result combinatorially (e.g., insert an edge in a graph if this point is on the boundary of two polygons), then an exact number type needs to be adopted. If the intermediate computations will only affect the final result quantitatively, then an inaccurate number type such as the floating point types provided by the hardware is adequate.

- Since the size of arbitrary-precision rationals expands on demand, the maximum as well as the average depth of the expression trees used to evaluate predicates will suggest whether a rational number type is appropriate. The trouble is that the time taken by an elementary operation (such as addition or multiplication) is no longer constant as when a primitive number type is abandoned, but would depend on the size of the two operands.

6.4 Algebraic Degree

As happens so often, the geometric problem can already be investigated by looking at one dimension. We can even ignore a geometric interpretation and consider only an arthmetic expression. The expression is parsed by the compiler to generate the machine equivalent to the human readable form. For example, if the expression $2+3\times 4$ is parsed into a tree, the binary multiplication operator will have two children, 3 and 4, and the binary addition operator will have two children, 2 and the result of the multiplication operator. In a geometric system one can argue that the quality of an algorithm can be evaluated based on the maximum depth of the predicate expressions [18, 3].

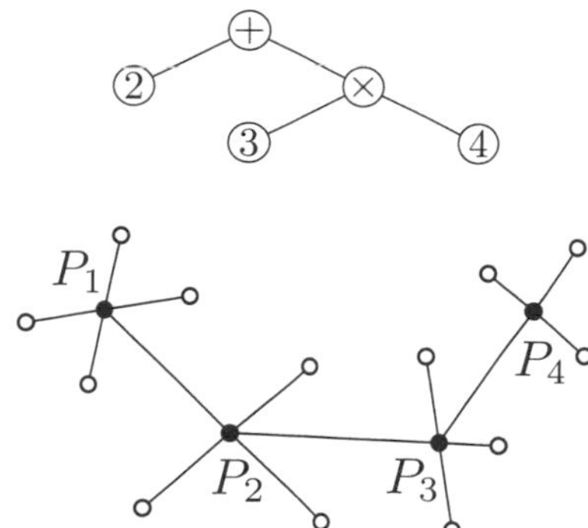

In a geometric setting consider a system that uses the points $P_1 \ldots P_4$ and suppose that we choose rational numbers as the number type to represent the points. Points marked with a circle are read as input, whereas those marked with a bullet ($P_1 \ldots P_4$) are constructed. P_1 is the intersection of two segments. P_2 lies at the intersection of two segments with P_1 defining the endpoint of one of the segments, and so on.

Just as we can talk about the algebraic degree of a polynomial, we can also talk about the algebraic degree of a point. Points read from the input have degree 0, P_1 has degree 1, P_2 has degree 2, etc. If the coordinates of the points are represented by rational numbers and if all input points are provided using x- and y-coordinates defined by numerators and denominators such that no more than a fixed number of bits are used to define each, then the denominator of P_j will in general need more bits than the denominator of P_i for $i < j$. A system may choose to calculate the greatest common divisor of the numerator and the denominator after each intersection computation and simplify the fraction (to reduce the number of bits) by dividing the two numbers by the GCD, but doing so will consume time and will, in general, not reduce the size of the numerator and denominator. No recipe is known to determine how often one should invoke a GCD routine. Various frequencies should be attempted and the program profiled to determine which one yields a suitable time and space profile.

Regardless, a subset of simplifications will reveal that the numerator and the denominator are relatively prime and that there is then no avoiding the increase of the size of the description of points in each successive generation. It is for this reason that it is crucial to consider carefully whether new points are being constructed from other points with the smallest "generation"—or that they have a low algebraic degree. Maintaining a small algebraic degree ensures that both the time and the storage the system needs will be minimized, but also, if only floating point variables are used, ensures that the chance of "corruption" of the lowest order bits, and the ones that potentially determine the outcome of predicates, remains minimized.

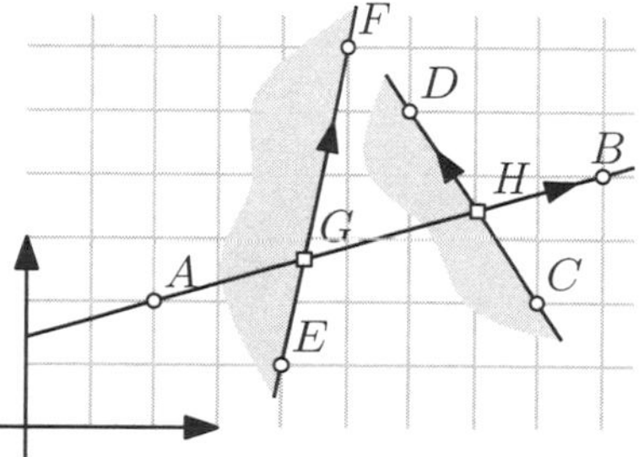

This theme is the topic of Chapter 7, but lest this issue appear to be of mere theoretical interest, an example is warranted. Consider clipping a segment AB in the plane by the two positive (i.e., left) halfspaces of CD and EF. In theory the resulting clipped segment does not depend on the order of the two clipping operations. The code below uses the type **float** to compute the final intersection point directly (Gd) by intersecting AB and EF, and also computes the ostensibly identical point indirectly (Gi) by first computing AH then intersecting AH and EF.

```
int main()
{
  const Point_E2f A(2,2), B(9,5);
  const Point_E2f C(8,2), D(6,5);
  const Point_E2f E(4,1), F(5,6);
```

```
    const Segment_E2f AB(A,B), CD(C,D), EF(E,F);

    const Point_E2f Gd = intersection_of_lines(AB, EF);
    const Point_E2f H = intersection_of_lines(AB, CD);

    const Segment_E2f AH(A,H);
    const Point_E2f Gi = intersection_of_lines(AH, EF);

    // assert( Gd == Gi ); // fails

    print( Gd.x() );
    print( Gi.x() );

    print( Gd.y() );
    print( Gi.y() );
}
```

After finding that the coordinates differ, we print the sign bit, the exponent, and the mantissa of the two x-coordinates then those of the y-coordinates (see § 7.5).

```
0 10000001 00011010000000000000000
0 10000001 00011010000000000000001
0 10000000 10000100000000000000000
0 10000000 10000100000000000000000
```

And so we see that the direct computation of the x-coordinate leads to a mantissa with a long trail of zeros, whereas the indirect computation leads to an ending least-significant bit of 1. This single-bit difference suffices for the equality operator to conclude that the two points are not equal. Performing the same computation using a combination of indirect steps only reduces the quality of the resulting floating point coordinates.

6.5 Repercussions from Changing the Number Type

Building a geometric system based on a definition for a point such as the one used above

```
template<typename NT> class Point_E2
```

makes it easy to experiment with different number types before choosing one. It is important to observe that in this case the repercussions resulting from changing a single identifier are significant. The dependency tree for such a system will show that changing a token in a (presumably single) file will force a recompilation of all or most files. If one anticipates the need to change number type during development—perhaps for comparison purposes—it can be useful to anticipate generating and maintaining different object files and different executables in different directory hierarchies. Switching the number type would then not force a recompilation of the entire system.

The ability to delay one's decisions concerning such basic details of a geometric system is discussed again in Part III, where the choice between Cartesian and homogeneous coordinate representations can also be made after a system has been designed and developed.

6.6 Exercises

6.1 You may wish to look ahead at Chapter 7 before answering this question, but you do not need to do so to tackle it.

Do floating point operations on your machine obey

a. transitivity (associativity) of addition,

b. transitivity (associativity) of multiplication,

c. and distributivity of multiplication over addition?

If you choose, for instance, three random floating point numbers a, b, and c in the interval $[0, 1)$, how likely is the equality $a + (b + c) = (a + b) + c$ to hold? Write a program that iterates a large number of times and that selects random numbers to determine whether your floating point processor is prone to failing each of these tests. If it does fail for the type **float**, determine the rate of success.

6.2 Implement a class for a rational number starting from the following code:

```
class Rational {
    int numerator;
    int denominator;
    ...
};
```

Determine the set of operators that must be overloaded on Rational for it to be usable for geometric objects in Point_E2.

6.3 Solve Exercise 6.2 and then add a (nonmember) function for computing the greatest common divisor of the numerator and the denominator. Make it possible for the client of Rational to normalize the two numbers by dividing by the GCD.

6.4 Complete Exercise 6.3 and then implement a system that generates line segments as described in § 6.4 and determine how quickly the size of the numerator and the denominator increases when the input points are constrained to lie on an integer grid (with coordinates, say, in the range $0 \cdots 255$). Look ahead in Chapter 7 to see examples for generating random points and segments. Does the depth of the expressions that your code can handle increase significantly if you use **long** instead of **int**?

6.5 The three medians of a triangle should intersect in one point. Likewise the three points of intersection of three lines with a fourth line should be colinear. Yet one cannot rely on machine built-in types to make these conclusions correctly. Read § 7.3 and confirm that your implementation for a rational number in Exercise 6.2 does indeed make it possible to make the conclusions reliably.

6.6 After reading Chapter 7, devise an additional example for a computation (in any dimension) that cannot be performed reliably if one instantiates

```
typedef Point_E2<double> Point_E2d;
```

but that is reliable if one instantiates instead geometric objects from a rational number type:

```
typedef Point_E2<Rational> Point_E2q;
```

7 Numerical Precision in Geometric Computing

Programmers in any discipline know that they cannot rely on floating points. A comparison of two floating point numbers that ought to be equal will, in general, not generate the answer expected. This problem is compounded when implementing geometric systems. The objective of this chapter is to convince the reader how flawed a geometric system that relies on floating point computation can become and to point to alternatives that do not succumb to numerical problems. The problems range from a system crashing to its producing an incorrect output. A system is said to be *robust* when it successfully completes its computation for any input and produces a correct output.

This material in this chapter and that in Chapter 6 intertwine and should be read concurrently.

7.1 One-Dimensional Analogue

After using floating point variables even briefly, programmers quickly discover that they should not use a test such as the following:

```
float a = ..;
float b = ..;
if (a==b)
  ..
else
  ..
```

The trouble with such a construct is that the underlying machine will perform the comparison between two floating point variables by checking the equality of every bit representing the variable, but after some nontrivial computations, the likelihood that the two are identical at every bit is so slim that the **else** part in this construct is almost surely going to be called many times when the "then" part should be executed instead.

Calculator users are also accustomed to the lack of precision of floating point numbers. On many such handheld machines, evaluating, say, $(\sqrt{5})^2$ will not yield 5, but another number with a minuscule error.

7.2 The Importance of Precision

Minuscule sacrifices in precision are wholly adequate in most science and engineering applications. In geometric computing the lack of precision can also

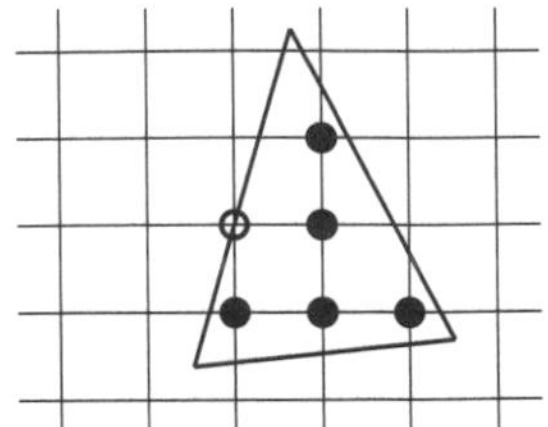

frequently be tolerated, but the lack of precision is a bothersome source of trouble if combinatorial information is going to be constructed from geometric data. Intuitively, combinatorial information derived from geometric data ("do these two points coincide?") is far more brittle: A small change in the input data generates a (qualitatively as well as quantitatively) different output. For example, if a point lies on one of the segments defining the boundary of a triangle, it is enough to move one of the vertices of the segment infinitesimally for the result of the point-inside-a-triangle test to change from true to false.

A natural reaction to the previous argument is that such a difference is likely inconsequential. A standard operation in computer graphics is *scan conversion*—the generation of the points on a grid (or "pixels") that are inside a triangle (Chapter 20). The objective is to produce a raster, or pixelated, image that conveys the shape of the triangle. If the pixels inside the triangle are displayed in one color and the background is displayed in another, it would indeed be inconsequential if one pixel on the grid is marked in black instead of white or vice versa.

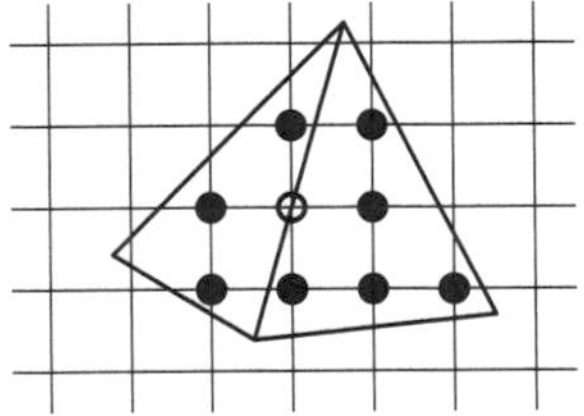

But consider a case where *two* triangles sharing an edge are displayed in, for example, two shades of gray. If the common edge passes by a point on the grid, it would be noticeable if this point is colored in the background color. The problem is compounded in an animation of these two triangles in which the four vertices defining the triangles are moving along trajectories in the plane. Such flickering of pixel colors will be perceptible.

To ensure that the rasterization or the scan conversion of the interior of each triangle in a moving mesh is correct, it suffices in this case to rasterize the boundary segments consistently, by always starting, for instance, from the endpoint with smaller y-coordinate. Since the rasterization algorithm is deterministic, the same outcome will be produced when the common segment is rasterized with either of its adjacent triangles (see § 20.4).

Even if not implemented correctly, such flickering pixels would be distracting and unaesthetic, but the problem can also become more serious. In their various incarnations, the precision-based troubles we encounter have their basis in the inadvertent breaking of some geometric truth. The truth broken in the previous example, for instance, is that all the points in the interior of the quadrilateral of points are shaded in gray (on the assumption that the two non-common vertices lie on opposite sides of the common edge). Precision-based errors lead to the inadvertent breaking of some geometric theorem also when the programmer of the geometric system is not even aware of the existence of the theorem in question. The difficulty of the precision-based problems in their general setting can be seen when we consider that two entirely distinct computation paths A and B may lead *in theory* to a single point. Because the two computation paths use different operations, the lowest-order bits representing the floating point number will differ and a test for, say, whether the two points coincide will fail (return false). How often the various floating point numbers are prone to generating an incorrect outcome is the topic of the next section.

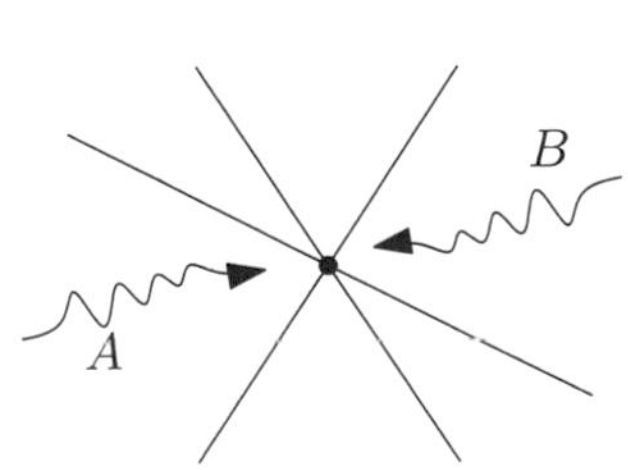

7.3 Examples of Precision Difficulties

We look at synthetic ways of devising examples of geometric problems in the plane that break some elementary theorem. We consider whether two points that ought to coincide are indeed coincident, whether three points that ought to be colinear are indeed so, and whether four points that ought to be cocircular also are so.

The reason these examples are described early in this discussion is that beginning designers of geometric systems will often argue that their systems would work properly, or would be robust, if they are fed only *nondegenerate* inputs—or inputs in which no two points are coincident, no three points are colinear, and no four points are cocircular (or equivalent degeneracies in higher dimensions). But it is impossible to guarantee in general that these conditions are satisfied, because the two, three, or four points may have been generated from entirely different computation paths and it is difficult or impossible to determine that these objects ought to be coincident, colinear, or cocircular.

Two Coincident Points

The three medians in a triangle meet at a point, or are concurrent, but suppose we compute the intersections of two different pairs of the three medians, would the use of floating point arithmetic conclude that the two intersection points have equal coordinates?

Whether the two points coincide depends both on the quality of the floating point arithmetic and on the choice of the coordinates of the triangle. In an attempt to evaluate the merit of floating point calculations on one machine, we measure the success rate of such a test. Three random points are chosen at random inside a square and the equality of the intersection points is evaluated. This leads to the following code:

```
template<typename T>
int
find_intersecting_medians(int trials)
{
  int non_degenerate_triangles = 0;
  int medians_meet_at_one_point = 0;

  while(trials) {

    Point_E2<T> P0, P1, P2;
    set_to_random(P0);
    set_to_random(P1);
    set_to_random(P2);

    if( oriented_side(P0, P1, P2) != ON_ORIENTED_BOUNDARY ) {

      non_degenerate_triangles++;

      T half( 0.5 );
      Point_E2<T> median01 = affine_combination(P0 , P1, half);
      Point_E2<T> median12 = affine_combination(P1 , P2, half);
      Point_E2<T> median20 = affine_combination(P2 , P0, half);

      Line_E2<T> L0( P0, median12 );
      Line_E2<T> L1( P1, median20 );
      Line_E2<T> L2( P2, median01 );
```

```
        Point_E2<T> P01 = intersection(L0, L1);
        Point_E2<T> P12 = intersection(L1, L2);

        if( P01 == P12 )
            ++medians_meet_at_one_point;

        --trials;
    }
  }
  return medians_meet_at_one_point;
}
```

Type	Size in bytes	Smallest number representable	Success out of 1,000,000
float	4	1.17549×10^{-38}	78,176
double	8	2.22507×10^{-308}	145,570
long double	16	3.3621×10^{-4932}	145,831

Running the above code using **double** on one million points yielded a confirmation that medians meet at a point for just 14.5% of the random triangles. One's intuition might suggest that using lower precision would result in substantially less accuracy and higher precision more accuracy. That is the case for **float**: (7.8%), but the advantage gained from doubling the number of bits is insignificant. The type **long double** succeeds at nearly the same ratio as **double**.

Still, a true skeptic would (justly) consider that the above example is contrived. If we know in advance that the intersections of distinct pairs of medians should coincide, why would we want to trip, so to speak, the machine? It would be, one could argue, simpler to compute the intersection of just one pair and use the resulting point as the triangle median. The trouble is that geometry is awash with theorems, some we are familiar with, some are obscure, but (probably) most are not even yet known. In this case the coincidence of the three medians just happens to be a well-known theorem.

Three Colinear Points

Now consider that we create a line passing through two random and noncoincident points and then generate three points on the line by intersecting it with three lines that are not parallel to it. How often does floating point arithmetic lead the program to conclude that the three points are indeed colinear? The following code, which closely emulates code in LEDA [67], generates the ratio:

```
template<typename NT>
int
colinear_points(int trials)
{
  int points_are_colinear = 0;
  int i = trials;

  while(i--) {
    Line_E2<NT> L;
    set_to_random<NT>(L);
```

```
        Line_E2<NT> L0, L1, L2;
        do { set_to_random<NT>(L0); } while(L.is_parallel(L0));
        do { set_to_random<NT>(L1); } while(L.is_parallel(L1));
        do { set_to_random<NT>(L2); } while(L.is_parallel(L2));

        Point_E2<NT> P0 = intersection(L, L0);
        Point_E2<NT> P1 = intersection(L, L1);
        Point_E2<NT> P2 = intersection(L, L2);

        if(oriented_side(P0,P1,P2) == ON_ORIENTED_BOUNDARY)
            ++points_are_colinear;
    }

    return points_are_colinear;
}
```

Type	Size in bytes	Smallest number representable	Success out of 1,000,000
float	4	1.17549×10^{-38}	234,615
double	8	2.22507×10^{-308}	256,827
long double	16	3.3621×10^{-4932}	256,644

In other words, floating point produces an incorrect result at least three times out of four!

Four Cocircular Points

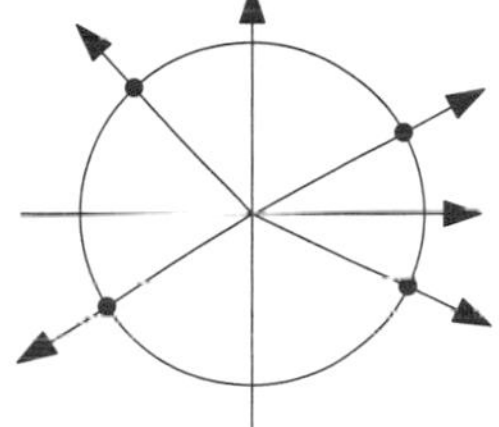

To conclude, we consider generating quadruples of random points on the unit circle and testing whether the four points are cocircular. Four random vectors are generated and normalized. Interpreting the resulting vectors as points, we determine how often the incircle predicate will conclude that the quadruples of points are indeed cocircular. The code follows.

```
template<typename T>
int
cocircular_points(int trials)
{
    int points_are_cocircular = 0;
    int i = trials;

    while(i--) {
        vector<Vector_E2<T> > V(4);
        set_to_random(V[0]);

        do {
            set_to_random(V[1]);
        } while(V[1] == V[0]);
        do {
            set_to_random(V[2]);
        } while(V[2] == V[0] || V[2] == V[1]);
        do {
            set_to_random(V[3]);
        } while(V[3] == V[0] || V[3] == V[1] || V[3] == V[2]);

        vector<Point_E2<T> > P(4);
        for(int i=0;i<4;++i)
            P[i] = Point_E2<T>(0,0) + Direction_E2<T>(V[i]);

        if(inside_circle(P[0], P[1], P[2], P[3]) == COCIRCULAR)
            ++points_are_cocircular;
    }
    return points_are_cocircular;
}
```

Since **double sqrt(double** X) and **float** sqrtf(**float** X) are the only functions available in the standard library, only these two types are tested and summarized in the table below.

Type	Size in bytes	Smallest number representable	Success out of 1,000,000
float	4	1.17549×10^{-38}	11
double	8	2.22507×10^{-308}	0

7.4 A Visual Example

The above experiments make it clear that one should not rely on floating point arithmetic. But since the floating point equality tests in the three experiments succeed only if all 32, 64, or 128 bits resulting from the determinant computation are equal, one could argue that the tests are overly stringent and could suspect that relying on floating point arithmetic would not be visually perceptible. The following experiment illustrates that at some magnification level, the lack of precision can make an image incorrect.

We generate a set of points with coordinates $(1, y)$ for $y \in [-1, 1]$. We calculate the reflection of each point from a given "center of interest" not lying on the line $x = 1$, and then construct a line segment from each sample point and its reflection. If we magnify the area around the center of interest (shown as a circle), all lines should coincide. The adjacent figure shows a square viewport of side 2×10^{-15}.

The example is of course contrived because if we wish to ensure the segments look coincident in a graphical application, we could draw two segments—each passing by the point in question. The second-generation (reflected) set of points could be avoided in this case, but there are many cases in which it is impossible to avoid increasing the generation level of geometric data. This issue, the algebraic degree of geometric objects, was discussed in § 6.4.

7.5 Bit-Level Reliability of Floating Point Numbers

The Single-Precision Floating Point Type in IEEE 754

We examine how floating point numbers are encoded in the IEEE 754 standard [54]. A floating point number is represented using a sign bit, a base-2 exponent, and a base-2 mantissa—in that order. The exponent of a single-precision float is encoded as an unsigned number offset by 127 (so one needs to subtract 127 to obtain the actual exponent); the sign bit is set to true for negative numbers; and the mantissa uses the remaining 23 bits of the 32 bits allocated. The latter consists of 1 followed by a fractional part, but because the integer part is always normalized and set to 1, that highest-order bit is not stored.

The code below reveals the structure of a single-precision floating point number captured by the type **float**. Each **float** is cast to characters and the latter's bits are printed to mirror **float**'s structure.

```
void binary_print(int n, char c)
{
  cout << ((c & 0x80) ? 1 : 0);
  if(n>=2) cout << "␣";
  cout << ((c & 0x40) ? 1 : 0)
   << ((c & 0x20) ? 1 : 0)
   << ((c & 0x10) ? 1 : 0)
   << ((c & 0x08) ? 1 : 0)
   << ((c & 0x04) ? 1 : 0)
   << ((c & 0x02) ? 1 : 0)
   << ((c & 0x01) ? 1 : 0);
}
```

```
void print(float x)
{
  char *ix =
   reinterpret_cast<char*>(&x);

  int n = 4;
  while(n--)
  binary_print(n, ix[n]);
  cout << endl;
}
```

Because all numbers below are of the form $1 \times 2^{\pm i}$, the fractional part of their mantissa is indeed zero, signaling that the mantissa is 1. The exponent of 1.0f, for example, is 127 since the exponent is 2^0 or 1.

```
int main()
{
  float x;
  x =  4.0f ; print(x);      0 10000001 00000000000000000000000
  x =  2.0f ; print(x);      0 10000000 00000000000000000000000
  x =  1.0f ; print(x);      0 01111111 00000000000000000000000
  x =  0.5f ; print(x);      0 01111110 00000000000000000000000
  x =  0.25f; print(x);      0 01111101 00000000000000000000000
  cout << endl;
  x = - 0.25f; print(x);     1 01111101 00000000000000000000000
  x = - 0.5f ; print(x);     1 01111110 00000000000000000000000
  x = - 1.0f ; print(x);     1 01111111 00000000000000000000000
  x = - 2.0f ; print(x);     1 10000000 00000000000000000000000
  x = - 4.0f ; print(x);     1 10000001 00000000000000000000000
  cout << endl;
}
```

We confirm that the encoding of numbers of the form $1 + 2^{-i}$ does indeed have a single bit of decreasing order. The outcome confirms that the smallest fraction that can be added to 1.0f without disappearing is $1/2^{23}$; the bits of $1 + 1/2^{24}$ are indistinguishable from those of 1.

Constructing Increasing Floats

We can now manually construct 16 consecutive four-byte sequences to obtain an array of integers. Casting in the opposite direction from elements of the array one_bit confirms that the resulting floating point numbers are indeed in strictly increasing order.

```
x = 1.0625000f; print(x); // 1 + 1/16                        0 01111111 00010000000000000000000
x = 1.0312500f; print(x); // 1 + 1/32                        0 01111111 00001000000000000000000
x = 1.0156250f; print(x); // 1 + 1/64                        0 01111111 00000100000000000000000
x = 1.0078125f; print(x); // 1 + 1/128                       0 01111111 00000010000000000000000
...                                                          ...
x = 1.00000047683715820312f; print(x); // 1 + 1/2^21         0 01111111 00000000000000000000100
x = 1.00000023841857910156f; print(x); // 1 + 1/2^22         0 01111111 00000000000000000000010
x = 1.00000011920928955078f; print(x); // 1 + 1/2^23         0 01111111 00000000000000000000001
x = 1.00000005960464477539f; print(x); // 1 + 1/2^24         0 01111111 00000000000000000000000
```

```
int one_bit[] =
{
  0x3F800000,      0 01111111 00000000000000000000000   1
  0x3F800001,      0 01111111 00000000000000000000001   1.00000011920928955078125
  0x3F800002,      0 01111111 00000000000000000000010   1.0000002384185791015625
  0x3F800003,      0 01111111 00000000000000000000011   1.00000035762786865234375
  0x3F800004,      0 01111111 00000000000000000000100   1.000000476837158203125
  0x3F800005,      0 01111111 00000000000000000000101   1.00000059604644775390625
  0x3F800006,      0 01111111 00000000000000000000110   1.0000007152557373046875
  0x3F800007,      0 01111111 00000000000000000000111   1.00000083446502685546875
  0x3F800008,      0 01111111 00000000000000000001000   1.00000095367431640625
  0x3F800009,      0 01111111 00000000000000000001001   1.00000107288360595703125
  0x3F80000a,      0 01111111 00000000000000000001010   1.0000011920928955078125
  0x3F80000b,      0 01111111 00000000000000000001011   1.00000131130218505859375
  0x3F80000c,      0 01111111 00000000000000000001100   1.000001430511474609375
  0x3F80000d,      0 01111111 00000000000000000001101   1.00000154972076416015625
  0x3F80000e,      0 01111111 00000000000000000001110   1.0000016689300537109375
  0x3F80000f       0 01111111 00000000000000000001111   1.00000178813934326171875
};
```

The Orientation Predicate Under Floats

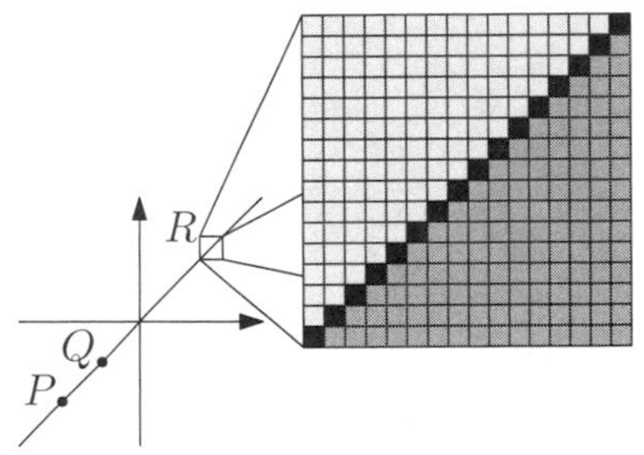

We now evaluate the orientation predicate on three nearly colinear points and use the floating point numbers thus constructed for the x- and y-coordinates of the third point. The experiment will reveal how unpredictable geometric computation using floating point numbers can be [57].

If the first two points P and Q are taken on the line $x = y$ and the bits of the two coordinates of the third point R are drawn from the table one_bit, we obtain a two-dimensional array of points with the diagonal aligned with the line $x = y$. The orientation predicate should conclude that the points on the diagonal are colinear with P and Q and that the set of points is cleanly split into those on the right- and on the left-hand side of the line PQ.

Invoking the orientation predicate on the three points (P, Q, R) where $R = (1 + \epsilon_x, 1 + \epsilon_y))$ should report a left turn if $\epsilon_x < \epsilon_y$, a right turn if $\epsilon_x > \epsilon_y$, and should conclude that the three points are colinear if $\epsilon_x = \epsilon_y$. These three values are encoded in three shades of gray and the result displayed in a matrix that mirrors a magnification of the discrete space of the type **float**. Depending on the values of P and Q, the table of orientation flags varies. The correct table is indeed generated if $P = (-5, -5)$ and $Q = (0, 0)$. If the two points are instead $(-5, -5)$ and $(-2, -2)$, a few points on the left are signaled to be on the right and vice versa. For the pair $(-6, -6)$ and $(-5, -5)$, some points are signaled to be colinear even though they are not, a large region on the right is flagged to be on the right, and each region is no longer even convex. The last case shows that the increasing density of the sample points that are not colinear but that may be flagged as such.

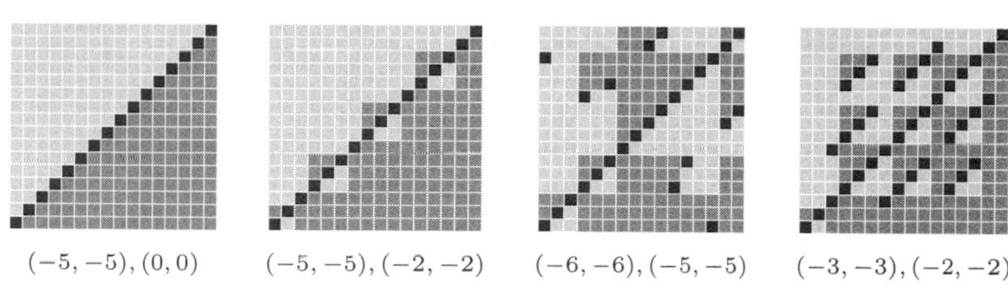

As Exercise 7.9 suggests, readers can compile the program "orientation-robustness" accompanying this text on their machines to seek tables even more unusual than the ones shown above.

7.6 Solutions to Precision Problems

Different fields will adopt different solutions depending on the application at hand.

The first and the third solutions below require that we abandon the assumption that computation is performed on real numbers. In the first case, integer arithmetic is used and in the third, either rational or algebraic numbers are used. The first method is used predominantly in hardware implementations of algorithms (such as the scan conversion of triangles) where the objectives of speed and exactness must be simultaneously satisfied. The third solution, exact arithmetic, is used in CAD and computational geometry applications. In that case, making an incorrect decision could lead to a program producing an incorrect result or crashing. The second method, epsilon geometry, is the one adopted in computer graphics. It balances the need for speed with the need for precision, but, as we shall see, it is very difficult to use that method to write code that never produces incorrect results.

Using Integer Arithmetic

With integer arithmetic, we abandon many of the operations discussed in Chapter 5. We will not attempt to implement, for instance, a function for the intersection of two line segments. This solution is sufficient for one important application: rendering on the raster devices that are widely used currently as a display technology.

Avoiding floating point arithmetic has the additional benefit of efficiency. The operation of rasterizing, or *scan converting*, segments is performed so often at the lower level of many rendering algorithms that improving the performance of this one step would percolate to affect that of many others.

Chapter 19 discusses Sproull's refactoring [102] of Bresenham's classical algorithm [21] as an example for how one may systematically redesign an algorithm to rid it of floating point.

Efficiency Through Epsilon Geometry

Epsilon geometry is a generalization to two and higher dimensions of the programming idiom that was discussed at the beginning of this chapter and that is widely used outside geometric applications. Programmers quickly observe they should not write

```
float a = ...;
float b = ...;
if( a == b ) {
  ...
}
```

because the Boolean condition is very unlikely to evaluate to true. An easy solution is to write

```
const float EPSILON = 1e−8;
float a = ...;
float b = ...;
if( fabs(a − b) < EPSILON ) {
  ...
}
```

The new comparison is favored over floating point equality on the premise that two numbers that are "sufficiently close" should be considered identical. But how should one determine the value of EPSILON, the maximum separation between two floating point numbers before they are considered equal? Should the comparison be relative to the size of a and b or remain in absolute terms?

There is unfortunately no firm foundation on which to base a geometric system built on comparisons with an epsilon. Nevertheless, this approach is widely adopted in real-time computer graphics and whenever efficiency is more important than accuracy.

This is another motivation for writing generic geometric classes as discussed in Chapter 6 rather than writing concrete classes that seal the data type to, say, **double**.

The decision to make coordinate comparisons based on some small epsilon is too important to be left for scattered use throughout a system. If it is adopted, it should be included in the design. The design would include a data type Epsilon_double that encapsulates the strategy for identifying points that are sufficiently close. The use of generic classes makes it possible to choose easily between Epsilon_double and **double** until the last stages of an implementation.

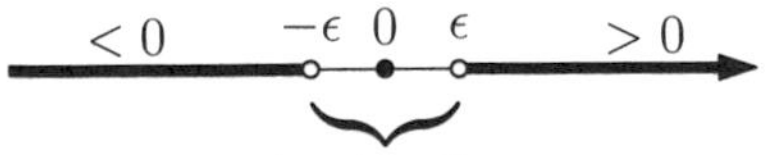

< 0 $\quad -\epsilon \;\; 0 \;\; \epsilon \quad$ > 0

treat as 0

If an epsilon is chosen in absolute terms (i.e., as a constant irrelevant of the magnitude of the numbers being compared—surely an inadequate choice in many cases), then the comparison operators would need to be defined for the new number type.

```
static const double EPSILON = 1e−7;
class Epsilon_double
{
  double d;
public:
  Epsilon_double(double d = 0.0) : d(d) {}

  bool operator==(const Epsilon_double& d2) const
  { return (fabs(d2.d − d) < EPSILON); }
```

```
    bool operator!=(const Epsilon_double& d2) const
    { return !operator==(d2); }

    bool operator<(const Epsilon_double& d2) const
    { return d - d2.d < -EPSILON; }
    bool operator<=(const Epsilon_double& d2) const
    { return d - d2.d <= -EPSILON; }
    bool operator>(const Epsilon_double& d2) const
    { return d - d2.d > EPSILON; }
    bool operator>=(const Epsilon_double& d2) const
    { return d - d2.d >= EPSILON; }
    ...
};
```

Notice that even though Epsilon_double is based on the assumption that two floating point numbers will only rarely coincide, we still need to aim to a modicum of consistency by ensuring that the Boolean operators are defined symmetrically for the new number type. Consistency can be confirmed by simply including a test routine such as the following:

```
void test_epsilon_double()
{
    Epsilon_double a(2.0);
    Epsilon_double b(2.0 + EPSILON);

    assert( (a < b) != (a == b) );
    assert( (b > a) != (a == b) );
}
```

In this case we choose for the $\pm\epsilon$-interval surrounding a given number to be open: $a \neq a+\epsilon$. The alternative, assuming a closed interval, is equally suitable.

Achieving this level of symmetric predicates is as far as tests based on epsilon will allow. Transitivity of equality is clearly unattainable, which gives one indication why systems based on epsilons breach basic properties of numbers.

Accuracy Through Rational Numbers

If the assumption that the countable set captured by floating point variables in a machine is an adequate representation of real numbers is inherently flawed, one may wonder whether it is necessary to use real numbers in the first place. If algebraic numbers (which arise as a solution to an algebraic equation) appear as part of the input of a problem, then our only options are either to use a floating point approximation or to use manipulations on algebraic numbers. But in yet many other cases, the input may be represented with rational numbers.

Often also few digits of precision are needed in the output of an algorithm. The intermediate computations, on the other hand, can in general not be implemented reliably using primitive data types. If the problem considered consists of linear sets (the focus of this text), then implementing a rational number type guarantees predicate accuracy. At the outset this appears to be simply a matter of implementing a class Simple_rational.

```
class Simple_rational {
    int numerator;
    int denominator;
    ...
};
```

But as is clear from Exercise 6.4, such an implementation would be inadequate. After multiple generation levels of geometric objects, the precision (32 bits in this case) available in any constant-sized number type will be insufficient—even after normalizing the rational numbers used. The answer lies in introducing a new class Rational that stores two integers of arbitrary precision. As more precision is needed by a BigInteger object, additional storage is allocated. The type mpq_class mentioned earlier (§ 6.1) is an example of such an extended-precision rational number.

```
class Rational {
    BigInteger numerator;
    BigInteger denominator;
    ...
};
```

Using rational numbers may result in a dramatic increase in execution time. The use of floating point filters makes it possible to develop systems that are both accurate and efficient [37, 90].

Yet, as discussed in § 29.8, filters are occasionally ineffective and restructuring an algorithm so that its predicates are dependent via brief expressions on the input data is needed. The BSP trees described in Chapters 28 and 29 differ from the classical presentation to attain such brief expressions.

Considerably more can be said about exact computation in particular and robustness in general. The interested reader is referred to the expository papers [120, 52, 91] and the references therein.

7.7 Algebraic Vs. Synthetic Geometry

There are two fundamentally different ways for doing geometry: *Synthetic* geometry is the art of constructing statements by building on a minimal set of self-evident statements (axioms and postulates). *Algebraic* geometry, on the other hand, is the art of solving geometric problems using algebraic manipulation on (numerical) representations of these objects.

The former has its origin with Euclid's development in about 300 B.C. of a geometry, named today after him. The latter has its origins with the introduction of Cartesian coordinates by Fermat and by Descartes in 1637 [95] and the introduction of homogeneous coordinates (see Chapter 12).

But synthetic geometry was not replaced by algebraic geometry. That the two strategies for building geometry can not only coexist, but can complete each other, was not understood in the early 19th century. A rivalry arose, for instance, between Gergonne and Poncelet [19], the first as an advocate of analysis and the second an advocate of synthesis. Looking at what we now know in geometric computing would suggest that synthesis took the back seat—computation is essentially analytic, but it is conceivable that the other face of geometry will some day fully reveal itself. The main strategy for resolving predicates today relies on their coordinates. Determining whether two points are equal, for example, is reduced to a computation on their coordinates. This is the best one can do if both points are read as input, but if one of the points is constructed from data that includes the other point, a proof may exist that the two points ought to be equal. Effective ways for encoding known proofs and for determining whether a proof of equality exists may one day make it possible to skirt many of the numerical problems that currently plague implementations of geometric algorithms.

7.8 Documenting Degeneracies

If the details of handling the unlikely events, but the ones that do arise in practice and that cause geometric software to fail, are so important, the reader will wonder why these details are seldom reported as part of the description of an algorithm in the literature. The trouble is that these details are unglamorous. The algorithm developer will have doubtless spent many hours finding these special cases, implementing them, debugging, and then finding that some more have been missed. The developer who does not realize that reporting on such an experience will save others who attempt the same task significant effort is likely to assume that finding the special cases is merely a detail of debugging.

The crucial observation is that special cases should be documented as intrinsic parts of an algorithm. If they are, the test cases that the programmer implementing the algorithm will write will *in particular* include these special cases.

Many precision-rooted troubles that do not arise in the one-dimensional version of a problem are a particular source of difficulty in higher dimensions. The contrast between the computation of Boolean operations on regular sets in one or higher dimensions (discussed in Chapter 30) is an example of the ease of implementing the one-dimensional problem compared to higher dimensions.

Someone moving from traditional programming to geometric programming who wishes to design and implement robust systems has to undergo a basic shift in thinking. This shift was already outlined in Chapter 2; predicates need to be designed to have ternary, not binary, outcomes, and the programmer needs to design and implement algorithms always concerned about all three possible outcomes at a geometry-based branching statement—even if most often only two cases will handle the three potential outcomes.

7.9 Exercises

7.1 Exercise 6.1 asked you to confirm whether several numerical problems that appear not to be geometric would be handled correctly on your floating point processor. Determine for each subproblem a geometric computation that would resolve to an identical calculation and that would thus fail if you concluded so in the numeric instance.

7.2 Write a program that generates a postscript file (see Appendix A) confirming the claim in § 7.4.

7.3 Implement a system for computing the convex hull of a set of points in the plane [83] using **double** precision and generate a postscript drawing of the hull. How many (random) points can your implementation handle before you notice errors in the convex hull?

7.4 Repeat the previous exercise for the Delaunay triangulation [63].

7.5 a. A *simple polygon* is one whose edges intersect only at vertices. Devise and implement a method for generating random simple polygons of n vertices. Your implementation should make it possible to generate either integer or floating point coordinates.

b. Read Chapters 20 and 22 and then implement a system for counting the number of grid points inside a simple polygon using floating point coordinates.

c. The area of a simple polygon whose vertices have integer coordinates is $\frac{1}{2}b + c - 1$, where b is the number of grid points that lie on the boundary of the polygon and c is the number of grid points inside the polygon. (For a proof of this equation due to Pick, see Coxeter [27].) Implement a system that uses this formula as well as your implementation for part b to compute the area of polygons whose vertices have integer coordinates.

7.6 If the vertices of a simple polygon of n vertices are $P_i(x_i, y_i), i = 0 \cdots n-1$, then the area of the polygon is given by

$$\frac{1}{2}\sum_{i=0}^{n-1} x_i y_{i+1} - x_{i+1} y_i.$$

Indexing is done modulo n, so when $i = n-1, i+1 = 0$ [74]. Implement a function that calculates the area of a simple polygon.

7.7 Implement Exercises 7.5 and 7.6. What is the ratio of success between the floating point and the exact computations?

7.8 The program labeled "precision" accompanying this text is an interactive version of the precision problems encountered in § 7.5 [57]. Experiment with different input values and provide those that give the least reasonable behavior for floating point arithmetic.

7.9 The figures in § 7.5 [57] were chosen among those generated by the program "orientation-robustness" accompanying this text. Choose different coordinates for the two constant points P and Q in an effort to generate tables that show even more unusual behavior for floating point arithmetic than the one documented earlier.

7.10 The tables shown in § 7.5 may seem to suggest that the behavior of floating point is entirely chaotic. Determine whether that is the case by generating a large set of images while gliding one of the two points P and Q. Begin at the following excerpt from the program "orientation-robustness" accompanying this text.

```
for(float P = −10; P<= −1; P += 1.0)
  for(float Q = P+1; Q <= 0; Q += 1.0)
    float_orientation_robustness(
      "psout/orientation−robustness", P, P, Q, Q);
```

Use a utility to convert the resulting set of vector images to a raster representation, and then use another utility to generate an animation showing the result of the glide. Can you detect any pattern?

7.11 If consecutive vertices of a regular polygon (§ 26.3) are used to define a set of vectors, then the vectors should always sum to the zero vector. Generate a large number of random-sized regular polygons of $n = 3 \cdots 7$ sides and determine how often such vectors sum to the zero vector. Is the outcome for a square significantly better than that for a triangle? Does the outcome change significantly if triangles take an arbitrary orientation (rotation)? Does it for squares?

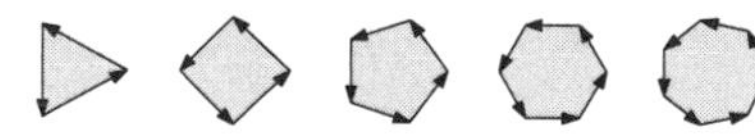

Part II

Non-Euclidean Geometries

8 Computational Spherical Geometry in One Dimension

Before studying spherical objects, we spend a moment looking at the objects one can define on the circle. We assume that the circle in question is a unit circle centered at the origin and that S^1 denotes the set of points on such a canonical circle. What objects can one define for the geometry of the circle, and what operations could be implemented on these objects? These are the topics of this chapter. Note that we usually say a "sphere" in any dimension, except for $d = 1$, for which we say "circle" [9].

8.1 A Point

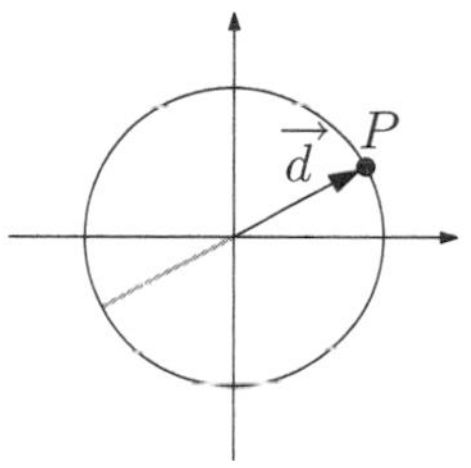

We start by defining a class for a *point on a circle* Point_S1. A library for circular objects may be designed by building on the notion of a direction in the Euclidean plane (§ 1.3). At the price of a little repetition, we choose here to reduce the coupling between the two components, which in any case leads to a clearer exposition and ensures that no level of indirection will remain in compiled code. The outline of a class for a point in S^1 may look as follows:

```
template<typename NT>
class Point_S1
{
  NT _x, _y;
public:
  Point_S1() : _x(1), _y(0) {}
  Point_S1(const NT& x, const NT& y) : _x(x), _y(y) {}
  Point_S1(const Point_E2<NT>& source, const Point_E2<NT>& target)
    : _x(target.x() − source.x()), _y(target.y() − source.y()) {}
  ...
};
```

The vector used to create the spherical point is not normalized, as doing so limits the possible number types, makes subsequent equality tests less reliable (Chapter 7), and unnecessarily wastes time. To determine whether two circular points are equal, we check that the two corresponding vectors are linearly dependent *and* that their inner product is positive. This last condition is the crucial one that distinguishes the geometry of S^1 from that of the projective line P^1 (§ 11.1).

```
  bool operator==(const Point_S1<NT>& p) const {
    return (this == &p) ||
      determinant(_x, _y, p._x, p._y) == 0 &&
      inner_product(_x, _y, p._x, p._y) > 0;
  }
```

8.2 Point Orientation

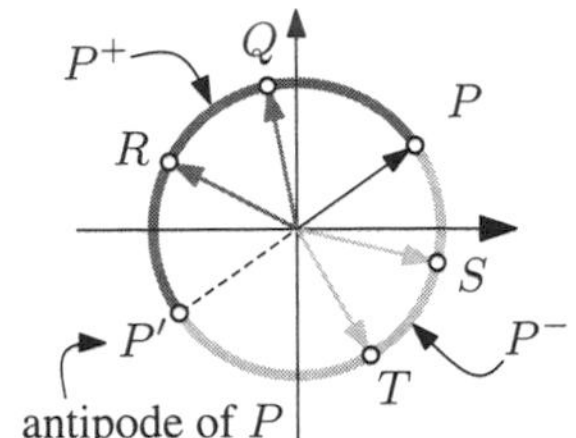

As discussed in § 2.5, a point in the Euclidean line E^1 divides the line into a positive and a negative side. A sidedness predicate returns *three* values, to signal the relative orientation of a query point.

A point in S^1 is somewhat trickier. At first sight it would appear that a point does *not* divide S^1 into two halves—only one section results from cutting the circle at a point. Yet as will become evident in § 28.3, it is natural to divide the circle into positive and negative parts with respect to a point. For that the notion of the *antipode* of a point is needed. The antipode of a point P with direction $\overrightarrow{d}$ is the point P' defined by the direction $\overrightarrow{-d}$. The pair P and P' divide the circle into two parts. The positive halfspace P^+ is the open halfcircle counterclockwise from P—containing the points Q and R in the figure—and the negative halfspace P^- is the clockwise halfcircle—containing the points S and T.

```
const Point_S1<NT> antipode() const { return Point_S1(-_x, -_y); }
```

Yet the point orientation predicate will not distinguish between P' and P; the third set in the classification of points in S^1 depending on their orientation with respect to P consists of the set $\{P, P'\}$. It is the responsibility of the client software to subsequently test whether the query point is the antipode.

```
template<typename NT>
Oriented_side
oriented_side(
        const Point_S1<NT>& p1,
        const Point_S1<NT>& p2)
{
  NT d = cross_product( p1.get_Direction_E2(), p2.get_Direction_E2() );

  return enum_Oriented_side(d);
}
```

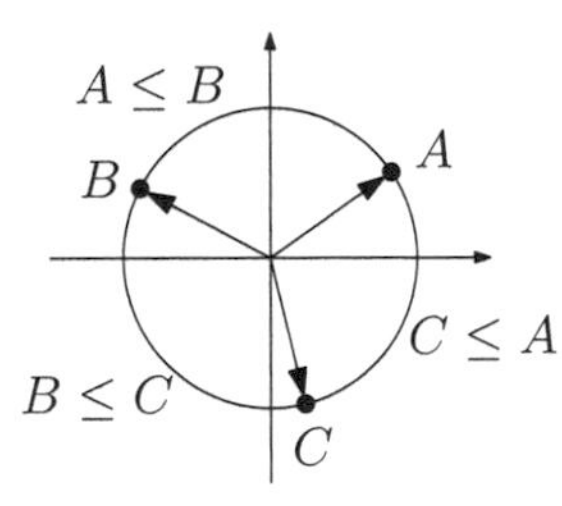

One property that we expect from point orientation on the Euclidean line—transitivity—is not satisfied for point orientation on S^1, for three points A, B, and C may be ordered such that $A \leq B$ and $B \leq C$, yet $A \leq C$ does not hold. Point orientation remains antisymmetric. If $A \leq C$ does not hold, it must be the case that $C \leq A$ does hold, but the lack of transitivity means that a set of points in S^1 cannot be sorted in absolute terms, but only in relation to a given point.

8.3 A Segment

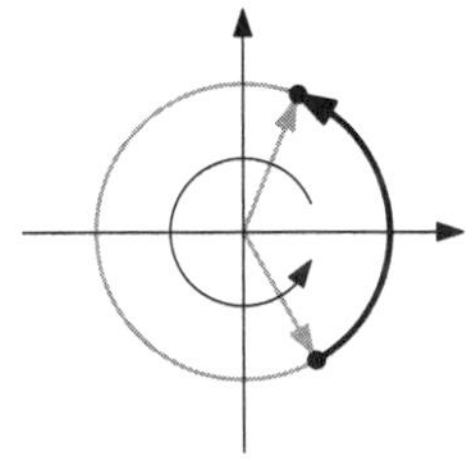

The next natural object to define is that of a circular segment. Since two distinct circular points could define two segments on the circle, we may choose to add a Boolean flag to identify which of the two segments is intended, but it is more elegant to impose a counterclockwise orientation on the circle and to define the segment as directed from a source to a target point.

```
template<typename NT>
class Segment_S1
{
  Point_S1<NT> _source, _target;
public:
  Segment_S1(
          const Point_S1<NT>& source = Point_S1<NT>(1,0),
          const Point_S1<NT>& target = Point_S1<NT>(0,1))
     : _source(source), _target(target)
  {}

  Point_S1<NT> source() const { return _source; }
  Point_S1<NT> target() const { return _target; }
};
```

Several options are possible in a library for objects in S^1. Can the source and the target of a segment coincide? Can the source and target of a segment be antipodal points? Can a segment be larger than half a circle? If the source and target coincide, would the segment defined have degenerated to a single point or to the full circle?

We leave it possible for a segment to have coincident endpoints; the segment then degenerates to a single point. We also leave it possible for the two endpoints to be antipodal, a case we will need in Chapters 28 and 29. But to simplify intersection operations, we disallow a segment from being larger than half a circle.

Three operations may be anticipated: detecting whether a point is among those defined by a segment, determining the intersection of two segments, and interpolating between two points on the circle. The last operation would be useful if one wishes, for instance, to animate a particle moving along a segment.

8.4 Point-Segment Intersection

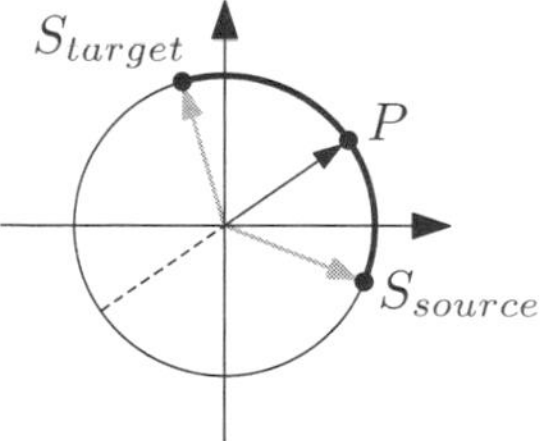

Rather than return a Boolean flag identifying whether a point intersects a segment, it is equally easy to return three values; the point may be inside the segment, outside, or on the boundary (coinciding with one of the endpoints). If a point is in the positive halfspace of the source of a segment and in the negative halfspace of the target, it is inside. If the query point coincides with either segment endpoint, it lies on the boundary

```
template<typename NT>
Set_membership
classify(
      const Segment_S1<NT>& segment,
      const Point_S1<NT>& P)
{
  const Point_S1<NT>& source = segment.source();
  const Point_S1<NT>& target = segment.target();

  Oriented_side side_of_source = oriented_side(source, P);
```

```
  Oriented_side side_of_target = oriented_side(target, P);

  if(
    side_of_source == ON_POSITIVE_SIDE &&
    side_of_target == ON_NEGATIVE_SIDE )
     return INSIDE_SET;
  else if(
        side_of_source == ON_ORIENTED_BOUNDARY &&
        inner_product(source.x(), source.y(), P.x(), P.y()) > 0
        ||
        side_of_target == ON_ORIENTED_BOUNDARY &&
        inner_product(target.x(), target.y(), P.x(), P.y()) > 0)
     return ON_SET_BOUNDARY;
  else
     return OUTSIDE_SET;
}
```

8.5 Segment Intersection and Interpolation

Intersecting two circular segments consists of making four containment tests. If all four fail, the two segments do not intersect. The function return type is a pair consisting of a Boolean flag in addition to the segment resulting from the intersection

```
template<typename NT>
std::pair<bool,Segment_S1<NT> >
intersection(
        const Segment_S1<NT>& seg1,
        const Segment_S1<NT>& seg2)
{
  if(seg1.contains(seg2.source()) && seg1.contains(seg2.target()))
     return std::make_pair(true, seg2);
  else if(seg2.contains(seg1.source()) && seg2.contains(seg1.target()))
     return std::make_pair(true, seg1);
  else if(seg1.contains(seg2.source()) && seg2.contains(seg1.target()))
     return std::make_pair(
        true, Segment_S1<NT>(seg2.source(), seg1.target()));
  else if(seg2.contains(seg1.source()) && seg1.contains(seg2.target()))
     return std::make_pair(
        true, Segment_S1<NT>(seg1.source(), seg2.target()));
  else
     return std::make_pair(false, Segment_S1<NT>());
}
```

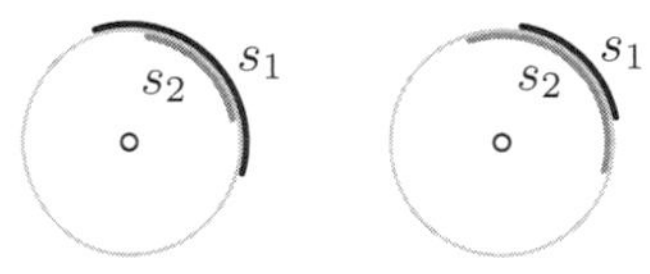

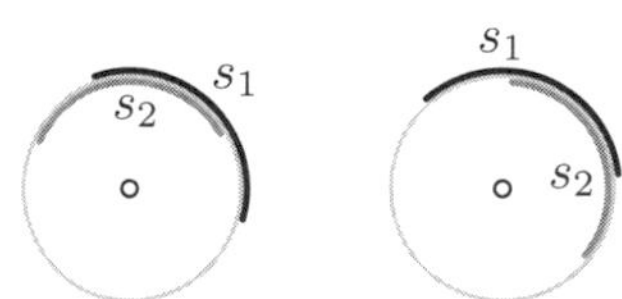

Consider that we wish to generate an animation of a spherical point along a spherical segment. The point lies initially at the source of the segment and reaches the target of the segment at the conclusion of the animation. We also would like the speed of the point to be uniform; the rate with which the point traverses angles at the center of the circle should be constant. Unlike intersection, writing a function to interpolate between two spherical points requires the use of trigonometric functions.

```
template<typename NT>
Point_S1<NT>
interpolate(
        const Point_S1<NT>& P0,
        const Point_S1<NT>& P1,
        const NT& t )
{
  const NT zero(0.0);
  const NT unity(1.0);

  assert(zero <= t && t <= unity);

  NT sourceAngle = atan2(NT(P0.y()), NT(P0.x()));
  NT targetAngle = atan2(NT(P1.y()), NT(P1.x()));

  if(targetAngle > sourceAngle) {
    NT angle = sourceAngle * (unity-t) + targetAngle * (t);
    return Point_S1<NT>(std::cos(angle), std::sin(angle));
  }
  else {
    NT angle =
      (M_PI + sourceAngle) * (unity-t)
      + (targetAngle + M_PI) * (t);
    return Point_S1<NT>(-std::cos(angle), -std::sin(angle));
  }
}
```

If the functions atan2, cos, and sin are defined through overloading for the types **float**, **double**, and **long double**, the above definition would adequately invoke the appropriate set of functions corresponding to the number type to which interpolate is instantiated. If a different number type is used, an error would be issued since the three trigonometric functions would not be resolved. (An alternative, overloading three concrete definition for interpolate, would cause the error to be issued when that function itself would not be resolved.)

One operation, *splitting* a segment by a point, is delayed to Chapter 28 when it will be needed to represent regular sets in S^1 and to perform Boolean operations on such sets.

8.6 Exercises

8.1 The implementation of Point_S1 (§ 8.1) could have stored an instance of Direction_E2 (§ 1.3).

```
template<typename NT>
class Point_S1
{
  Direction_E2<NT> d;
  ...
};
```

The class Point_S1 would then act as a wrapper that delegates messages it receives to the Direction_E2 class.

Yet a system developed on such a foundation could be paying a price for the extra level of indirection. Implement two prototype systems and compare the assembly language generated by your compiler to determine what price, if any, would be paid by the indirection. Do the compiler's various optimization stages have any effect?

8.2 The function interpolate defined in § 8.5 would be adequate if it is used to visualize a moving particle, since spherical objects then need in any case to be converted to a floating point type. But if the number type used is a rational type (such as mpq_class) and if the result of the interpolation would be subsequently used for other than visualization, then the function needs to be defined for the number type used.

Interpolation may be performed without having to use trigonometric functions by determining a number of points that are sufficiently close to the circle and performing linear interpolation between the closest two.

Sketch a design that would allow you to define interpolation in the first quadrant on Gmpq by recursively dividing that quadrant by two until there are 256 points on the interval $[(1, 0), (0, 1))$. Your design should cache the generated points between invocations.

8.3 Implement a prototype for your solution to Exercise 8.2.

9 Computational Spherical Geometry in Two Dimensions

Studying spherical geometry before classical (Chapter 11) and oriented (Chapter 14) projective geometries provides considerable insight to these geometries. Assigning coordinates to spherical objects also makes the assignment of homogeneous coordinates to projective objects more intuitive.

But spherical geometry is also interesting in its own right as many geometric systems are implemented on the sphere. Such systems will most naturally be implemented not using two spherical coordinates (latitude and longitude), but by operating on vectors in Euclidean 3D space (Chapter 3). Systems using spherical geometry could indeed be implemented by operating on vectors without abstracting the structure of the sphere, but such systems are inherently flawed. The notions that need to be abstracted are repeatedly defined within the system; the operations risk being multiply defined; and debugging the system must proceed in an ad-hoc manner. These flaws are removed by encapsulating objects and predicates in a software layer for spherical geometry, the topic of this chapter.

At the time of this writing, the CGAL (see Chapter 18) kernel exclusively contains Euclidean geometry objects, but an extension package for spherical geometry that includes a number of algorithms not discussed here has been designed and implemented [92] for LEDA [67].

9.1 Spherical Geometry Objects

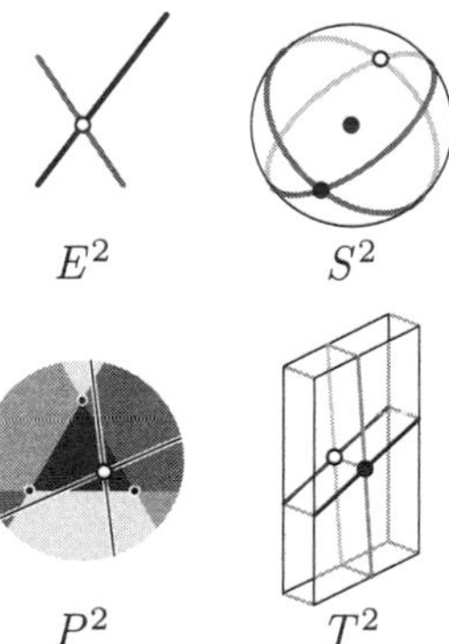

A distinguishing characteristic of the four geometries (E^2, Chapter 1; S^2, this chapter; P^2, Chapter 11; and T^2, Chapter 14) is the number of points in which two lines intersect. In Euclidean geometry two distinct lines intersect in a point unless they are parallel; in spherical and oriented projective geometries two distinct lines intersect in two points (but see § 9.5); and in projective geometry two distinct lines intersect in one point.

As with the geometry of a circle, a point on the sphere, or a *spherical point*, will be represented by a direction in three dimensions. If an observer situated at a spherical point starts moving in an arbitrary direction on the surface of the sphere, the path taken will eventually reach the starting point—and so as far as the observer is concerned, the path taken is a straight line. Thus, either a "spherical line" or a "spherical circle" is an adequate name for such an oriented path. The first considers the sphere as experienced by a minuscule creature walking on its surface; the second considers the line as seen by a distant observer. Either name is adequate; here we use the latter.

The analogy with the Euclidean plane goes further: just as an oriented line in the Euclidean plane divides the plane to left and right halfplanes, the line on the sphere also divides the sphere into left and right hemispheres.

A spherical circle is also the intersection of a plane passing by the center of the sphere with the surface of the sphere. The resulting set of points and circles can for convenience be thought to lie on a sphere of unit radius, but an identical geometry would result if a sphere of an arbitrary radius is used.

Adopting oriented spherical circles means that the points on a given spherical circle also define a second spherical circle with the opposite orientation. An oriented plane will be used to construct a spherical circle and the right-hand rule will apply: If the thumb of the right hand points to the normal of the oriented plane defining the circle, the fingers are aligned with the orientation of the circle.

9.2 A Spherical Point

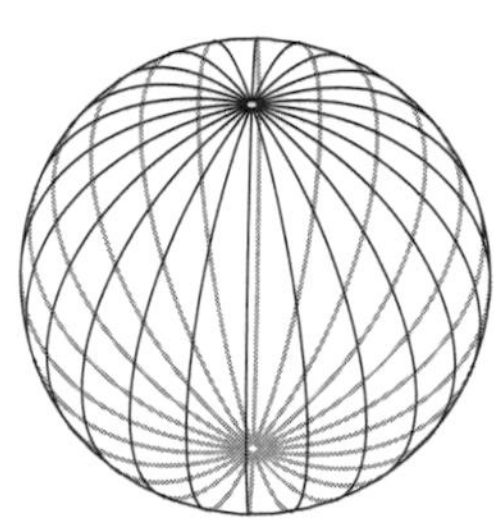

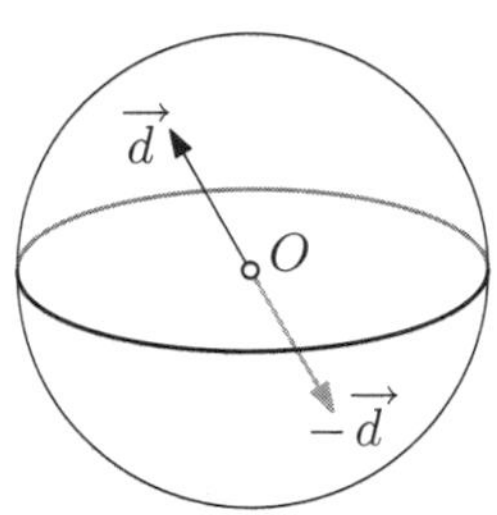

It is tempting to parameterize a spherical point using two angles θ and ϕ that measure longitude and latitude. After choosing two antipodal points as poles, a "Greenwich line" can be declared and an equator deduced as the perpendicular bisector of the two poles. But building a library for geometric components on such a parameterization is flawed. One problem is that a discontinuity occurs when an observer crosses the Greenwich line or passes by either pole. Another is that rates of change of the coordinates for someone traveling on the surface of the sphere will vary greatly. Perceived speeds near the poles are skewed compared to those near the equator. Adopting a direction in Euclidean space as the parameterization for a spherical point avoids these problems.

A spherical point is described by a direction $\overrightarrow{d}$ in E^3. The *antipode* of this point is defined as the spherical point with direction $-\overrightarrow{d}$. The classes for spherical objects have the suffix S2 to identify them as lying on the two-dimensional sphere.

```
template<typename NT>
class Point_S2
{
  NT _x, _y, _z;
public:
  Point_S2() : _x(1), _y(0), _z(0) {}
  Point_S2(const NT& x, const NT& y, const NT& z) : _x(x), _y(y), _z(z) {}
  Point_S2(const Direction_E3<NT>& D) : _x(D.x()), _y(D.y()), _z(D.z()) {}
  Point_S2(const Point_E3<NT>& source, const Point_E3<NT>& target)
    : _x(target.x() - source.x()),
      _y(target.y() - source.y()),
      _z(target.z() - source.z())
  {}

  Direction_E3<NT> get_Direction_E3() const {
    return Direction_E3<NT>(_x,_y,_z);
  }
```

```
    NT x() const { return _x; }
    NT y() const { return _y; }
    NT z() const { return _z; }

    const Point_S2<NT> antipode() const { return Point_S2(-_x, -_y, -_z); }

    Point_S2 operator-() const { return antipode(); }
    ...
};
```

Equality of Two Spherical Points

As with directions in Euclidean 3D space (§ 3.2), here also two points represented by vectors $\overrightarrow{v_1}$ and $\overrightarrow{v_2}$ are equal if there exists a *positive* k such that $\overrightarrow{v_1} = k\overrightarrow{v_2}$. Also as with directions in E^3, there is no need to normalize the vectors before determining equality.

```
    bool operator==(const Point_S2<NT>& p) const {
        return (this == &p) ||
            are_dependent(_x, _y, _z, p._x, p._y, p._z) &&
            inner_product(_x, _y, _z, p._x, p._y, p._z) > 0;
    }
```

9.3 A Spherical Circle

Because we know that the oriented plane carrying a spherical circle must pass by the origin, it would be wasteful to represent a spherical circle using an oriented plane. Storing the normal to the plane suffices.

```
template<typename NT>
class Circle_S2
{
    NT _x, _y, _z;
public:
    Circle_S2() : _x(1), _y(0), _z(0) {}
    Circle_S2(const NT& x, const NT& y, const NT& z) : _x(x), _y(y), _z(z) {}
    ...
};
```

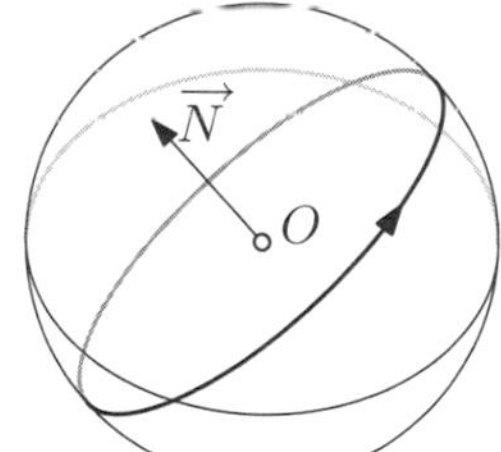

Determining the Spherical Circle Carrying Two Spherical Points

Two distinct and nonantipodal spherical points define a spherical circle. A portion of the circle is traced by the shortest path on the sphere between the two points. Reversing the order of the points defines a circle coincident with the first, but with the opposite orientation.

```
template<typename T>
class Circle_S2
{
    ...
```

```
    Circle_S2(const Point_S2<NT>& p1, const Point_S2<NT>& p2)
    {
      cross_product(
                p1.x(), p1.y(), p1.z(),
                p2.x(), p2.y(), p2.z(),
                _x, _y, _z);
    }
};
```

Finding an Orthonormal Basis for a Spherical Circle

It is frequently useful to know two orthogonal points $\overrightarrow{B_1}$ and $\overrightarrow{B_2}$ on a given spherical circle $\overrightarrow{c}$. Parameterizing a point on a spherical circle, for instance, requires knowing such a basis. The first spherical point $\overrightarrow{B_1}$ can be determined by calculating the cross product of $\overrightarrow{c}$ with an arbitrary other vector $\overrightarrow{L} \neq \overrightarrow{c}$. Since the numerical quality of the outcome is maximized if $\overrightarrow{L}$ is itself orthogonal to $\overrightarrow{c}$, we choose $\overrightarrow{L}$ to be the least dominant (§ 3.7) among $\{\overrightarrow{X}, \overrightarrow{Y}, \overrightarrow{Z}\}$.

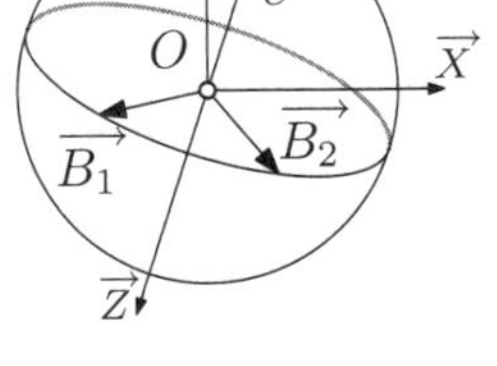


```
  Point_S2<NT> base1() const {
    Dominant_E3 least_dom = least_dominant(_x, _y, _z);
    Vector_E3<NT> least = get_Vector_E3<NT>(least_dom);

    Vector_E3<NT> b1 = cross_product(least, Vector_E3<NT>(_x,_y,_z));
    return Point_S2<NT>(b1.x(), b1.y(), b1.z());
  }
```

The second point B_2 can then be calculated as the cross product $\overrightarrow{c} \times \overrightarrow{B_1}$.

```
  Point_S2<NT> base2() const {
    NT b2x,b2y,b2z;
    Point_S2<NT> b1 = base1();
    cross_product(_x,_y,_z, b1.x(), b1.y(), b1.z(), b2x, b2y, b2z);
    return Point_S2<NT>(b2x, b2y, b2z);
  }
```

9.4 A Spherical Segment

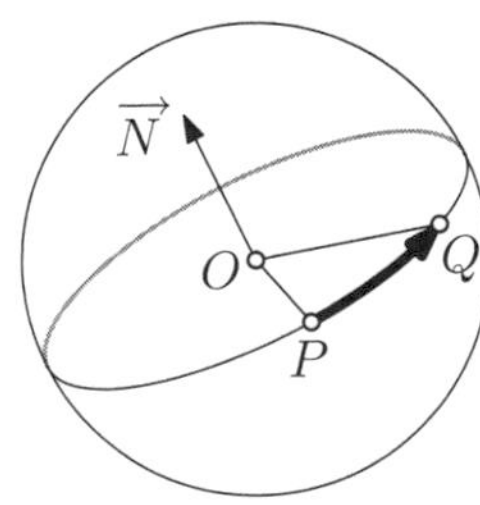

A spherical segment is defined by an ordered pair of spherical points and consists of one set of points traversed when moving from the source point to the target point. That much is evident, but otherwise the design of a spherical segment leaves many options open.

If two antipodal points are allowed to define a segment, for instance, or if the longer segment on the circle defined by two points is intended, then a spherical circle must be stored in each spherical segment object. The orientation of the circle would indicate which of the two segments is meant and, in the case of antipodal points, would specify the circle on which the segment lies.

Introducing a circle also acts as a cache that stores factors in a determinant when calculating the orientation of a third point with respect to the segment.

But numerical precision is one difficulty, for if one computes a circle from a pair of points, testing whether one of the two points lies on the circle may fail. Since precision needs in any case to be treated carefully, we choose the convenience of storing the circle.

```
template<typename NT>
class Segment_S2
{
  Point_S2<NT> _source, _target;
  Circle_S2<NT> _circle;
public:
  Segment_S2() : _source(), _target() {}
  Segment_S2(
        const Point_S2<NT>& source,
        const Point_S2<NT>& target)
    : _source(source), _target(target)
  {
    assert(_source!=_target && _source!=-_target);
    Direction_E3<NT> D = cross_product(
                         _source.get_Direction_E3(),
                         _target.get_Direction_E3());
    _circle = Circle_S2<NT>(D);
  }
  Segment_S2(
        const Point_S2<NT>& source,
        const Point_S2<NT>& target,
        const Circle_S2<NT>& circle)
    : _source(source), _target(target), _circle(circle)
  {
    assert(_circle.contains(_source) && _circle.contains(_target));
  }
  ...
};
```

A constructor whose parameters are two nonantipodal points can deduce the circle the two points define. Computing its orientation as a cross product will capture the shorter of the two segments. A second constructor makes it possible to represent segments described by two antipodal points. It is also useful for segments that are nearly coincident or antipodal, since the reliability of a cross product will diminish the closer the two points are to being dependent vectors.

In practice it is simpler to disallow spherical segments that are larger than half a circle, but it is also possible to make such segments legal and pay a price in the increased complexity needed to implement functions such as intersection operations. This is not unlike the decision taken by many geometric and graphical libraries to handle only triangles and disallow other (convex) polygons. The simplicity of the library is offset by constraining the application programmer.

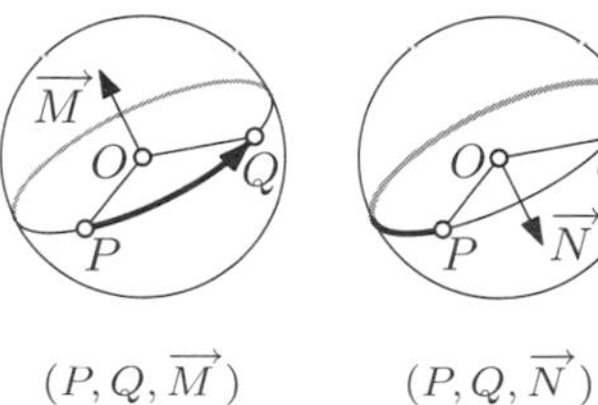

$(P, Q, \overrightarrow{M})$ $(P, Q, \overrightarrow{N})$

A degenerate segment defined by two coincident points is not necessarily also associated with a spherical circle. In practice a system that handles spherical objects will in any case need to confirm whether degeneracy has occurred and suitably handle such events, perhaps by replacing segments by points.

Computing whether a spherical point P_3 lies in the positive halfsphere defined by two points P_1 and P_2 is simply a matter of computing a determinant.

```
template<typename T>
Oriented_side
oriented_side(
        const Point_S2<T>& p1,
        const Point_S2<T>& p2,
        const Point_S2<T>& p3)
{
  T d = determinant(
            p1.x(), p1.y(), p1.z(),
            p2.x(), p2.y(), p2.z(),
            p3.x(), p3.y(), p3.z());

  return enum_Oriented_side(d);
}
```

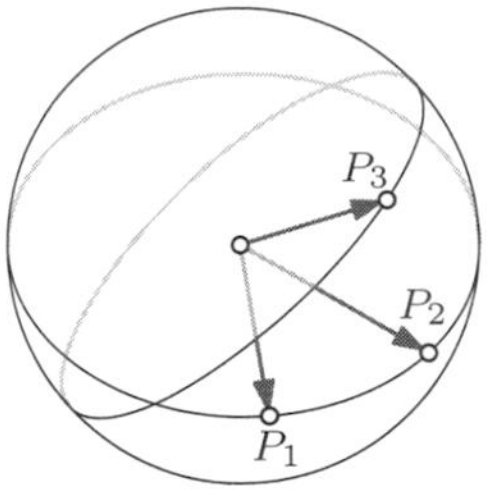

Additional fundamental objects that can be defined in S^2 are a spherical triangle and a spherical polygon. A convex spherical polygon can be defined as one that does not include two antipodal points—a definition Berger [10] credits to Cauchy. As with the other spherical objects that we have so far encountered, restricting spherical polygons to be convex simplifies its implementation, as will be seen when looking at intersections in S^2.

9.5 Intersections

Incidence of a Spherical Point and Circle

Determining whether a point is incident to a circle reduces to asking whether the two corresponding directions are orthogonal. In that case their inner product vanishes.

```
template<typename T>
class Circle_S2
{
  ...
  bool contains(const Point_S2<NT>& p)
  {
    return (dot_product(_x,_y,_z, p.x(),p.y(),p.z()) == 0);
  }
};
```

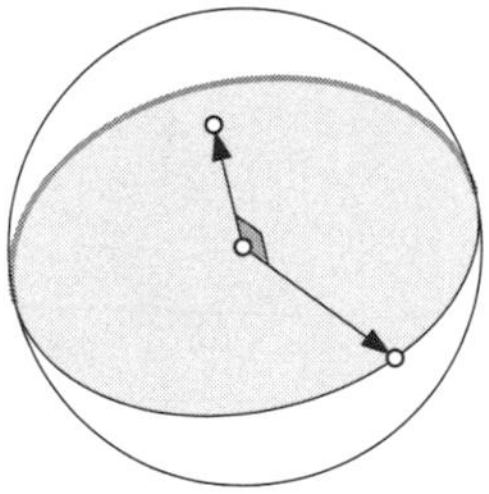

Determining the Points of Intersection of Two Spherical Circles

Two noncoincident spherical circles intersect in two spherical points. The cross product of the normals of the two planes defining the two circles determines one point. Its antipode is the other point of intersection.

Yet we will see momentarily in Chapter 11 the elegance that results from extending Euclidean geometry and declaring that no two lines are parallel. Rather than encumbering our source code with **if** statements confirming that two lines are not parallel before intersecting them in E^2, the projective plane P^2 will make it possible to develop a system in which two distinct lines always intersect at exactly one point. By contrast, a function that determines the intersection of two spherical circles in S^2 could be made to be rather awkward. The client would always be sent back two points and would be asked to choose between them. But there is a way to return to the elegance of one point per intersection [104, Chapter 6]. We abandon commutativity, which is in any case a small price to pay. We declare *the* point of intersection of two spherical circles $\overrightarrow{c_1} \cap \overrightarrow{c_2}$ to be the cross product $\overrightarrow{c_1} \times \overrightarrow{c_2}$. The antipodal point is returned by $\overrightarrow{c_2} \cap \overrightarrow{c_1}$. The client interested in both points would either call the intersection function two times or (more efficiently) compute the antipodal point of the first point obtained.

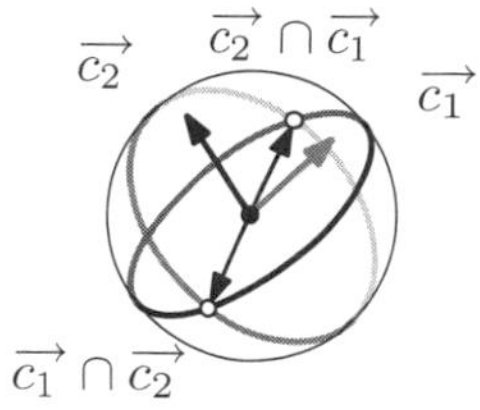

9.6 Exercises

9.1 The adjacent image shows a set of four spherical circles. Three spherical circles look reasonable, but one looks rather unusual. Without seeing the program that produced this image, suggest the reason two tones were produced at quick alternation.

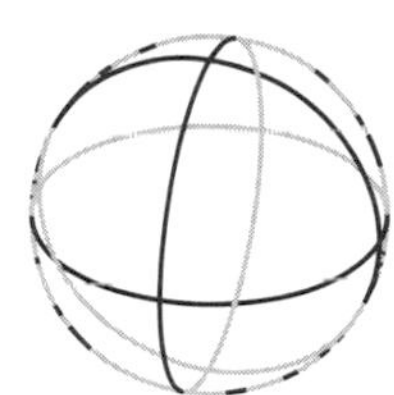

9.2 In the adjacent image the (minuscule) regions where curves overlap are striking. Mention how this overlap may have appeared this way and suggest a way to generate a more reasonable image.

9.3 Refer to Appendices B and C and implement a prototype for visualizing planet earth. Your system will maintain lines of longitude along the y axis of the viewing window and will provide an interface for rotating and for zooming. Draw the outline of continents after devising rudimentary coordinates for a reasonable number of vertices.

9.4 Read ahead on Platonic solids (§ 26.3) and then modify the start-up code labeled "spherical_model" such that for each vertex of a Platonic solid, the circle defined by the circular point coincident with that vertex is drawn. Generate one image for each solid and argue whether the strategy above should lead to a regular tiling of the sphere when starting from the vertices of Platonic solids.

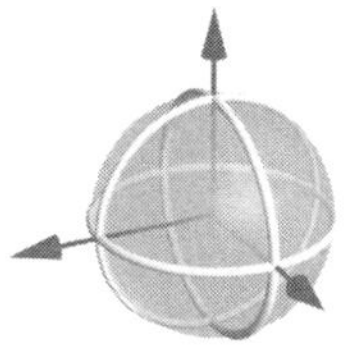

For many other examples of potential sphere partitions, see Wenninger's *Spherical Models* [116].

9.5 Read ahead on Platonic solids (§ 26.3) and on Euler operators (Chapter 27) and then implement the following system.

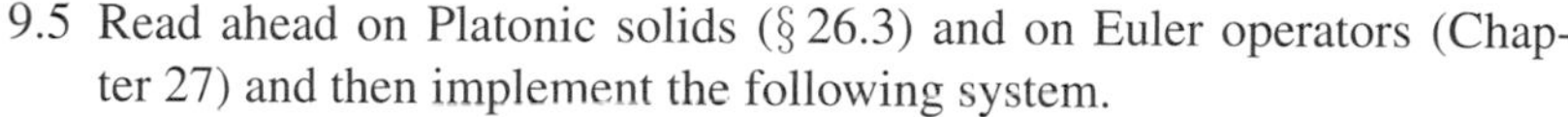

Choose a Platonic solid and implement the operation of either tapering a vertex or insetting a face. Apply the operation you chose on all the vertices (faces) of the solid and visualize the resulting solid as its vertices, edges, and faces project on S^2.

Visualize each face by implementing a recursive partitioning algorithm on each resulting spherical face. A face is divided if the angle for each of its edges (segments) is larger than a given threshold. Divide a triangle

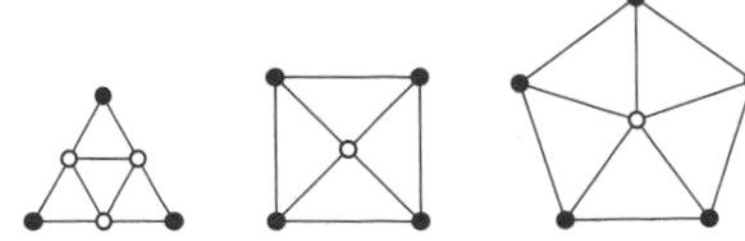

by inserting three additional vertices at the midpoint of each edge and divide any other regular polygon by inserting a point at the center of the polygon. Begin from the start-up code labeled "spherical_model."

9.6 Implement a function for the intersection of a spherical circle and a spherical segment in S^2.

9.7 Implement a function that returns a Boolean flag to signal whether two spherical segments in S^2 intersect.

9.8 Implement a function that finds the point of intersection of two spherical segments in S^2. Use a precondition that both segments are strictly smaller than half a spherical circle.

10 Rotations and Quaternions

Orthogonal matrices (§ 4.6) and orthogonal transformations (§ 4.7) would appear to perform rotations quite adequately. Yet we are not content. We suspect that an operation as elementary as rotation can be expressed more concisely (and hence more elegantly) than using a 2×2 matrix for E^2 or using a 3×3 matrix for E^3. It would also appear that the set of rotations is the natural set to study as the set of transformations that can be applied on objects in S^1 or on S^2. The magnitude of a vector in these two spherical geometries is insignificant and so a matrix that is also capable of scaling is surely excessive for capturing rotations. Rotations in the plane are simple and have been perfectly understood when complex numbers became an accepted notion, but rotations in space had to wait until the concise object that can capture them—the *quaternion*—was discovered in 1843 at the hand of one Irish man. Quaternions begin with a story involving Hamilton, a walk with his spouse, and a charming instance of defacing public property (a bridge) that has inspired historians of mathematics ever since.

10.1 Rotations in the Plane

How many ways are there to represent rotations in the plane? Chapter 4 discussed that rotations are a subset of objects captured by affine transformations, but here we seek rotation objects that we can apply on objects in the geometry of the circle S^1. A design using the basis vectors of the rotation would ensure during construction that the two unit vectors $\overrightarrow{u_1}$ and $\overrightarrow{u_2}$ are indeed orthogonal. Since directions in the plane are only normalized on demand, we can also not rely on directions being unit vectors; two explicit vector normalizations are needed.

```
template<typename NT>
class Rotation_S1
{
  Vector_E2<NT> _u1;
  Vector_E2<NT> _u2;
public:
  Rotation_S1() : _u1(1,0), _u2(0,1) {}
  Rotation_S1(const Vector_E2<NT>& u1, const Vector_E2<NT>& u2)
    : _u1(u1), _u2(u2)
  {
    assert( inner_product(_u1,_u2) == 0 );
    u1.normalize();
```

```
        u2.normalize();
    }
    ...
};
```

A more economical design is to store only one of the two vectors, say $\overrightarrow{u_1}$, and deduce the value of the second, $\overrightarrow{u_2}$, whenever needed. This is possible because the matrix M

$$M = [\overrightarrow{u_1}, \overrightarrow{u_2}] = \begin{bmatrix} u_1.x & u_2.x \\ u_1.y & u_2.y \end{bmatrix}$$

is proper orthogonal, and so $\overrightarrow{u_2}$ can be constructed by applying a quarter-turn counterclockwise rotation to $\overrightarrow{u_1}$. Rotations are multiplications in the complex plane. If a given vector is interpreted as lying in the complex plane, the desired rotation can be applied by multiplying by a unit complex number (one on the unit circle). The complex number needed to apply a $\frac{\pi}{2}$-rotation is $(0 + 1 \cdot i)$ and so $\overrightarrow{u_2} = (u_1.x + u_1.y \cdot i) \cdot (0 + 1 \cdot i) = -u_1.y + u_1.x \cdot i$.

An even more economical design is to use the rotation angle.

```
template<typename NT>
class Rotation_S1
{
    NT rotationAngle;
public:
    ...
};
```

But this design is unattractive because slow trigonometric functions would need to be invoked before the transformation is applied.

Such a long-winded approach is evidently not necessary: If we know that rotation is equivalent to multiplication in the complex plane, we might as well make do without a special rotation object in the first place. But this set-up is a useful prelude to quaternions and order-3 orthogonal transformations, which will be applied on S^2. In any case, we verify that the transformed point P' can be obtained from P by applying $P' = (P.x + P.y \cdot i) \cdot (r.x + r.y \cdot i)$ and that the equivalent 2×2 orthogonal matrix is indeed

$$\begin{bmatrix} r.x & -r.y \\ r.y & r.x \end{bmatrix}.$$

And so an instance of Rotation_S1 can conceivably act merely as a wrapper for a unit complex number.

```
template<typename NT>
class Rotation_S1
{
    std::complex<NT> unitz;
public:
    Rotation_S1() : unitz(1,0) {}
    Rotation_S1(const Direction_E2<NT>& d)
    {
        NT m = std::sqrt(d.x()*d.x() + d.y()*d.y());
```

```
    unitz = std::complex<NT>(d.x()/m, d.y()/m);
  }
  Rotation_S1(const NT& rx, const NT& ry)
  {
    NT m = std::sqrt(rx*rx + ry*ry);
    unitz = std::complex<NT>(rx/m, ry/m);
  }

  Point_S1<NT> rotate(const Point_S1<NT>& P) const
  {
    NT x = unitz.real() * P.x() - unitz.imag() * P.y();
    NT y = unitz.imag() * P.x() + unitz.real() * P.y();
    return Point_S1<NT>(x,y);
  }

  Segment_S1<NT> rotate(const Segment_S1<NT>& seg) const
  {
    return Segment_S1<NT>(rotate(seg.source()), rotate(seg.target()));
  }
  ...
};
```

Taking the ultra-purist approach of adding a layer of abstraction to complex numbers when the operations needed from that class could have easily been implemented has advantages. First, it illustrates how delegating the combination of rotations to complex multiplication (shown below) evidently discards some information: If the combined angle exceeds a full turn, only the remainder is stored and the rotation object maintains no recollection of the history. Second, we will encounter the same ideas again in the context of rotations on the sphere, and it is instructive to see most notions already present in the geometry of the circle. Last, such encapsulation gives us the freedom to use an alternative rotation scheme without making the changes perceptible from the client side.

```
template<typename T>
class Rotation_S1
{
  ...
  Rotation_S1<T> operator*(const Rotation_S1<T>& R2)
  {
  return unitz * R2.unitz;
  }
};
```

10.2 Rotations in Space

The rotation objects discussed in this section act equally well on spherical objects as they do on objects in Euclidean space. Yet the discussion is delayed to this point (rather than introduced in Chapter 4) to ensure that spherical objects and motions on a sphere are appreciated first.

Can transformations other than rotations be applied in spherical geometry? If it seems that it is too limiting for rotations to be the only transformation applicable, it is because they are not. A richer set of transformations can be applied on a geometry that has the same topology as spherical geometry, but because the transformations are distinct, the geometry shall also be considered distinct. This *oriented projective geometry* is discussed in Chapters 14 and 15.

The properties of quaternions can all be derived once we take faithfully the following two equations as those of a new type of imaginary number. This new object, dubbed a *quaternion* because it has four parts, will be to 3D what complex numbers are to 2D—a natural rotation object [31].

$$q = r + xi + yj + zk,$$
$$i^2 = j^2 = k^2 = ijk = -1$$

Once this definition has been accepted, a new algebra can be shown to apply 3D rotations:

$$ij = -ji = k, \qquad jk = -kj = i, \qquad \text{and} ki = -ik = j.$$

If $\overrightarrow{v}$ is a 3D vector $\overrightarrow{v}(x, y, z)$, a quaternion can be defined as the sum of a real scalar r with the inner product of $\overrightarrow{v}$ and the new imaginary basis (i, j, k):

$$q = [r + \overrightarrow{v} \cdot (i, j, k)]. \tag{10.1}$$

It then makes sense to talk about the *real part* of a quaternion, r, and its *imaginary part*, $\overrightarrow{v} \cdot (i, j, k) = xi + yi + zk$. A quaternion with a zero real part it termed a *pure imaginary* quaternion.

Using the above identities, two quaternions $q_1 = r_1 + x_1 i + y_1 j + z_1 k$ and $q_2 = r_2 + x_2 i + y_2 j + z_2 k$ can be added:

$$q_1 + q_2 = (r_1 + r_2) + (x_1 + x_2)i + (y_1 + y_2)j + (z_1 + z_2)k$$

and multiplied:

$$\begin{aligned} q_1 * q_2 = &(r_1 r_2 - x_1 x_2 - y_1 y_2 - z_1 z_2) \\ &+ (r_1 x_2 + r_2 x_1 + y_1 z_2 - y_2 z_1)\, i \\ &+ (r_1 y_2 + r_2 y_1 + z_1 x_2 - z_2 x_1)\, j \\ &+ (r_1 z_2 + r_2 z_1 + x_1 y_2 - x_2 y_1)\, k. \end{aligned} \tag{10.2}$$

As with complex numbers, it is useful to define the *conjugate* of a quaternion by negating the imaginary parts:

$$q^* = [r - v.(i, j, k)] = r - xi - yj - zk.$$

The product of a quaternion by its conjugate is a real-valued scalar:

$$\begin{aligned} q * q^* = &(rr + xx + yy + zz) \\ &+ (rx - rx + yz - yz)i \\ &+ (ry - ry + zx - zx)j \\ &+ (rz - rz + xy - xy)k \\ = &(r^2 + x^2 + y^2 + z^2). \end{aligned}$$

And so if the *norm* of a quaternion is defined as

$$N(q) = r^2 + x^2 + y^2 + z^2,$$

the multiplicative inverse of a quaternion becomes $q^{-1} = q^*/N(q)$, with the norm (not its square root) as the denominator. As can be easily verified, the left and right multiplicative inverses of a quaternion are equal.

If the norm of a quaternion q is 1, we say that q is a *unit quaternion*. Dealing with unit quaternions is convenient. Their products, their conjugates, and their inverses are all also unit quaternions.

Storing the Imaginary Component as a Vector

If one finds writing either Eq. (10.2) or its implementation tiring, delegating the computation to vectors as expressed in Eq. (10.1) (due to Gibbs [46]), can simplify both:

$$\begin{aligned} q_1 =& [r_1 + v_1 \cdot (i, j, k)] \\ q_2 =& [r_2 + v_2 \cdot (i, j, k)] \\ q_1 * q_2 =& [r_1 r_2 - v_1 \cdot v_2, [r_1 v_2 + r_2 v_1 + v_1 \times v_2] \cdot (i, j, k)] . \end{aligned}$$

A quaternion object is then implemented by encapsulating a vector, and the quaternion product is expressed as vector dot and cross products.

```
template<typename T>
class Quaternion_v
{
  T _r;
  Vector_3<T> _V;
  ...
  Quaternion_v<T>
  operator*(const Quaternion_v<T>& q2) const
  {
  T newr = _r * q2._r – dot_product(_V, q2._V);
  Vector_3<T> newv = q2._V * _r + _V * q2._r + cross_product(_V, q2._V);
  return Quaternion_v<T>(newr, newv);
  }
```

Rotation Using Quaternions

Just as with complex numbers, an arbitrary nonzero quaternion can be used to perform rotations in 3D. But unlike complex numbers, the operation requires the use of both the quaternion and its inverse.

If we construct a quaternion from a (not necessarily unit) vector $\overrightarrow{v}$ in 3D and, for simplicity, write the conversion as $v = \overrightarrow{v}$ where v is a pure imaginary quaternion, the result of the product

$$v' = q^{-1} v q$$

effects the rotation of $\overrightarrow{v}$ into $\overrightarrow{v'}$, represented as the pure imaginary quaternion v'.

Finding the Quaternion Corresponding to an Angle-Axis Pair

Rotation by an angle θ about the 3D axis $\overrightarrow{v}(x, y, z)$ can be effected by the quaternion

$$q = \cos\frac{\theta}{2} + \sin\frac{\theta}{2}\overrightarrow{v} \cdot (i, j, k). \tag{10.3}$$

An arbitrary nonzero quaternion is suitable for performing rotation because determining the conjugate will require dividing by the norm. Since the number of vectors that are mapped is typically large for each rotation, it is useful to avoid constantly dividing by the norm and restricting one's attention to unit quaternions. In that sense the class Rotation_S2 completely mirrors the connection between Rotation_S1 and complex numbers. Rotation_S2 is merely a wrapper for a unit quaternion.

```
template<typename T>
class Rotation_S2
{
    Quaternion<T> _unitq;
    ...
};
```

The convenience of rotation using a unit quaternion stems from the conjugate being equal to the inverse. Hence, for a unit quaternion we can write

$$v' = q^* v q.$$

Finding the Linear Transformation Corresponding to a Quaternion

The linear transformation corresponding to a quaternion can be found by the somewhat tedious (since it involves 64 multiplications) process of converting the two quaternion products to a matrix-vector multiplication, discarding the imaginary quantities in the process.

The rotation of $\overrightarrow{v}(a, b, c)$ using $q(r, x, y, z)$ is effected by

$$(0, x', y', z') = (r, -x, -y, -z)(0, a, b, c)(r, x, y, z).$$

Expanding, the requisite linear transformation matrix is found to be

$$M = \begin{bmatrix} 1 - 2y^2 - 2z^2 & 2xy + 2rz & 2xz - 2ry \\ 2xy - 2rz & 1 - 2x^2 - 2z^2 & 2yz + 2rx \\ 2xz + 2ry & 2yz - 2rx & 1 - 2x^2 - 2y^2 \end{bmatrix}.$$

Finding the Angle and Axis of a Quaternion

Given a quaternion $q(r, x, y, z)$, it is easy to find the angle θ and the axis of rotation $\overrightarrow{v}$ from Eq. (10.3).

$$\theta = 2\cos^{-1} q_r$$

$$\overrightarrow{v} = \left(\frac{q_x}{\sin(\cos^{-1} q_r)}, \frac{q_y}{\sin(\cos^{-1} q_r)}, \frac{q_z}{\sin(\cos^{-1} q_r)} \right)$$

Quaternion Interpolation in S^3

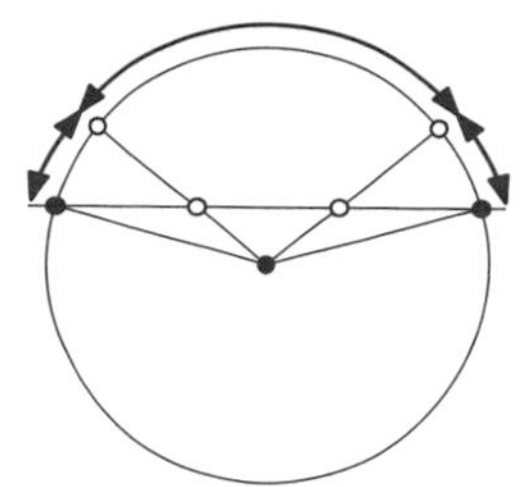

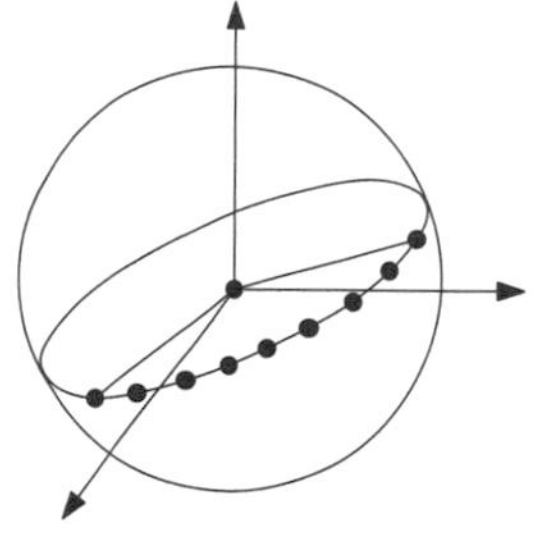

Interpolating linearly on the arc between two spherical points does not lead to linear interpolation on a sphere, or *spherical linear interpolation*—for otherwise trisecting a segment, for example, would lead to trisecting an angle [103]—the instance of the problem on the circle S^1 (Chapter 8). Likewise, interpolating spherical points in S^2 cannot be done by cutting across the sphere, but must be performed on the surface of the sphere.

Interpolating quaternions also needs to be done on a sphere. The sphere in question is the unit sphere S^3 embedded in Euclidean 4D [10]. The 4D points representing quaternions are constrained to lie on a sphere since any two quaternions related by a positive scalar multiple represent an identical orientation. In that quaternions share the constraint with directions in the plane, but the correspondence with orientation in 2D falters: Not all points in S^3 represent a unique orientation, because the quaternions q and $-q$ capture the same orientation. This can be seen by extracting the (oriented) axis and the (oriented) angle of rotation. If both the axis and the angle are reversed, the resulting rotation remains unchanged.

If $q_1 q_2 = \cos\theta$, spherical linear interpolation [97, 46] is determined by

$$slerp(q_1, q_2, u) = \frac{\sin(1-u)\theta}{\sin\theta} q_1 + \frac{\sin u\theta}{\sin\theta} q_2.$$

10.3 Exercises

10.1 a. Multiply the two quaternions

$$q_1 = 1 + 0i + 1j + 0k \text{ and } q_2 = 0 + 1i + 0j + 1k.$$

b. Find $1/q_3$ for a quaternion q_3

$$q_3 = a + bi + cj + dk; \qquad a^2 + b^2 + c^2 + d^2 = 1.$$

10.2 Even though it is not possible to trisect an angle using a ruler and a compass, it is possible to trisect a segment. Given a pair of points in the plane, show the steps needed to generate two additional points that divide the given segment into three segments.

10.3 Since rotations in the plane are a special type of rotation in space, one can use quaternions for the latter. Develop a class Rotation_S1 that acts as a wrapper for a unit quaternion specialized to this application.

10.4 The start-up code labeled "orbits" shows the trajectory of a planet "earth" and an orbiting "moon," but the motion is incomplete. Modify the code while incrementally implementing the following:

- The north-south axis of the earth is parallel to the rotation axis.
- The moon orbits around the earth.
- The earth spins around its own north-south axis.
- The axis of the earth tilts during rotation.

The objective is to animate all motions simultaneously, but not necessarily that the generated motions are faithful to reality. In particular, circular planetary trajectories are adequate for our purposes.

10.5 The start-up code labeled "slerp_on_s1" shows the effect of linear interpolation between two distinct and nonantipodal points on S^1. Modify the code to perform spherical interpolation instead. After completing the implementation, study the location of the interpolated point for values 0.25 and 0.75 and explain your observations.

10.6 The start-up code labeled "rotation-group" *slerps* a wire sphere between two orientations. Modify the code such that pressing a key from "a" to "k" animates a tetrahedron from some original position to any of the 11 positions to which it can be moved such that it rotates to a new orientation, but then rests coinciding with the original position (see § 26.3).

10.7 The motion shown in the code labeled "rotation-group" shows quaternion interpolation in S^3. Show two additional animations that exhibit interpolation in S^1 and S^2.

10.8 The start-up code labeled "euler-angles" accompanying Chapter 4 makes it possible to position an object in 3D using Euler angles.

Solve Exercise 10.6, but use interpolation over Euler angles instead. The objective is to be convinced how undesirable the result of interpolating rotation using Euler angles can be.

Make sure that the rotation for each individual Euler angle traverses the shortest path about its own axis (see Exercise 4.6).

10.9 We saw in § 4.10 that three pairs of points define a unique affine mapping in E^2. We will also see in § 11.1 that four pairs of points define a unique projective mapping in P^2. It is easy to see that two pairs of equidistant points define a unique rotation on S^2.

Write a function that takes two pairs of equidistant points in S^2 and that generates the quaternion needed to rotate the first pair of points so that they coincide with the second pair.

Notice that the problem is overconstrained since the distances (or angles) between the points in each pair will, in general, not be equal under floating point.

10.10 Solve Problem 10.9 then choose one of the start-up programs and modify it to visualize a uniform-speed animation taking one pair of points to another pair.

11 Projective Geometry

Projecting a 3D scene on a 2D canvas or image is the process of constructing a mapping between points in the scene and points on the canvas. Such projective mappings, which are the basis of constructing *perspective* images, are the topic of this chapter.

Desargues and Pascal showed projective properties of figures. Dürer before them showed how an artist can construct accurate projections, but the first to set projective geometry as a science was Poncelet. Lying injured among Napoleon's defeated army in 1812, Poncelet's life was spared because he was wearing an officer's uniform [8]. Bored with his captivity, and with access to pen, paper, and candles, but to no references, he developed axiomatic projective geometry in a jail cell in Russia in 1812–1814.

11.1 The Projective Line

We start with the simplest possible projection—one that maps points on one line to points on another. We wish to define a mapping from points on a line l to points on a line m through a *center of perspectivity* E not incident to either l or m (Figure 11.1). The image of a point A on l is the point A' on m. A' is the intersection of the line EA and m.

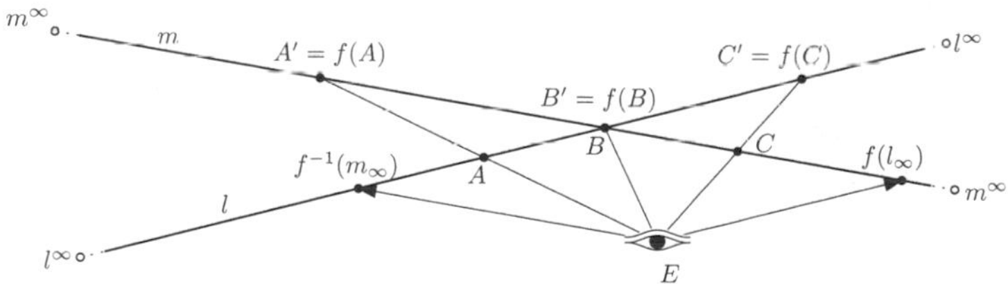

Figure 11.1
Projection between two lines

But consider the point on l at the intersection of the line passing by E and parallel to m and consider also the point on m at the intersection of the line passing by E and parallel to l. We could certainly study the mapping by removing the two points from l and m. With the two points removed, there is a one-to-one correspondence between the points remaining on l and m. But we are interested in projective mappings to construct images (1D images in this case). Having a point in the image map to no point in the scene, or vice versa, is inconvenient. An implementation would have frequently to test whether a point can be mapped before performing the mapping.

The mapping would be more elegant if we studied it without removing the two points and declared instead that either point does map. The basis of projective geometry lies in declaring that each of the two lines is augmented with an *ideal point*, or a point at infinity, that completes the mapping. Thus, each line will consist of a set of affine points in addition to a single additional ideal point. Each such line is an instance of the projective line P^1.

By augmenting the two lines with the points l^∞ and m^∞—and keeping the two points we contemplated removing—the mapping is bijective. Because the perspectivity maps an affine point to an ideal point, we refer to the mapping as a projective mapping. If the two lines l and m become parallel, the mapping would be affine: Each affine point would map to an affine point and the two ideal points would map to one another.

Finding the Center of Perspectivity

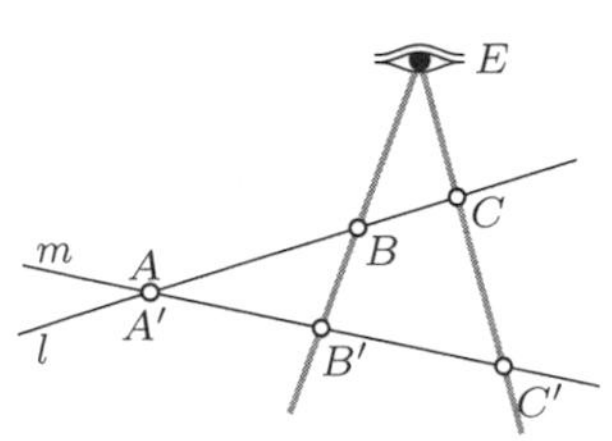

If the two lines and the center of perspectivity are known, then the mapping can be easily found, but what if we reverse the problem and ask for the center of perspectivity given two lines? We would first need to establish the number of pairs of points needed to uniquely define a projective mapping. We saw in Chapter 4 that precisely two pairs of points define an affine mapping. To see that three pairs of points are needed, consider the case when three distinct points are determined on two given lines. We take the liberty of displacing one of the two lines through translations and rotations until the mapping and the center of perspectivity are evident. If the three points A, B, C map to A', B', and C', we can position m such that A and A' coincide and determine E by finding the intersection of BB' and CC'.

This mapping satisfies all our constraints. The three points are mapped to each other through projection from E, the mapping is bijective, and the two ideal points on the two lines project to one point each. That the projectivity between two sets of three distinct points is unique is called *the fundamental theorem of projective geometry*. To establish that theorem, we can be more rigorous by not taking the liberty to move one of the two lines. Given three points on two lines, we proceed instead to find *two* projective mappings. The combination of the two mappings would yield the requisite projective transformation [79]. Choose a third line passing by A' distinct from ABC and $A'B'C'$. Also choose an arbitrary center of perspectivity on AA' distinct from both points. Project once (light projectors) on the new line. Choose the second center of perspectivity at the intersection of $B'B''$ and $C'C''$. Project a second time (dark projectors). By construction the second projection maps A' to itself. The composition of the two transformations is the requisite projective mapping.

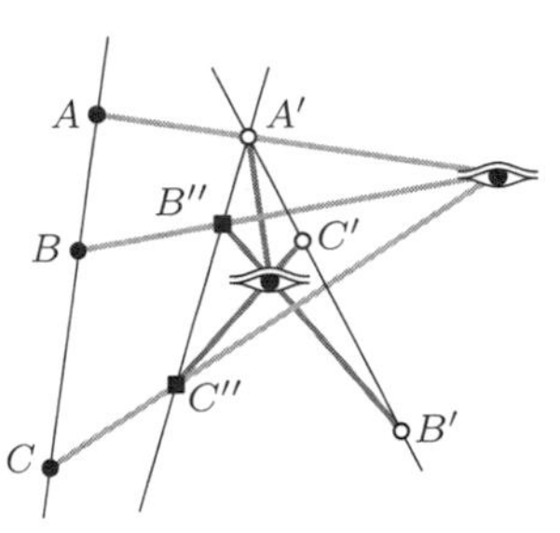

Suppose that the three points are not distinct and that, for instance, B is coincident with C. We might still declare that we have a perspectivity by positioning the eye E at $E = B = C$, but that perspectivity would be a degenerate one; it would map not just B' and C' to $B = C$ but all points on the line $B'C'$ to $E = B = C$. The mapping between A and A' could only be satisfied if A' also coincides with $B' = C'$. We would then have a degenerate mapping from *all* points on one line to a single point on the other. Such a mapping would, naturally, not be invertible. The characteristic of such degenerate mappings, as we shall see after introducing coordinates for points

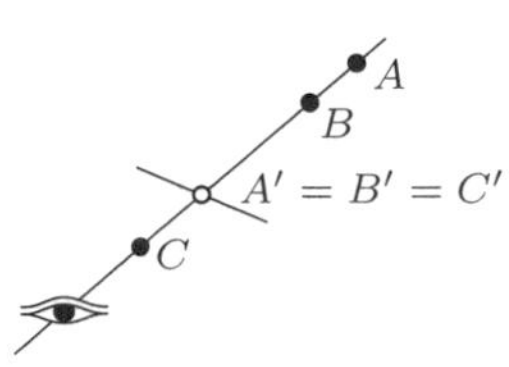

and matrices for transformations in Chapter 12, is that the determinant of the transformation matrix vanishes and the matrix is singular.

Nonseparability of the Projective Line

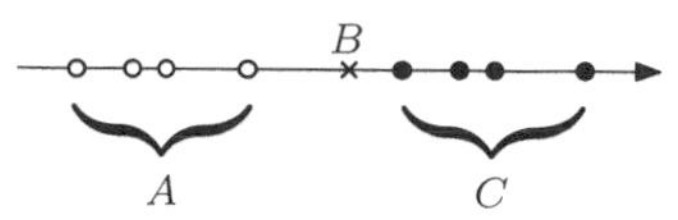

Notions that we take for granted in Euclidean geometry such as separation, sidedness, betweenness, and separability make no sense in projective geometry. The points in the Euclidean plane, for instance, can be classified as lying to the left, to the right, or on a directed line in that plane. In E^2 a point is also said to be on the left, to be on the right, or to be colinear with an ordered pair of points. Separation is exhibited on the Euclidean line as betweenness; a point B is said to lie between two other points A and C if it is impossible to repeatedly apply minuscule transformations to animate the motion of A until it coincides with C without passing by the separating point B. Affine transformations are incapable of changing this order; any nonsingular affine transformation will keep B between A and C.

In contrast, an inherent property of projective transformations is their ability to map an affine point to an ideal point and vice versa. We may position three distinct points A, B, and C in P^1 such that B would appear to separate A and C. In other words, it would appear that a particle at A could not be animated to reach C without passing by C. The futility of using B as a separator, or of attempting to divide P^1 into points on one side or the other (in addition to the separator itself), becomes evident when we consider the projection depicted.

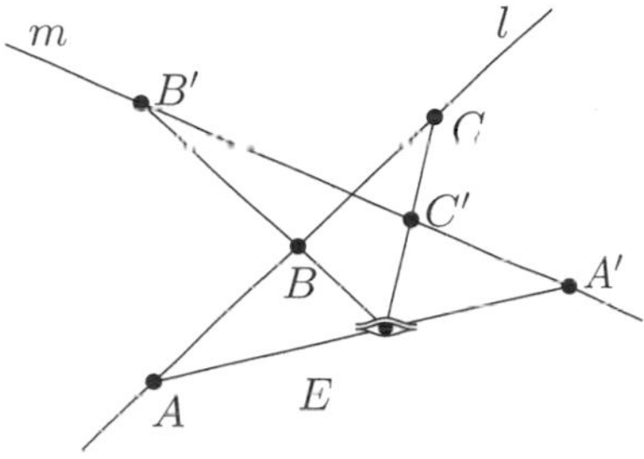

A projection from l to m has now mapped A to A' and C to C'. B' no longer separates A' and C'. But B also does not separate A and C. It will become more evident in Chapter 12 that the projective line is more accurately modeled as a circle, but we already see that that is the case because there is only one ideal point on each projective line. Cutting a projective line at one projective point B yields one connected piece of the circle, not two. It would still be possible to go from A to C by passing by the ideal point of l.

Would it perhaps be possible to restrict the class of projective transformations to disallow those that modify the betweenness relationships? It is reasonable to ask this question, for example, because it is reasonable in Euclidean geometry to restrict transformations to those that do not include a reflection (the transformation matrix then has a positive determinant). If a set of points lies on the right of a separator line in E^2 before a transformation, they continue to lie on its right after transformation. But restricting projective transformations in P^1 to map the ideal point to itself would result in precisely the class of affine transformations. The requisite construction for performing the projection would position the two "projective" lines parallel to one another. *Two* pairs of points now suffice to determine the center of perspectivity and an affine transformation is obtained!

Projection Does Not Preserve Ratios

Affine mappings (Chapter 4) preserve ratios; the ratio of the lengths of two segments does not change after an affine transformation. The ratio of lengths is

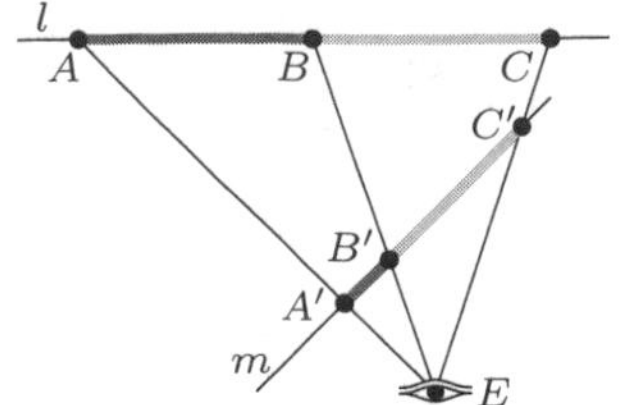

another attribute of affine transformations that is not preserved under projective transformations. The figure illustrates that even though the lengths of AB and BC are equal before projection, they are, in general, not equal after projection.

A Matter of Prepositions

The issue of prepositions was discussed in § 2.5, but it is worth recasting it in the context of projective geometry. Most of the time we consider a given space on its own. The projective line P^1, the projective plane P^2, and the projective space P^3 can all be studied independently as a standalone set of points. In that case we say that a point is *in* the corresponding set. The point models a creature of dimension equal to that of the space—the creature is unable to comprehend any higher dimension. By contrast, if we embed, say, P^1 in P^2 or in P^3, then we refer to a point *on* P^1. The creature is now either two- or three-dimensional, but is limited in its motion to one dimension. The same distinction can be made by using different labels [79]. We could refer to the *intrinsic* projective line when we think of P^1 on its own and to the *extrinsic* projective line when we think of it as lying in P^2 or P^3.

11.2 The Projective Plane

The two projective lines have so far been embedded in a plane to perform the mapping. Likewise when we now move one dimension higher to map between two projective planes P^2, the two planes will be embedded in three-dimensional space.

This embedding provides us with the first chance to argue that what we are doing is useful. Since 3D objects are, in general, modeled as *faceted* objects and since the projection of an object can be examined as the projection on a canvas, or *image plane*, of many facets, the projection of a 3D object can be studied as many instances of projecting between two projective planes. In each instance the plane carrying a facet is referred to as the *object plane*. As with projections between lines, the aim is to establish a bijective mapping between points on the object plane and points on the image plane.

Mapping of a Segment and a Line

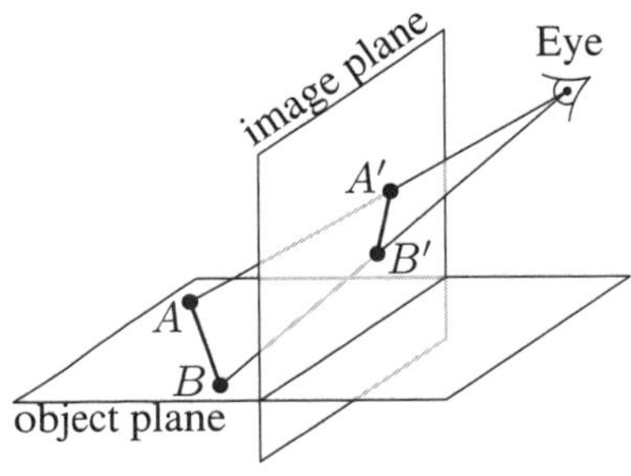

Consider a segment AB lying on the object plane. To determine the projection of AB on the image plane as seen through E, we construct the two *lines* EA and EB. The intersections of the two lines with the image plane determine the two projection points A' and B'. We also know that the projection of all points lying on the line (segment) AB are colinear on the image plane; they lie at the intersection of the plane EAB with the image plane. Thus, the projection of the segment AB is also a segment, $A'B'$, and the projection of the line AB is the line $A'B'$. No constraints are imposed on the position of either A or B. In particular, points A and B may lie on opposite sides of the image plane.

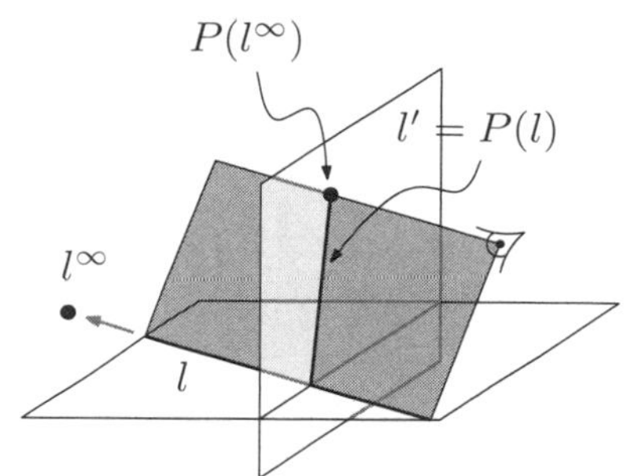

Once again the map between a line l and its projection $P(l)$ has a property that distinguishes it from the affine maps discussed in Chapter 4. The point l^∞, the ideal point of l, maps to an affine, or nonideal, point $P(l^\infty)$.

The projection of l is determined by constructing the plane in 3D that carries both E and l. The intersection of that plane (shaded in the figure) with the image plane determines $l' = P(l)$.

To find $P(l^\infty)$ we determine the line parallel to l and passing by E, and then find its intersection with the image plane. Observe that the mapping is bijective; l has only one ideal point.

A few special cases merit attention. If E lies on the image plane, the projection is singular: All points on l map to a point that coincides with E. If E lies on the object plane, $P(l)$ coincides with the intersection of the two planes. If E lies on l, the projection is again singular, and the line $P(l)$ collapses into the one point where l meets the image plane.

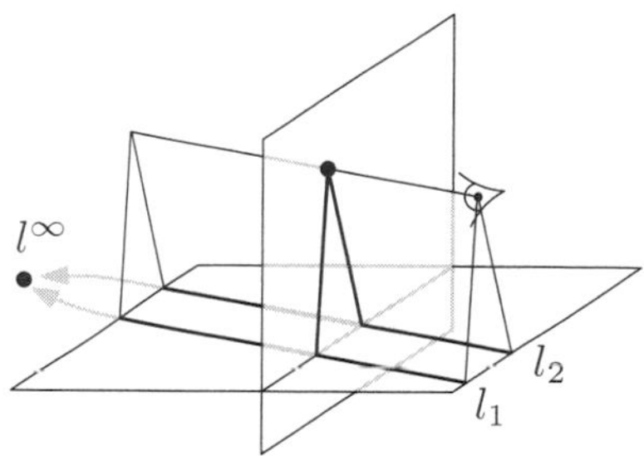

The fact that these cases mirror those in § 11.1 is no accident. All details are identical because we are now in a linear subspace of 3D space.

But if two parallel lines l_1 and l_2 are projected, the above method for finding the image of the ideal point of each line would produce the same point on the image plane. We must conclude that if $P(l_1^\infty)$ coincides with $P(l_2^\infty)$, it is because l_1^∞ and l_2^∞ are also coincident.

Now consider a set of parallel lines S_1 in the object plane and likewise another two sets S_2 and S_3. From the argument above the lines in S_1 meet at one ideal point S_1^∞. Similarly, those in S_2 and S_3 meet in S_2^∞ and S_3^∞.

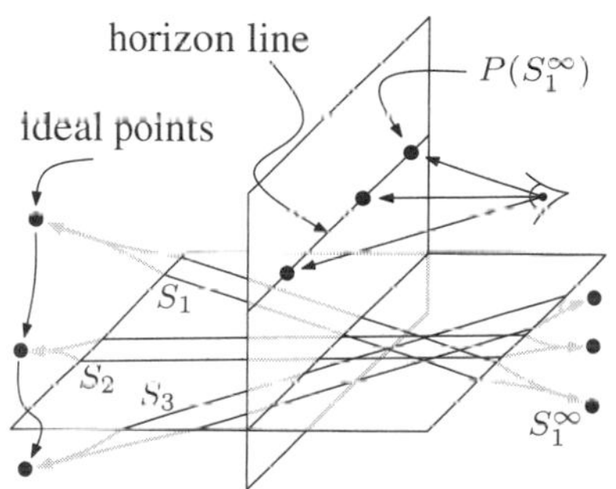

The projections of S_1^∞, S_2^∞, and S_3^∞ through E are $P(S_1^\infty)$, $P(S_2^\infty)$, and $P(S_3^\infty)$. Because the three sets of parallel lines are coplanar, we must conclude that the projections of their ideal points ($P(S_1^\infty)$, $P(S_2^\infty)$, and $P(S_3^\infty)$) are colinear. Construct the plane passing by E and parallel to the object plane. The line defined by $P(S_1^\infty)$, $P(S_2^\infty)$, and $P(S_3^\infty)$ is determined by the intersection of that plane with the image plane.

Because the line thus defined is akin to what we commonly perceive as a horizon, it is called the *horizon line*. Just as an ideal point on the projective object plane maps to an affine point on the projective image plane, so does *the* ideal line on the object plane map to an affine line on the image plane.

It is necessary for the image plane also to be a projective plane. To see that that is the case, consider a point on the plane parallel to the image plane and passing by the center of perspectivity E. The image of that point is an ideal point on the image plane.

Notice also that a point on the opposite side from the image plane of the plane parallel to the object plane and passing by E would project onto a point on the opposite side from the object plane of the plane parallel to the object plane and passing by E.

Projection Does Not Preserve Parallelism

Perhaps the most prominent feature with which we recognize perspective images is that it destroys parallelism. When surrounded by tall buildings, we experience their sides as intersecting rather than as parallel lines. We also perceive railroad tracks as meeting at a point.

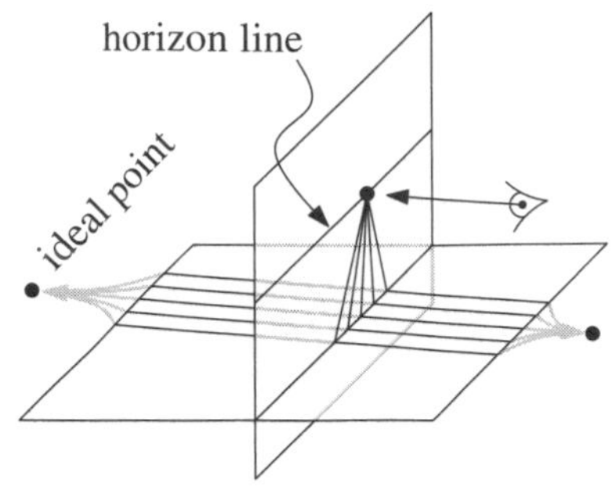

As long as the image plane is not parallel to the set of parallel lines (i.e., to the object plane), the projection of the lines will meet at an (affine) point on the object plane. Projective geometry is one possible model of the physical world. Lines that are parallel in a Euclidean model remain parallel in a projective model, but, unlike in a Euclidean model, they intersect at a point.

Nonseparability of the Projective Plane

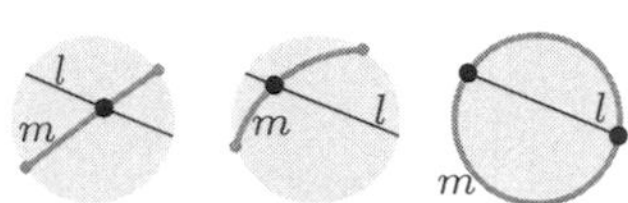

Consider two intersecting lines l and m in the projective plane and imagine someone traveling from the intersection point along the line l. The traveler eventually reaches the ideal point of l and emerges on the opposite side of m. This means that it is possible in the projective plane to go from one side to the other of some line m without crossing the line. A line m does in fact not have two, but only one side.

Imagine constructing the projective plane P^2 out of some rubberlike material. Now proceed to push the material, along with the two lines l and m, until the line m reaches the ideal line of the plane P^2. The point of intersection of l and m has been dragged all along and now that point is an ideal point. The one-sidedness of m is now evident: All points of P^2 lie on that one side.

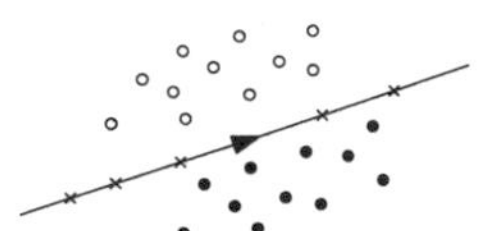

The implication is that the projective plane is not separable. A line in the Euclidean plane divides the points in the plane into three sets: those lying on either side of the line as well as those lying on the line.

Because by contrast a line in P^2 has only one side, the projective plane is said to be nonseparable. A given line on P^2 divides the points in P^2 into *two* sets: those that lie on the line and those that do not. And so it is possible to implement a predicate that determines whether a point is incident to a line in the projective plane, but not whether two points lie on the same side of the line.

Topology of the Projective Plane

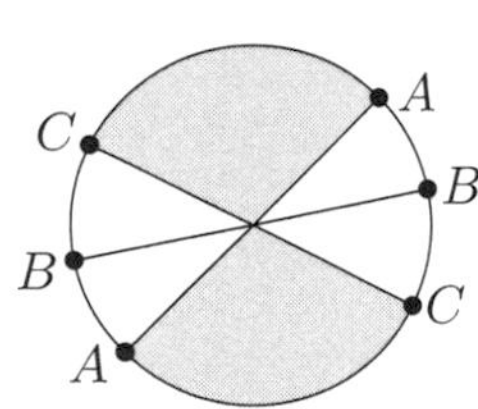

Now suppose that we do attempt to build the projective plane out of some fabric, what would the resulting object look like? We would take a circular piece of fabric and stitch together opposite points. In doing so we would be performing physically what we earlier decided to do logically: to declare opposite points identical. But after joining the two points marked A together, and likewise for B and C, we will no longer be able to proceed; the projective plane cannot be embedded in 3D space without self-intersecting. The structure obtained after connecting three opposite points is the familiar Möbius strip. Indeed, all we have to do to complete the projective plane from a Möbius strip is to stitch a band at the boundary of the strip and pull until the entire boundary is a single point [49, 87].

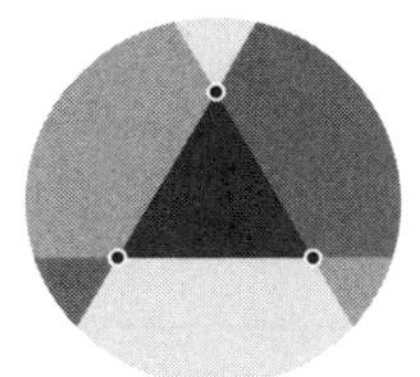

Another diagram [28, 15] illustrates the topology of the projective plane as well as hints to its connections with spherical and oriented projective geometries. Consider three points and the three lines they define in the projective plane P^2. Each pair of points defines *two* segments, an *internal*, entirely affine, segment, and an *external* segment that includes an ideal point. The three lines partition P^2 to four triangles, shown in different shades of gray. In addition to the internal, entirely affine, triangle, there are three external triangles. Each such triangle includes one (not two) segment on the ideal line.

Difficulty in Defining Projective Segments

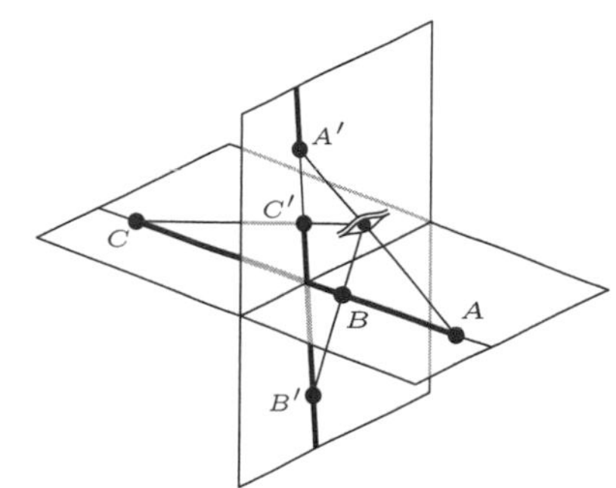

Computing in the projective plane holds other problems. We earlier looked at the projection of a segment AB from one plane to another. But the notion of a segment itself poses a difficulty. To see the difficulty, it suffices to repeat a diagram discussed under P^1, but after embedding it in P^2.

The two points A and C project to the points A' and C'. But even though all points of the segment AC are affine, not all points on the segment $A'C'$ are affine.

Suppose that we define a class Segment_P2. Consider also that we define a class Transformation_P2 (sketched in § 12.6). We would encounter a difficulty after applying a transformation on a given segment; there will be no way to determine which of the two potential projective segments resulted from the transformation. We continue this argument in § 12.1.

11.3 The Projective Space

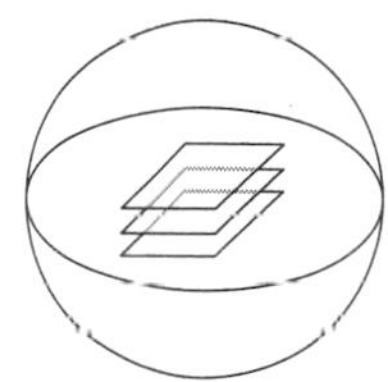

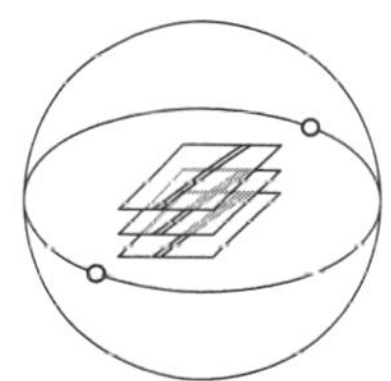

It is once again easiest to think of the projective space P^3 as consisting of the set of points in the Euclidean space E^3 in addition to the points on a sphere at infinity. A point may be either an affine point or an ideal point. A plane may be either affine; it then includes one ideal line, or ideal; and then all lines on it are ideal. A set of parallel planes meet at one ideal line.

A line may be either affine or ideal. An affine line consists of a set of affine points in addition to an ideal point—the two points at infinity being identified. An ideal line lies on the ideal plane. The topology of the line remains equivalent to a circle.

There are an infinite number of projective planes embedded in projective space, each of which has the structure of the canonical projective plane P^2. So, for example, two lines lying in a projective plane intersect at an affine point if they are not parallel and at an ideal point if they are. Likewise, two projective planes always intersect: at an affine line if they are not parallel and at an ideal line if they are.

A plane and a line not lying in the plane always intersect at a single point. If either the plane or the line is ideal, the point of intersection is ideal. The intersection point is also ideal if the line and the plane are affine and parallel.

Before tackling the algebra in Chapter 12, we spend a moment discussing construction operations. Aside from determining points, lines, and planes directly through coordinates, we may wish to construct them from other geometric objects. A point in P^3 is determined from the intersection of three planes or from the intersection of a line and a plane. A line in P^3 is determined by either joining two points or intersecting two planes. A plane is determined either by joining three points or by joining a point and a line.

Adding projections to rigid-body and affine transformations, we obtain Table 11.2.

Transformation	**Rigid body**	**Affine**	**Projective**
Preserves	colinearity parallelism distances and angles	colinearity parallelism	colinearity
Examples	rotation translation	rotation translation scale shear	rotation translation scale shear projection

Figure 11.2
Properties preserved by transformations

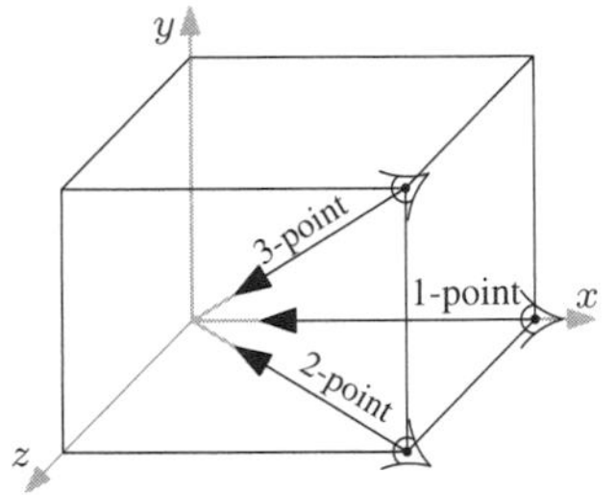

11.4 One-, Two-, and Three-Point Perspective

Architects often use one-, two-, and three-point perspective renderings of city scenes and interior design scenes. Because most lines in such scenes are parallel to one of the three orthogonal axes, it is convenient to classify them into one of these three categories of perspective.

One-Point Perspective

If the view direction is parallel to, say, the x-axis, all lines parallel to y and all lines parallel to z will remain parallel in the image. Lines parallel to x will have a vanishing point. The resulting images similar to those in Figure 11.3 are termed *one-point perspective* images.

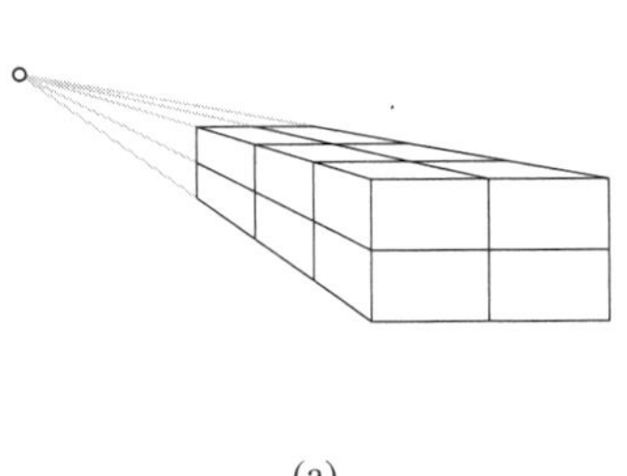
(a)

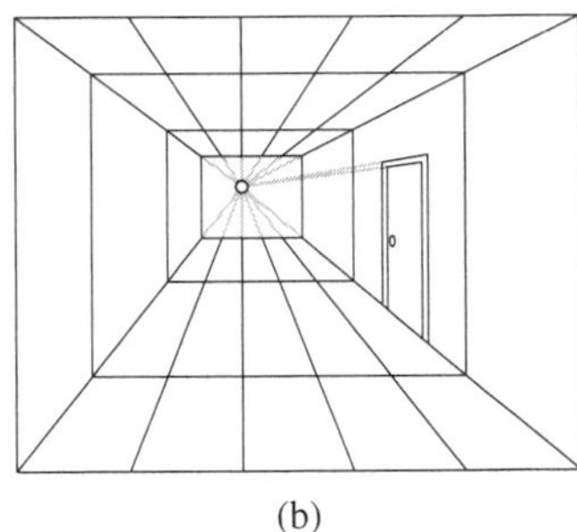
(b)

Figure 11.3
Examples of one-point perspective

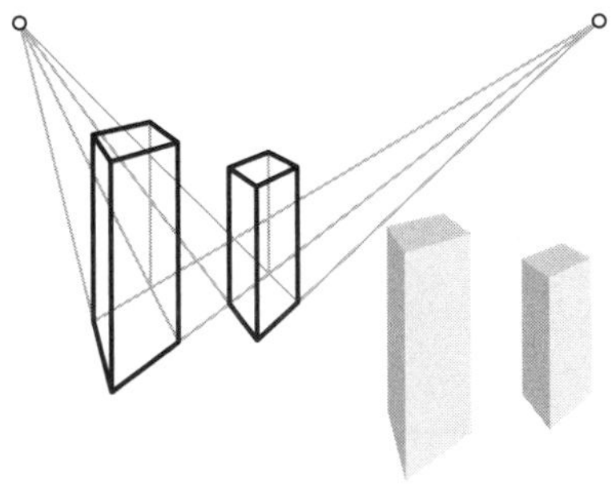

Two-Point Perspective

If the view direction is parallel to one plane, say the xz-plane (and therefore the view plane is parallel to the y-axis), then all lines parallel to y will remain parallel in the image, but lines parallel to x and lines parallel to z will each have one vanishing point.

When rendering such a scene, the artist will start by declaring a line as the *vanishing line* of the ground plane (and of all planes parallel to it) as well as two vanishing points on that line.

Three-Point Perspective

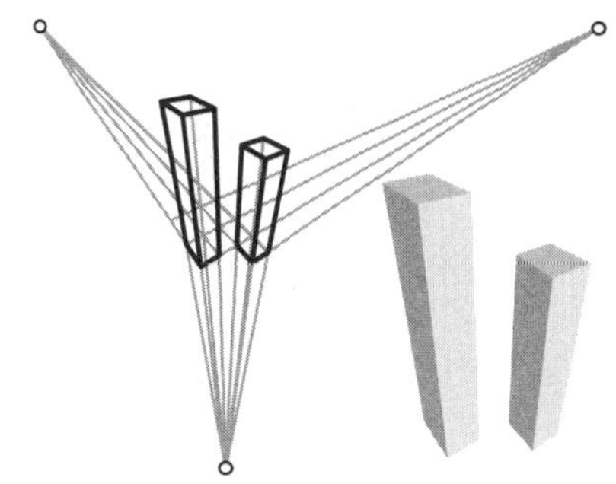

Three-point perspective images are used to accentuate the height of a building and to amplify its height compared to the height of the observer. In this case a third vanishing point is chosen as the point where all vertical lines meet. Since the vertical lines are not coplanar with the ground plane, the third vanishing point should not lie on the first chosen vanishing line.

In this case the view direction is not parallel to any of the three major axes and each set of lines parallel to one of the axes will have a vanishing point.

11.5 Exercises

11.1 One way to generate a perspective drawing of a grid of equal-sized squares is to fix the nearest corner of the grid and then to choose two vanishing points for the grid lines and a third for the diagonals in addition to the size of one side of the nearest square. The four points are shown as bullets in Figure 11.4. Implement a system that draws such a grid and that makes it possible for the user to move the four points to generate a different perspective.

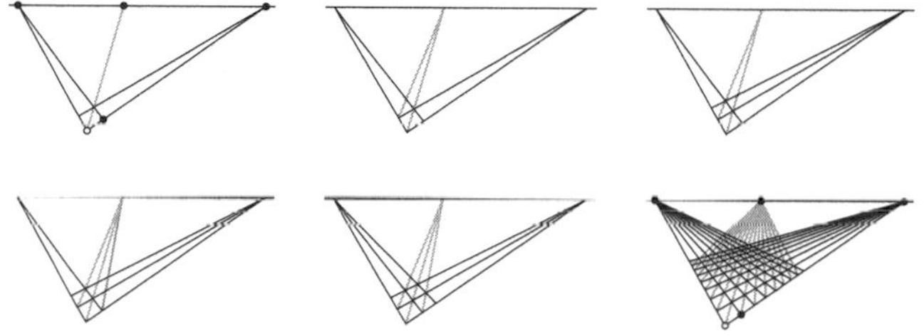

Figure 11.4
Perspective of a regular grid

11.2 Using Euclidean geometry objects, implement a drawing system that renders a three-point perspective image consisting of two "buildings." The system would maintain the drawing while making it possible for the user to drag any of the three vanishing points.

11.3 Does the ideal plane in P^3 include any lines that are not ideal? Explain.

11.4 Draw a sketch showing the point of intersection in P^3 of

1. an affine plane and an affine line that are not parallel.
2. an affine plane and an affine line that are parallel.
3. the ideal plane and an affine line.
4. an affine plane and an ideal line.

11.5 This perspective drawing of a parallelepiped is incorrect. Mention the property of perspective that is not respected and that reveals this drawing is a false perspective. Sketch a plausible correction.

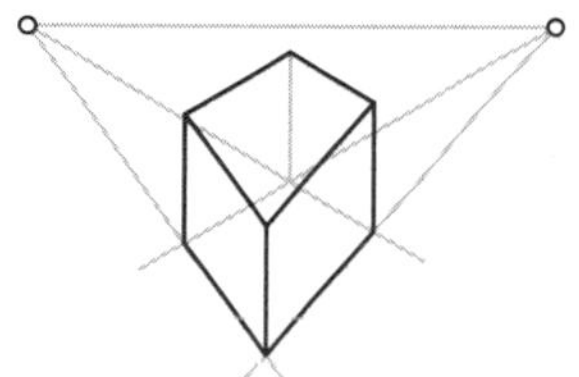

11.6 We saw how to project a set of parallel lines on an image plane. Consider the converse problem. You are given a set of lines that meet at a point as well as the location of an observer. Describe how you would construct a set of parallel lines that project to the given set of intersecting lines.

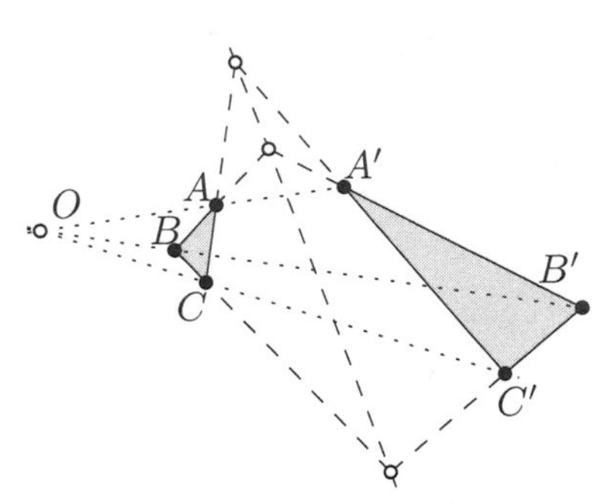

11.7 Desargues, an architect from Lyon, showed the following theorem in 1648 [106]. If in two triangles ABC and $A'B'C'$ the lines AA', BB', and CC' meet in one point, then the point of intersection of AB and $A'B'$, that of AC and $A'C'$, and that of BC and $B'C'$ are colinear.

Develop a system that visually confirms Desargues's theorem by allowing the user of your program to manipulate any of the six vertices interactively. The theorem holds in the projective plane, but your implementation will use only Euclidean objects and Cartesian coordinates. To simplify confirming visually that the three intersection points are colinear, draw a degenerate triangle.

Exercise 12.9 is a sequel implementation in the projective plane using homogeneous coordinates.

11.8 Is it the case that any projective transformation that preserves ratios is an affine transformation? Justify.

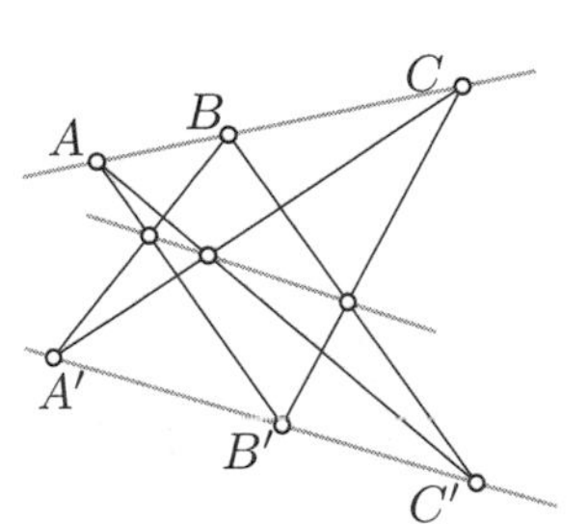

11.9 Pappus of Alexandria established in the fourth century B.C. the following theorem.

> *If the six vertices of a hexagon lie alternately on two lines, the three points of intersection of pairs of opposite sides are colinear* [27].

Develop a system that visually confirms Pappus's theorem by allowing the user to manipulate any of the six vertices interactively.

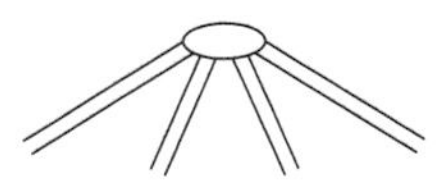

11.10 The figure shows a false (i.e., incorrect) perspective of a traffic circle. One possible mistake is that the two lines representing each street are parallel and should thus meet at a vanishing point, yet they are shown as parallel in this drawing.

1. Under what condition(s) would the drawing in fact be correct (while each two lines remain parallel and ignoring the circle itself)?
2. Modify the drawing such that each two parallel lines do meet at one vanishing point while paying special attention to the relationship between all vanishing points.

12 Homogeneous Coordinates for Projective Geometry

Chapter 1 presented Euclidean geometry alongside analytic geometry in the Euclidean plane using Cartesian coordinates. Such a familiar mapping needed no introduction. By contrast, Chapter 11 illustrated how projective spaces differ from affine spaces but gave no hint to how one could perform computation in projective spaces. Such an accompanying analytic geometry requires assigning coordinates to points and other objects—the topic of this chapter. Homogeneous coordinates have been discovered four times independently—by Bobillier, Plücker, Feuerbach, and Möbius [19]. Chapter 13 discusses Möbius's version, which remains interesting in its own right. The first known notes by Möbius on barycentric calculus date from 1818 [29, p. 49]. It is worth quoting Boyer's impression.

> *The year 1827 is of considerable importance in the history of analytic geometry in Germany [...]. It is sometimes said that Descartes arithmetized geometry, but this is not strictly correct. For almost two hundred years after his time coordinates were in essence geometric. [...] The arithmetization of coordinates took place not in 1637 but in the crucial years 1827–1829.* [19, p. 242]

12.1 The Projective Line

The main observation needed to assign coordinates to points on the projective line P^1 relies on embedding the line in a plane and observing that there is a one-to-one correspondence between points on P^1 and *lines* passing by a point not incident to P^1.

Consider embedding the projective line P^1 in a plane parameterized by (x, w) and colocating P^1 with the line $w = 1$. Each point in P^1 can be captured by a line passing by the origin.

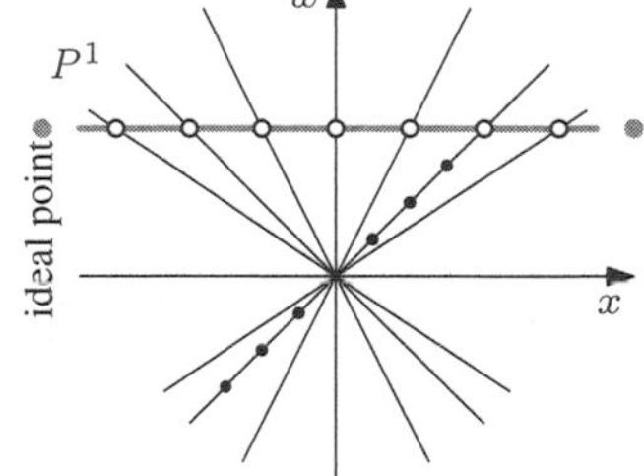

The mapping function is that of intersection. A line is mapped to its intersection with the projective line. The mapping is one-to-one because each line passing by the origin intersects P^1 in exactly one point. Because the line $w = 0$ intersects P^1 at its ideal point, the ideal point is represented by the line $w = 0$.

Homogeneous Coordinates of Points

To determine the coordinates of a point A on P^1, we choose the coordinates of an arbitrary point lying on the line passing by A and the origin. The existence of multiple representations for one element in a set is already familiar from the elements of $\mathbb{Q}$; a rational number can be captured by an infinite number of quotients. Likewise, the *homogeneous coordinates* of any point on the line passing by the origin and a point on P^1 are suitable coordinates.

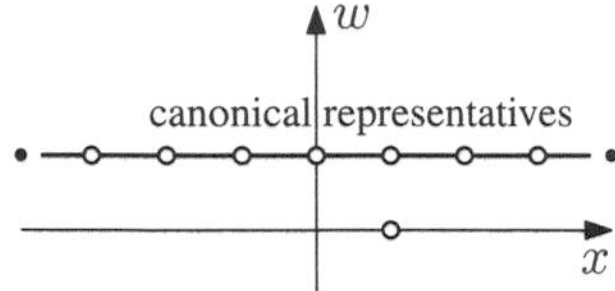

Just as it is often desirable to find a canonical representation for a rational number—with the constraints that the denominator be positive and that the numerator and the denominator be relatively prime (have no common divisor other than unity), it is also often desirable to find a canonical representative for each element on the projective line. The canonical representative of affine points will be a point on the line $w = 1$. The canonical representative of the ideal point will be the point $(1, 0)$. Thus, the canonical representative for an affine point can be found by using $(x/w, 1)$ for (x, w) and the canonical representative of the ideal point is found by using $(1, 0)$ for $(x, 0)$ with $x \neq 0$. Determining a canonical representative is referred to as a *normalization* step.

If a point on the projective line is represented by the tuple (x, w), an arbitrary other tuple (kx, k), $k \neq 0$, is an equally suitable representative. This is true for both affine ($w \neq 0$) as well as for the ideal ($w = 0$) point. Notice in particular the lack of restriction on the sign of k; it may be positive or negative.

Point Equality on the Projective Line

Determining whether two pairs (x_1, w_1) and (x_2, w_2) represent the same point amounts to checking whether the determinant

$$\begin{vmatrix} x_1 & x_2 \\ w_1 & w_2 \end{vmatrix}$$

vanishes, or whether $x_1 w_2 = x_2 w_1$. We can now sketch a class for a point on the projective line using homogeneous coordinates.

```
template<typename NT>
class Point_P1
{
  NT _x, _w;
public:
  Point_P1(const NT& x = 0, const NT& w = 1) : _x(x), _w(w) {}

  const NT hx() const { return _x; }
  const NT hw() const { return _w; }

  bool operator==(const Point_P1<NT>& p) const {
    return determinant(_x, _w, p._x, p._w) == 0;
  }
};
```

The term *homogeneous* hints at several properties of homogeneous coordinates:

1. The ideal point is treated no differently than any affine point. Indeed, as was seen in Chapter 11, projective transformations may map the ideal point to an affine point and an affine point to the ideal point.

2. No special meaning is attributed to the w value within projective geometry; there is homogeneity between x and w.

3. As we shall shortly see, a homogeneous equation in the first degree (e.g. $Xx + Yy + Ww = 0$ in P^2 or $Xx + Yy + Zz + Ww = 0$ in P^3, for given constants X, Y, Z, and W) represents a linear subspace. This is also true in P^1, but the outcome is rather trivial: $Xx + Ww = 0$ for a given pair of constants X and W is still just a point.

Because projective geometry acts as a kind of augmented Euclidean geometry with a special topology, it is clear that we can easily map points from Euclidean to projective spaces. But since the set of projective points is a superset of the set of Euclidean points, the converse mapping is by necessity ill-defined. Were we to be studying projective geometry on its own, we would have no need to perform a mapping back to Euclidean spaces and the difficulty of mapping would not arise.

Yet in computing we are usually interested in projections when implementing visualization applications. The image we render is best described as a subset of the Euclidean plane E^2. For this reason we shall attribute special status to the w variable and will refer to it as the *homogenizing* variable. We will always use a pair (x, w) with the knowledge that we will eventually map the outcome of the computation from projective to Euclidean space. If one is to remain theoretical and not develop systems, one would have no need ever to move between Euclidean and projective spaces. In practice one needs to do such moves and distinguishes between x and w.

Segments and the Nonseparability of the Projective Line

We can now revisit the nonseparability of the projective line discussed in §11.1. Suppose that in analogy with the classes for points and segments in Euclidean (Point_E1, Segment_E1) and spherical (Point_S1, Segment_S1) geometries, we attempt to define in addition to Point_P1 a class Segment_P1 that captures a segment (or simply an interval) on the projective line.

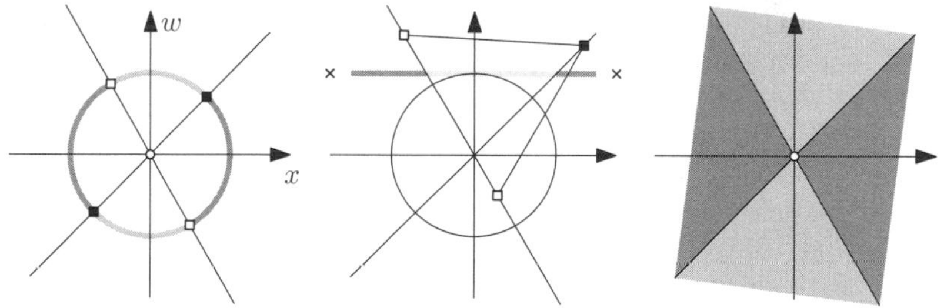

Figure 12.1
The nonseparability of the projective line makes it impossible to define objects of type Segment_P1.

With homogeneous coordinates, the points on the projective line are represented using a nonzero vector in the Euclidean plane. We look at the question of nonseparability in the three different ways shown in Figure 12.1. On the left of the figure we think of the vectors as normalized such that the magnitude of each vector is unity, but signs are also irrelevant and so any two antipodal points also capture the same point. Thus, the projective point that can be represented by either of the two squares and the projective point that can be represented by either of the two boxes define *two* segments, each of which is shown using two (same-shade) arcs on the circle.

Consider instead that we initially do not normalize the vectors representing the two projective points (still shown as a box and a square in the figure at the center). Two segments can join one point to the other. If we now consider a normalization step in which we project on the line $w = 1$, either of the two distinct segments (in the two shades) is potentially the segment that we intended to capture. One crucial distinction between the two segments is now evident: One of the two segments includes the ideal point on the projective line (marked as a cross on either end of the line) whereas the other consists of affine points only.

Since any two points that are colinear with the origin capture the same projective point, we can also consider the mapping between a line in the plane and a point in the projective line. As shown on the right of Figure 12.1, two such lines in the plane, which represent two points in the projective line, define not one, but two segments on the projective line.

Also because the projective line is topologically a circle, a point on the projective line does not split it into two parts. Even after cutting the projective line at a point R, we would be able to move unimpeded between two arbitrary other points A and B on the line. This suggests that it makes no sense to attempt implementing a predicate such as

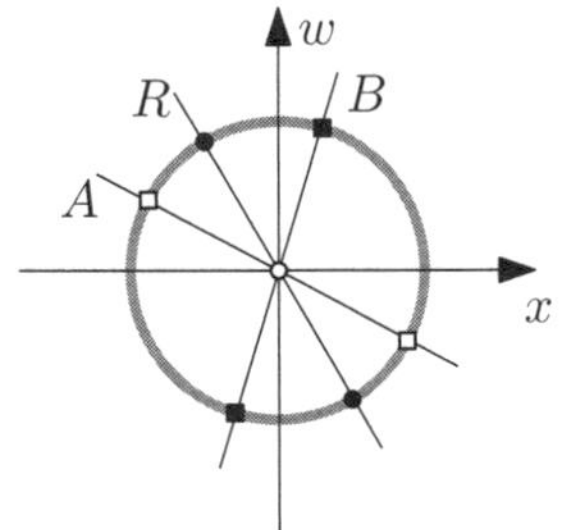


```
template<typename NT>
bool
are_on_same_side(Point_P1<NT> R, Point_P1<NT> A, Point_P1<NT> B);
```

that would determine whether both A and B lie on the same side of R; R does not split the projective line into two parts.

Just as we embed P^1 in E^2 to assign coordinates, we will also embed P^2 in E^3 and P^3 in E^4. The main objective from looking at P^1 (even though P^1 is unlikely to arise in practice) is to make it easier to be convinced that it will also not make sense to attempt defining predicates for reporting whether two points lie on the same side of a hyperplane in P^2 or P^3. The trouble is that one-dimensional geometries do not make it clear that a point is also a hyperplane, but the notion that a point in one-dimensional geometries acts as both a point and a hyperplane will be evident in a much more pragmatic way in Chapters 28 and 29.

12.2 The Projective Plane

We move one dimension higher and seek a way to assign coordinates to projective points and lines on the projective plane P^2. If P^2 is embedded in three space and if an arbitrary point O not incident to the plane is chosen, then the lines passing by O are in one-to-one correspondence with the points on P^2. Also, the planes passing by the same point are in one-to-one correspondence with the lines on the projective plane. We can verify that the topology of the projective plane discussed in Chapter 11 is indeed respected. There is only one plane parallel to the projective plane and passing by O; that plane captures the ideal line. But there is an infinite number of lines passing by O and parallel to the projective plane; each such line captures one ideal point.

Just as with the projective line, it is convenient to choose O at the origin and to position the projective plane at the plane $w = 1$ in a space parameterized by a triple of axes (x, y, w).

Homogeneous Coordinates of a Point in the Projective Plane

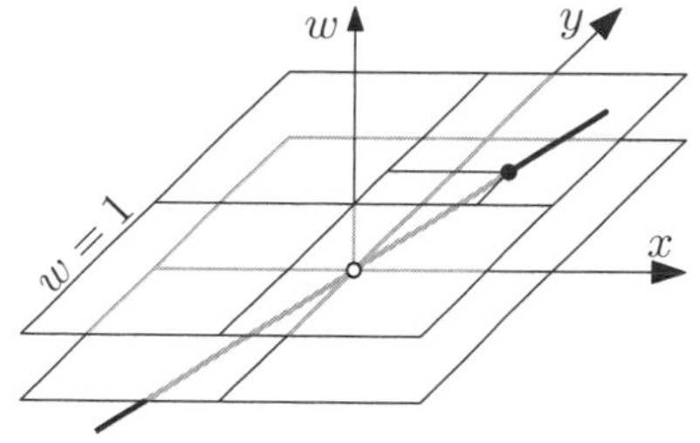

An arbitrary point on the line passing by a projective point A and the origin is a suitable representative for A. The canonical representative of an affine point with homogeneous coordinates (x, y, w) is $(x/w, y/w, 1)$. The canonical representative of an ideal point is $(x, y, 0)$, either such that $x^2 + y^2 = 1$, or more simply by setting either x or y to 1 and thus choosing a point on a square centered at the origin and lying in the xy-plane.

Point Equality in the Projective Plane

Two points (x_1, y_1, w_1) and (x_2, y_2, w_2) in P^2 are equal if there exists a $k \neq 0$ such that $x_1 = kx_2$, $y_1 = ky_2$, and $w_1 = kw_2$. If we think of the coordinates as those of vectors in 3D, then asking whether two points are not equal amounts to asking whether the two vectors are linearly independent, which can be verified by asking whether the matrix

$$\begin{bmatrix} x_1 & x_2 \\ y_1 & y_2 \\ w_1 & w_2 \end{bmatrix}$$

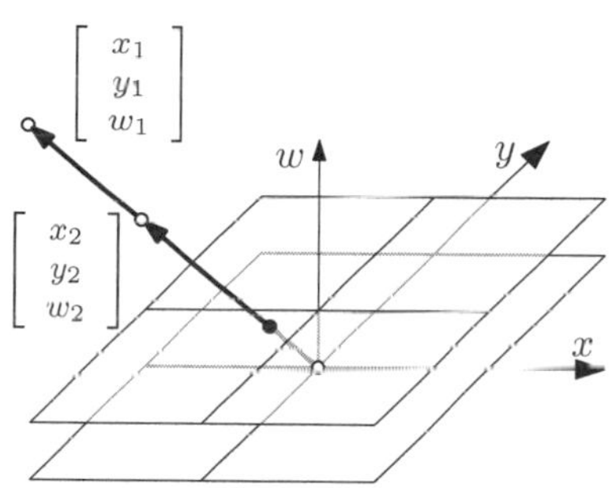

has rank 2. In turn, this condition holds if and only if all three determinants

$$\begin{vmatrix} x_1 & x_2 \\ y_1 & y_2 \end{vmatrix}, \quad \begin{vmatrix} x_1 & x_2 \\ w_1 & w_2 \end{vmatrix}, \quad \begin{vmatrix} y_1 & y_2 \\ w_1 & w_2 \end{vmatrix}$$

do not simultaneously vanish. Overloading the equality operator on a generic class Point_P2 leads to

```
template<typename NT>
bool
Point_P2<NT>::operator==(const Point_P2<T>& p) const
{
  return (this == &p) ||
        are_dependent(_x, _y, _w, p._x, p._y, p._w);
}
```

Homogeneous Coordinates of a Line in the Projective Plane

If the projective plane is embedded in 3D space, a line in P^2 can likewise be captured by a plane passing by the origin O [87]. That plane is in turn captured by its normal vector; because we know that it passes by the origin, the constant d in § 3.4 is zero. If the normal vector to the plane is $\overrightarrow{N}(X, Y, W)$, then a

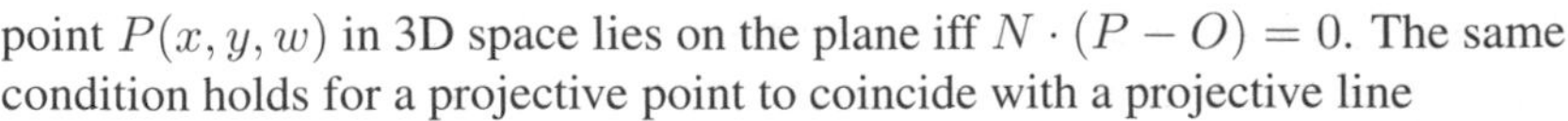

point $P(x, y, w)$ in 3D space lies on the plane iff $N \cdot (P - O) = 0$. The same condition holds for a projective point to coincide with a projective line

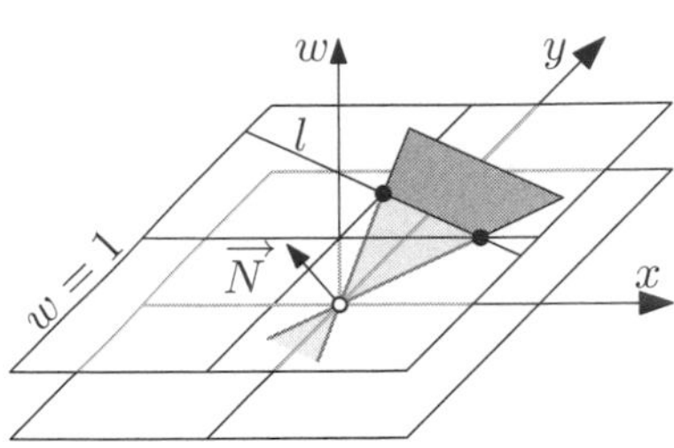

$$N \cdot (P - O) = 0 \implies L \cdot P = 0 \implies \begin{bmatrix} X & Y & W \end{bmatrix} \cdot \begin{bmatrix} x \\ y \\ w \end{bmatrix} = 0.$$

This homogeneous equation makes it evident that neither scaling the coefficients of the line (the normal vector in 3D) nor scaling the coordinates of the point matters. For the equation to be meaningful, neither the three line coefficients nor the three point coordinates can simultaneously vanish. If $X = 0$ and $Y = 0$ (but $W \neq 0$), the plane is parallel to the plane $w = 1$ and the projective line captured is the ideal line.

It is easy to derive the coefficients of the normal vector by appealing to the equation for the coplanarity of four points in 3D [Eq. (3.2)]. The condition for the point $P(x, y, w)$ to be coplanar with the three points $P_1(x_1, y_1, w_1)$, $P_2(x_2, y_2, w_2)$, and $O(0, 0, 0)$ is

$$\begin{vmatrix} x & 0 & x_1 & x_2 \\ y & 0 & y_1 & y_2 \\ w & 0 & w_1 & w_2 \\ 1 & 1 & 1 & 1 \end{vmatrix} = 0.$$

But we can be more direct and seek the condition for three vectors not to span 3D space (or for the three to lie in the same subspace), which leads to

$$\begin{vmatrix} x & x_1 & x_2 \\ y & y_1 & y_2 \\ w & w_1 & w_2 \end{vmatrix} = 0.$$

Expanding either way, we get

$$x \begin{vmatrix} y_1 & y_2 \\ w_1 & w_2 \end{vmatrix} - y \begin{vmatrix} x_1 & x_2 \\ w_1 & w_2 \end{vmatrix} + w \begin{vmatrix} x_1 & x_2 \\ y_1 & y_2 \end{vmatrix} = 0$$

$$\implies X = \begin{vmatrix} y_1 & y_2 \\ w_1 & w_2 \end{vmatrix}; \qquad Y = -\begin{vmatrix} x_1 & x_2 \\ w_1 & w_2 \end{vmatrix}; \qquad W = \begin{vmatrix} x_1 & x_2 \\ y_1 & y_2 \end{vmatrix}.$$

This allows us to sketch the implementation of a class Line_P2 for a projective line.

```
template<typename T>
class Line_P2
{
private:
    T _X, _Y, _W;              // Xx+Yy+Ww=0
public:
    Line_P2() : _X(1), _Y(0), _W(0) {}
    Line_P2(const Point_P2<T>& source, const Point_P2<T>& target)
    {
        _X = + determinant(source.hy(), source.hw(),  target.hy(), target.hw());
        _Y = − determinant(source.hx(), source.hw(),  target.hx(), target.hw());
```

```
    _W = + determinant(source.hx(), source.hy(), target.hx(), target.hy());
  }

  bool is_incident(const Point_P2<T>& p) const
  {
    return inner_product(_X, _Y, _W, p.hx(), p.hy(), p.hw()) == 0;
  }
  ...
};
```

The preceding code does not guard against the two points used to create a line not being distinct (defined by linearly dependent vectors). One could take a run-time penalty and throw an exception if the line is initialized to the zero vector.

```
struct Maldefined_Line_P2 : public std::runtime_error {
  Maldefined_Line_P2(const std::string& s) : std::runtime_error(s) {}
};

template<typename NT>
class Line_P2
{
  Line_P2(const Point_P2<NT>& P1, const Point_P2<NT>& P2)
  {
    ...

  NT zero(0);
  if( _X == zero && _Y == zero && _W == zero )
    throw Maldefined_Line_P2("");
  }
};
```

Alternatively, one could write an assertion that is subsequently turned off at the conclusion of development.

Notice that because the condition for two points to be coincident is precisely the test for the three coefficients to vanish simultaneously, we perform the test after computing the coefficients rather than call the equality operator before computing them.

Intersection of Two Projective Lines

Interpreted in Euclidean 3D space, constructing a line passing by two points is equivalent to finding the normal to two vectors. The converse problem, finding the point of intersection of two lines, is also answered by finding the normal to two vectors.

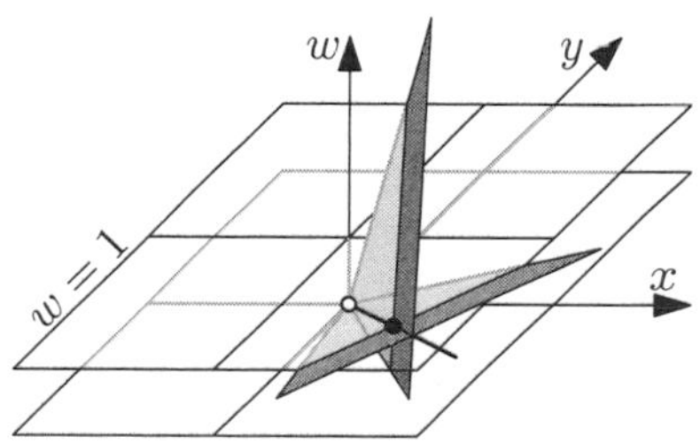

The first problem sought the line *joining* two points. The second seeks the *meeting point* of two lines. If we consider that the coefficients of the two lines are the normal vectors to the corresponding planes, then the meeting point is the line simultaneously orthogonal to the two planes—a computation reminiscent of computing the "cross product."

The coordinates of the point of intersection, up to a (positive *or negative*) factor, can be found by code such as the following. The assertion confirms that the projective point constructed is sensible (its three components do not simultaneously vanish, confirming that the two given planes were distinct).

```
template<typename T>
Point_P2<T>
intersection(const Line_P2<T>& l1, const Line_P2<T>& l2)
{
    T detx = + determinant(l1.Y(), l1.W(), l2.Y(), l2.W());
    T dety = − determinant(l1.X(), l1.W(), l2.X(), l2.W());
    T detw = + determinant(l1.X(), l1.Y(), l2.X(), l2.Y());

    assert((detx != 0) || (dety != 0) || (detw != 0));

    return Point_P2<T>(detx, dety, detw);
}
```

Duality

Observing that meeting two lines (intersecting them) and joining two points (constructing a line from them) are computationally identical leads us to an intriguing property of projective geometry, the duality of points and lines.

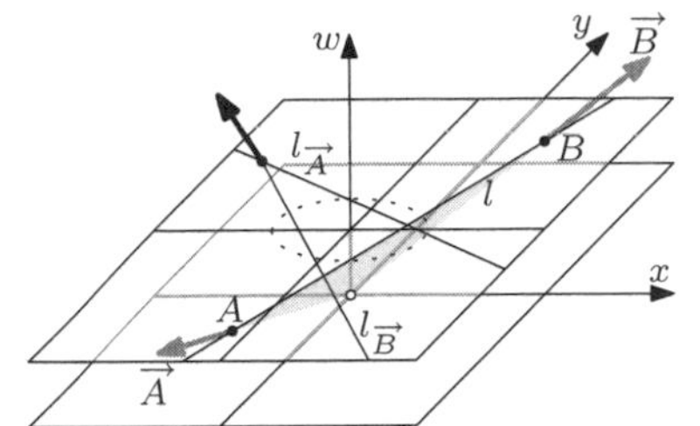

Suppose we are given two points A and B and are asked to form the line l passing by A and B. The procedure above (finding the plane carrying O, A, and B) has a dual interpretation. Instead of considering that A and B are two points, we take them instead to be two normal vectors $\overrightarrow{A}$ and $\overrightarrow{B}$. $\overrightarrow{A}$ in turn defines a line $l_{\overrightarrow{A}}$ that is the intersection of the plane $w = 1$ with the plane whose normal vector is $l_{\overrightarrow{A}}$—and likewise for $l_{\overrightarrow{B}}$. The two planes $l_{\overrightarrow{A}}$ and $l_{\overrightarrow{B}}$ intersect (or meet) at a point $l_{\overrightarrow{A}} \cap l_{\overrightarrow{B}}$. If we now interpret this point as a vector and assume that the vector is a normal to a plane, then the plane passes by A and B. The normal (and the plane) also precisely defines the line passing by A and B that we were seeking!

Computationally, duality makes it possible to solve two problems with one implementation. To do so one manipulates the problem statement, swapping the words "point" and "line" and swapping also "meet" and "join." So "find the line that joins two points" becomes "find the point that meets two lines." Exercise 12.4 pursues this theme.

Coordinates of Ideal Points and the Ideal Line

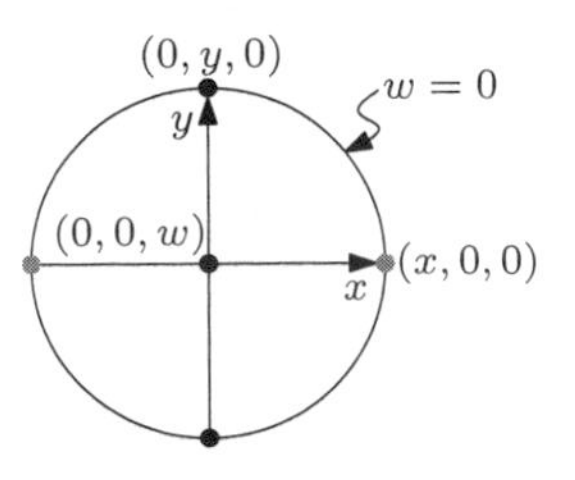

Homogeneous coordinates make no distinction between affine and ideal points or affine lines and the ideal line. If the ideal line is illustrated as some large circle encompassing the plane, the coordinates $(x, 0, 0)$ for any $x \neq 0$ capture one ideal point, namely, the one lying at the intersection of the x-axis and the ideal line. Likewise, the point infinitely far along the y-axis has coordinates of the form $(0, y, 0)$ for any $y \neq 0$. Just as the equation of the y-axis is $x = 0$ and that of the x-axis is $y = 0$, the ideal line has the equation $w = 0$.

12.3 The Projective Space

We proceed largely in analogy with the discussion on the projective line and plane.

A Point in Projective Space

Two projective points $P_1(x_1, y_1, z_1, w_1)$ and $P_2(x_2, y_2, z_2, w_2)$ are equal if the matrix

$$\begin{bmatrix} x_1 & y_1 & z_1 & w_1 \\ x_2 & y_2 & z_2 & w_2 \end{bmatrix}^T$$

has rank strictly less than 2, which holds if and only if the following six determinants simultaneously vanish:

$$\begin{vmatrix} x_1 & x_2 \\ y_1 & y_2 \end{vmatrix}, \quad \begin{vmatrix} x_1 & x_2 \\ z_1 & z_2 \end{vmatrix}, \quad \begin{vmatrix} x_1 & x_2 \\ w_1 & w_2 \end{vmatrix},$$

$$\begin{vmatrix} y_1 & y_2 \\ z_1 & z_2 \end{vmatrix}, \quad \begin{vmatrix} y_1 & y_2 \\ w_1 & w_2 \end{vmatrix}, \quad \begin{vmatrix} z_1 & z_2 \\ w_1 & w_2 \end{vmatrix}$$

A Plane in Projective Space

Four points are needed to define a hyperplane in affine 4D space, and the condition for the coplanarity of *five* points (one being the origin) is for the following determinant to vanish:

$$\begin{vmatrix} x & 0 & x_1 & x_2 & x_3 \\ y & 0 & y_1 & y_2 & y_3 \\ z & 0 & z_1 & z_2 & z_3 \\ w & 0 & w_1 & w_2 & w_3 \\ 1 & 1 & 1 & 1 & 1 \end{vmatrix} = 0.$$

or we ask more directly for the condition under which four vectors lie in the same hyperplane in 4D:

$$\begin{vmatrix} x & x_1 & x_2 & x_3 \\ y & y_1 & y_2 & y_3 \\ z & z_1 & z_2 & z_3 \\ w & w_1 & w_2 & w_3 \end{vmatrix} = 0.$$

This homogeneous form for the equation of a plane, which is credited, along with its analog in one lower dimension, to Cayley in 1843 by Boyer and Merzbach [20], identifies the factors of a plane $\alpha(X, Y, Z, W)$ on which a point $P(x, y, z, w)$ lies:

$$Xx + Yy + Zz + Ww = 0.$$

Hence, we can now write

$$X = + \begin{vmatrix} y_1 & y_2 & y_3 \\ z_1 & z_2 & z_3 \\ w_1 & w_2 & w_3 \end{vmatrix},$$

$$Y = - \begin{vmatrix} x_1 & x_2 & x_3 \\ z_1 & z_2 & z_3 \\ w_1 & w_2 & w_3 \end{vmatrix},$$

$$Z = + \begin{vmatrix} x_1 & x_2 & x_3 \\ y_1 & y_2 & y_3 \\ w_1 & w_2 & w_3 \end{vmatrix},$$

$$W = - \begin{vmatrix} x_1 & x_2 & x_3 \\ y_1 & y_2 & y_3 \\ z_1 & z_2 & z_3 \end{vmatrix}.$$

12.4 A Line in the Projective Space

The condition for four points $(x_i, y_i, z_i, w_i), i = 1, \ldots, 4$, in projective space to be coplanar can be stated as the vanishing of the following determinant:

$$\begin{vmatrix} x_1 & y_1 & z_1 & w_1 \\ x_2 & y_2 & z_2 & w_2 \\ x_3 & y_3 & z_3 & w_3 \\ x_4 & y_4 & z_4 & w_4 \end{vmatrix} = 0 \implies +x_1 \begin{vmatrix} y_2 & z_2 & w_2 \\ y_3 & z_3 & w_3 \\ y_4 & z_4 & w_4 \end{vmatrix} - y_1 \begin{vmatrix} x_2 & z_2 & w_2 \\ x_3 & z_3 & w_3 \\ x_4 & z_4 & w_4 \end{vmatrix}$$

$$+ z_1 \begin{vmatrix} x_2 & y_2 & w_2 \\ x_3 & y_3 & w_3 \\ x_4 & y_4 & w_4 \end{vmatrix} - w_1 \begin{vmatrix} x_2 & y_2 & z_2 \\ x_3 & y_3 & z_3 \\ x_4 & y_4 & z_4 \end{vmatrix} = 0.$$

To show that the final forms are elementary, we proceed to expand in detail:

$$\begin{aligned}
\implies & + x_1 \left[y_2 \begin{vmatrix} z_3 & w_3 \\ z_4 & w_4 \end{vmatrix} - z_2 \begin{vmatrix} y_3 & w_3 \\ y_4 & w_4 \end{vmatrix} + w_2 \begin{vmatrix} y_3 & z_3 \\ y_4 & z_4 \end{vmatrix} \right] \\
& - y_1 \left[x_2 \begin{vmatrix} z_3 & w_3 \\ z_4 & w_4 \end{vmatrix} - z_2 \begin{vmatrix} x_3 & w_3 \\ x_4 & w_4 \end{vmatrix} + w_2 \begin{vmatrix} x_3 & z_3 \\ x_4 & z_4 \end{vmatrix} \right] \\
& + z_1 \left[x_2 \begin{vmatrix} y_3 & w_3 \\ y_4 & w_4 \end{vmatrix} - y_2 \begin{vmatrix} x_3 & w_3 \\ x_4 & w_4 \end{vmatrix} + w_2 \begin{vmatrix} x_3 & y_3 \\ x_4 & y_4 \end{vmatrix} \right] \\
& - w_1 \left[x_2 \begin{vmatrix} y_3 & z_3 \\ y_4 & z_4 \end{vmatrix} - y_2 \begin{vmatrix} x_3 & z_3 \\ x_4 & z_4 \end{vmatrix} + w_2 \begin{vmatrix} x_3 & y_3 \\ x_4 & y_4 \end{vmatrix} \right] = 0 \\
\implies & + \begin{vmatrix} x_1 & w_1 \\ x_2 & w_2 \end{vmatrix} \cdot \begin{vmatrix} y_3 & z_3 \\ y_4 & z_4 \end{vmatrix} + \begin{vmatrix} y_1 & z_1 \\ y_2 & z_2 \end{vmatrix} \cdot \begin{vmatrix} x_3 & w_3 \\ x_4 & w_4 \end{vmatrix} \\
& - \begin{vmatrix} y_1 & w_1 \\ y_2 & w_2 \end{vmatrix} \cdot \begin{vmatrix} x_3 & z_3 \\ x_4 & z_4 \end{vmatrix} - \begin{vmatrix} x_1 & z_1 \\ x_2 & z_2 \end{vmatrix} \cdot \begin{vmatrix} y_3 & w_3 \\ y_4 & w_4 \end{vmatrix} \\
& + \begin{vmatrix} x_1 & y_1 \\ x_2 & y_2 \end{vmatrix} \cdot \begin{vmatrix} z_3 & w_3 \\ z_4 & w_4 \end{vmatrix} + \begin{vmatrix} z_1 & w_1 \\ z_2 & w_2 \end{vmatrix} \cdot \begin{vmatrix} x_3 & y_3 \\ x_4 & y_4 \end{vmatrix} = 0.
\end{aligned} \tag{12.1}$$

This equation makes it possible to interpret the original 4×4 determinant not as the condition under which four points are coplanar, but as the condition under which two lines L and M intersect (including the possibility of their meeting at an ideal point).

If L is defined by (P_1, P_2) and M by (P_3, P_4), we can discard the points and argue that the lines are instead defined by the six 2×2 determinants in Eq. (12.1). The condition for L to be *incident* to M can then be more conveniently written as

$$+L_{xw}M_{yz} + L_{yz}M_{xw} - L_{yw}M_{xz} - L_{xz}M_{yw} + L_{xy}M_{zw} + L_{zw}M_{xy} = 0.$$

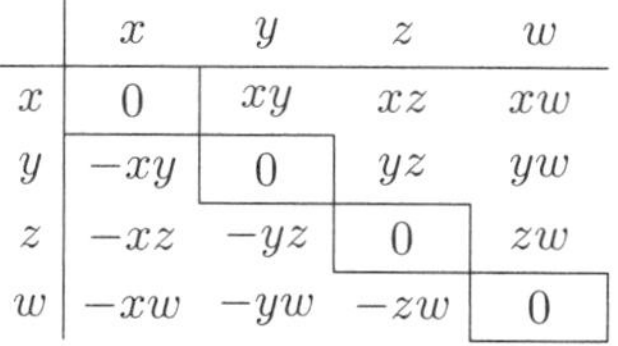

	x	y	z	w
x	0	xy	xz	xw
y	$-xy$	0	yz	yw
z	$-xz$	$-yz$	0	zw
w	$-xw$	$-yw$	$-zw$	0

Since $L_{xx} = 0$ and $L_{xy} = -L_{yx}$ (and likewise for the other point coordinates), the six numbers $L_{xy}, L_{xz}, L_{xw}, L_{yz}, L_{yw}, L_{zw}$ are the only six 2×2 determinants one can usefully derive from the point coordinates. These six numbers, which define a line L up to a constant scalar factor, are called its *Plücker coordinates* [94, 28].

That six numbers are used to define a line—even if restricted to affine ones—may seem to clash with the established notion that four suffice (say the coordinates of the intersection with two parallel planes). The six numbers are in fact doubly redundant. One redundancy is carried over from the homogeneity of the pair of points defining the line. Because multiplying the coordinates of one or both points by a scalar factor yields a different representation for the same points, the six determinants also continue to capture the same line under multiplication by a (nonzero) scalar. Each determinant is effectively the slope of the line in a given hyperplane in 4D. The second redundancy arises because a line is incident to itself:

$$\begin{aligned} &+ L_{xw}L_{yz} + L_{yz}L_{xw} - L_{yw}L_{xz} \\ &- L_{xz}L_{yw} + L_{xy}L_{zw} + L_{zw}L_{xy} = 0, \end{aligned}$$

which yields a second constraint on the six coordinates for each of the two lines:

$$+ L_{xw}L_{yz} - L_{yw}L_{xz} + L_{xy}L_{zw} = 0.$$

Recall from § 12.3 that all six determinants simultaneously vanish if and only if the two points coincide, but then we cannot hope in any case to define a line.

Coordinates of Ideal Points and the Ideal Plane

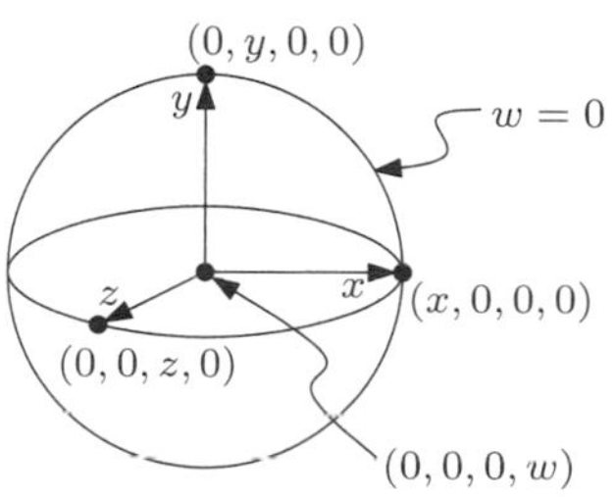

Intersecting the ideal plane $w = 0$ with any projective line yields an ideal point. The w-coordinate of the resulting point will be 0, signaling that it is an ideal point. The same ideal point would result from the intersection of the given line with another parallel to it—or also from the intersection of two distinct planes passing by the given line with the ideal plane. The intersection of the x-axis with the ideal plane is the ideal point with coordinates $(x, 0, 0, 0)$ for any $x \neq 0$, and so on.

12.5 Transformations in the Projective Line

Transforming Points

The set of projective transformations on projective points is the set of transformations achievable by multiplying a real-valued matrix M

$$M = \begin{bmatrix} p_{00} & p_{01} \\ p_{10} & p_{11} \end{bmatrix}$$

by the coordinates of a point represented as a vector:

$$P' = MP = \begin{bmatrix} p_{00} & p_{01} \\ p_{10} & p_{11} \end{bmatrix} \begin{bmatrix} x \\ w \end{bmatrix} = \begin{bmatrix} p_{00}x + p_{01}w \\ p_{10}x + p_{11}w \end{bmatrix}.$$

To see the power of projective transformations compared to affine transformations it suffices to examine the effect of a matrix such as

$$M = \begin{bmatrix} 0 & -1 \\ 1 & 0 \end{bmatrix}$$

on the three canonical points (§ 12.8) $P_0(0,1)$, $P_1(1,1)$, and $P^\infty(1,0)$. The three points map to $P_0' = MP_0 = (-1,0)$, $P_1' = MP_1 = (-1,1)$, $P_\infty' = MP_\infty = (0,-1)$. Since $(-1,0)$ coincides with $(1,0)$ on the projective line, the effect of M is to swap P_0 and P_∞. Transformations in the projective line (and in higher dimensions) will, in general, map affine points to ideal points and will map ideal points to affine ones.

Scale

We start by positing that either of the two transformation matrices

$$S_1 = \begin{bmatrix} k & 0 \\ 0 & 1 \end{bmatrix}, \qquad S_2 = \begin{bmatrix} 1 & 0 \\ 0 & 1/k \end{bmatrix}$$

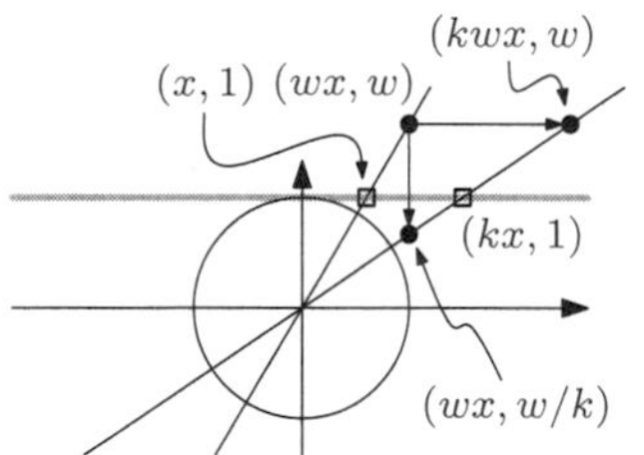

would effect a scale in the projective line P^1. We wonder what effect either matrix has on the set of points colinear with the origin if P^1 is embedded in the Euclidean plane. Let us start from the canonical representative $(x,1)$ of a projective point. This point will in practice have any coordinates (wx, w), $w \neq 0$. The effect of applying S_1 on that point is illustrated by the horizontal arrow in the figure, leading to a point with coordinates (kwx, w). In E^2 the effect of S_2 would have been different—illustrated by the vertical arrow in the figure—but the effect of S_2 is identical to that of S_1 if considered in the projective line P^1. As we can confirm by normalizing, the resulting canonical representative is $(kx, 1)$ regardless of whether S_1 or S_2 was applied.

Translate

We consider next the effect of the transformation matrix

$$T = \begin{bmatrix} 1 & t \\ 0 & 1 \end{bmatrix}.$$

Once again we take the arbitrary point in the plane to be (wx, w) such that its canonical representative is $(x, 1)$. The effect of T—as perceived in E^2—is then $(wx + wt, w)$. Normalizing, we obtain $(x + t, 1)$.

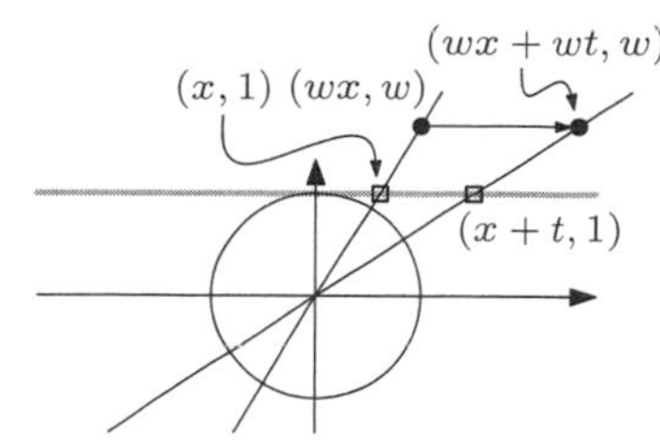

Perspective

If all we wanted to do was to scale and to translate, we would have been content to stay in Euclidean space—homogeneous coordinates by themselves have no obvious appeal (but see Chapter 16). The compelling reason for moving to homogeneous coordinates and to projective geometry is the ability to map between the affine point and the ideal point. Consider applying the matrix

$$P = \begin{bmatrix} 1 & 0 \\ p & 1 \end{bmatrix}$$

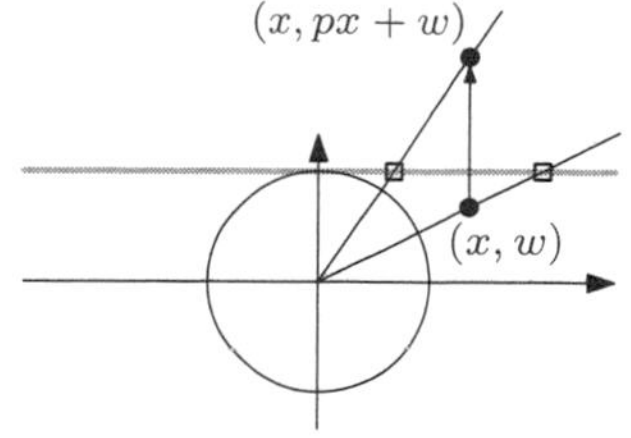

to a projective point. We can no longer take the point to have the coordinates (wx, w) since the implicit constraint that $w \neq 0$ presumes that the point is not ideal. Under scale and translation the ideal point maps to itself, but now it will map to an affine point.

We take the point to be the more general $(x, w), xw \neq 0$ instead. The outcome of applying P is the point $(x, px + w)$. Two interesting questions immediately come up: Which point maps to the ideal point and to which point does the ideal point map? To answer the first we write $px + w = 0$. This is simply the equation of a line in E^2 with slope $w/x = -p$. But that confirms that the ideal point does not map to itself—for if $w = 0$, it must be the case that $x = 0$, but we disallow the origin of E^2, or $p = 0$, and then we're just applying the identity map.

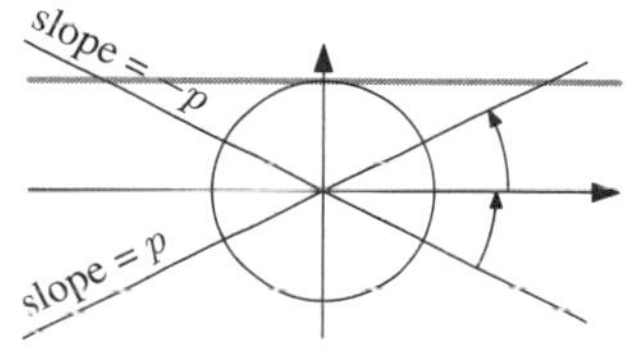

To answer the second question (find the image of the ideal point), we start from a point $(x, 0), x \neq 0$. Applying P, we get (x, px). That point is also the equation of a line in E^2—one with slope p.

12.6 Transformations in the Projective Plane

Transforming Points

A point $P(x, y, w)$ in the projective plane is likewise transformed by a matrix M

$$M = \begin{bmatrix} p_{00} & p_{01} & p_{02} \\ p_{10} & p_{11} & p_{12} \\ p_{20} & p_{21} & p_{22} \end{bmatrix},$$

leading to

$$P' = MP = \begin{bmatrix} p_{00} & p_{01} & p_{02} \\ p_{10} & p_{11} & p_{12} \\ p_{20} & p_{21} & p_{22} \end{bmatrix} \begin{bmatrix} x \\ y \\ w \end{bmatrix}.$$

Translation

Consider using the matrix

$$M = \begin{bmatrix} 1 & 0 & p_{02} \\ 0 & 1 & p_{12} \\ 0 & 0 & 1 \end{bmatrix}$$

to transform the point $P = [\ wx \quad wy \quad w\ ]^T$. P is a point on the projective plane, which is embedded in Euclidean three space for the purpose of performing the transformation.

If, following the transformation M, the projective point P'

$$P' = MP = \begin{bmatrix} 1 & 0 & p_{02} \\ 0 & 1 & p_{12} \\ 0 & 0 & 1 \end{bmatrix} \begin{bmatrix} wx \\ wy \\ w \end{bmatrix} = \begin{bmatrix} wx + wp_{02} \\ wy + wp_{12} \\ w \end{bmatrix}$$

is mapped back to the Euclidean plane, the coordinates are seen to be $(x + p_{02}, y + p_{12})$. Such a transformation matrix is termed a *translation* matrix because the effect it induces on points on the projective plane, when mapped to the Euclidean plane following division by w, is seen to be a translation. Observe that the coefficients p_{02} and p_{12} are determined irrespective of the value of w for a given point.

Rotation

One can likewise verify that the effect of the matrix

$$M = \begin{bmatrix} \cos\theta & -\sin\theta & 0 \\ \sin\theta & \cos\theta & 0 \\ 0 & 0 & 1 \end{bmatrix}$$

on E^3 or P^2 is a rotation about the w-axis or about the origin.

Uniform Scale

Either of the following two matrices effect a uniform scale on Euclidean points:

$$M_1 = \begin{bmatrix} s & 0 & 0 \\ 0 & s & 0 \\ 0 & 0 & 1 \end{bmatrix}, \qquad M_2 = \begin{bmatrix} 1 & 0 & 0 \\ 0 & 1 & 0 \\ 0 & 0 & 1/s \end{bmatrix}.$$

The effect of M_2 is division of a point's w-coordinate by s. Subsequently dividing by the value $w = 1/s$ produces the desired magnification by a scale s.

Nonuniform Scale

A nonuniform scale is effected if the two scales s_x and s_y are not equal:

$$M = \begin{bmatrix} s_x & 0 & 0 \\ 0 & s_y & 0 \\ 0 & 0 & 1 \end{bmatrix}.$$

Perspective

If the matrix M

$$M = \begin{bmatrix} 1 & 0 & 0 \\ 0 & 1 & 0 \\ p_{20} & p_{21} & 1 \end{bmatrix}$$

is used to transform the point $P = [\ x \quad y \quad w\ ]^T$, the resulting point is

$$P' = MP = \begin{bmatrix} 1 & 0 & 0 \\ 0 & 1 & 0 \\ p_{20} & p_{21} & 1 \end{bmatrix} \begin{bmatrix} x \\ y \\ w \end{bmatrix} = \begin{bmatrix} x \\ y \\ p_{20}x + p_{21}y + w \end{bmatrix}.$$

The resulting projective point P' is an affine point only if $(p_{20}x + p_{21}y + w) \neq 0$. But $(p_{20}x+p_{21}y+w) = 0$ is itself the equation of a plane that passes by the origin in E^3 and is therefore a projective line in P^2. Thus, points lying on that line map to the ideal line.

Conversely, a point on the ideal line has the form $(x, y, 0)$ and maps to $(x, y, p_{20}x + p_{21}y)$. One sees again that a necessary and sufficient condition for the matrix M above to map ideal points to (other) ideal points is for $p_{20} = p_{21} = 0$. If that is not the case, the ideal line maps to the affine line with points of the form $[x/(p_{20}x + p_{21}y), y/(p_{20}x + p_{21}y)]$. That this equation does not look like that of a line is not surprising; its form exhibits that the ideal point in the *direction* (x, y) maps to such a point.

Transforming Lines

We are given a line $L[X, Y, W]^T$ in P^2 and we wish to compute the line $L'[X', Y', W']^T$ resulting from applying the projective map M to L. If the set of points on L are represented by the point $P[x, y, w]^T$, then $L \cdot P = 0$, but let us avoid dot products and write $L^T P = 0$.

Projective transformations evidently preserve incidence: If a point P is incident to a line L before applying a transformation M, then it must be the case that $P' = T(P)$ is also incident to $L' = T(L)$, where the function T applies the transformation matrix M. We can write

$$L^T P = 0 \implies L^T M^{-1} P' = 0,$$

but since P' is incident to L', we know that $L'^T P' = 0$, so we have [94]

$$L'^T = L^T M^{-1} \implies L' = M^{-T} L.$$

If M is an orthogonal matrix, then its inverse is also its transpose

$$M^{-T} = M, \qquad L' = ML.$$

If M is an arbitrary (non-orthogonal) matrix,

$$M^{-T} \neq M$$

and so, compared to transforming a point, transforming a line also requires computing a matrix inverse.

12.7 Transformations in the Projective Space

Transforming Points

Predictably, a point $P(x, y, z, w)$ in P^3 is transformed by a matrix M:

$$M = \begin{bmatrix} p_{00} & p_{01} & p_{02} & p_{03} \\ p_{10} & p_{11} & p_{12} & p_{13} \\ p_{20} & p_{21} & p_{22} & p_{23} \\ p_{30} & p_{31} & p_{32} & p_{33} \end{bmatrix},$$

leading to

$$P' = MP = \begin{bmatrix} p_{00} & p_{01} & p_{02} & p_{03} \\ p_{10} & p_{11} & p_{12} & p_{13} \\ p_{20} & p_{21} & p_{22} & p_{23} \\ p_{30} & p_{31} & p_{32} & p_{33} \end{bmatrix} \begin{bmatrix} x \\ y \\ z \\ w \end{bmatrix}.$$

Types of Projective Transformations in Space

The arguments for the plane in § 12.6 apply equally well to the projective space. The matrix

$$\begin{bmatrix} 1 & 0 & 0 & p_{03} \\ 0 & 1 & 0 & p_{13} \\ 0 & 0 & 1 & p_{23} \\ 0 & 0 & 0 & 1 \end{bmatrix}$$

translates by the vector $[\ p_{03} \quad p_{13} \quad p_{23}\ ]^T$.

The matrices

$$M_\theta = \begin{bmatrix} 1 & 0 & 0 & 0 \\ 0 & \cos\theta & -\sin\theta & 0 \\ 0 & \sin\theta & \cos\theta & 0 \\ 0 & 0 & 0 & 1 \end{bmatrix},$$

$$M_\phi = \begin{bmatrix} \cos\phi & 0 & \sin\phi & 0 \\ 0 & 1 & 0 & 0 \\ -\sin\phi & 0 & \cos\phi & 0 \\ 0 & 0 & 0 & 1 \end{bmatrix},$$

$$M_\psi = \begin{bmatrix} \cos\psi & -\sin\psi & 0 & 0 \\ \sin\psi & \cos\psi & 0 & 0 \\ 0 & 0 & 1 & 0 \\ 0 & 0 & 0 & 1 \end{bmatrix}$$

effect a rotation with the angles θ, ϕ, and ψ about the x-, y-, and z-axes, respectively.

The matrix

$$\begin{bmatrix} s_x & 0 & 0 & 0 \\ 0 & s_y & 0 & 0 \\ 0 & 0 & s_z & 0 \\ 0 & 0 & 0 & s \end{bmatrix}$$

effects a scale by the factors $s_x/s, s_y/s, s_z/s$.

Finally, the matrix

$$\begin{bmatrix} 1 & 0 & 0 & 0 \\ 0 & 1 & 0 & 0 \\ 0 & 0 & 1 & 0 \\ p_{30} & p_{31} & p_{32} & 1 \end{bmatrix}$$

maps the ideal plane to the affine plane with points of the form

$$\left(\frac{x}{p_{30}x + p_{31}y + p_{32}z}, \frac{y}{p_{30}x + p_{31}y + p_{32}z}, \frac{z}{p_{30}x + p_{31}y + p_{32}z}\right).$$

Transforming Planes

If a plane in P^3 is given by $\pi[X, Y, Z, W]^T$ and we wish to compute the plane $\pi' = T(\pi)$ where the transformation T is captured by a matrix M, then, as in P^2, we use the transpose of the inverse of M:

$$\pi' = M^{-T}\pi.$$

The same equation is true for directions (normalized normal vectors), but we avoid declaring directions in projective geometry; we should be able to multiply a direction by an arbitrary nonzero constant, such as -1, and obtain the same direction in projective space, yet clearly a multiplication by -1 should also reverse the orientation of a direction. These issues will be solved in Chapter 14 when separability will be recovered for projective geometry.

12.8 Canonical Projective Points

Were we to be doing synthetic geometry in two or three dimensions (as Euclid would have), we would not put any particular significance to one point compared to the others. Assigning Cartesian coordinates to points in the plane requires establishing a point (the origin) and two vectors (the basis) and assigning a particular significance to the resulting three points (the origin and the two points resulting from its addition with the two basis vectors).

These three points figure prominently in the geometry that we subsequently practice. Yet often, as will be argued in Chapter 17, this is undesirable. But the three canonical points of E^2 or the four canonical points of E^3 are frequently helpful. The column vectors of a transformation in E^2, for instance, offer a direct reading of the vectors to which the basis vectors map.

A natural question to ask is to wonder whether there are any such canonical points in projective geometry. If there are, we also would wonder how many such points there are in each geometry. If there are $n + 1$ canonical affine points in an n-dimensional Euclidean geometry, the added power of projective transformations (modifying parallelism) already hints that there ought to be more than $n + 1$ canonical points in n-dimensional projective geometry.

The number of points is directly associated with the fundamental theorem of projective geometry (§ 11.1). There we noticed that $n + 2$ mappings between point pairs are needed in P^n. It is customary to take the points $(1, 0)$, $(0, 1)$, and $(1, 1)$ as the fundamental points in P^1; the points $(1, 0, 0)$, $(0, 1, 0)$, $(0, 0, 1)$, and $(1, 1, 1)$ as the fundamental points in P^2; and the points $(1, 0, 0, 0)$, $(0, 1, 0, 0)$, $(0, 0, 1, 0)$, $(0, 0, 0, 1)$, and $(1, 1, 1)$ as the fundamental points in P^3 [26, 79].

Just as with Euclidean transformations, the coordinates of the points to which $n + 1$ of the canonical points of a projective space map can be read directly as the column vectors of the projective transformation. The canonical points also serve in the opposite direction. We do not need to implement a routine that maps an arbitrary four points to another arbitrary four points; it is sufficient to implement a function that finds the mapping from the four canonical points to an arbitrary other quadruple. An arbitrary mapping can then be found by using the canonical mapping twice. Inverse and matrix composition would then complete the routine.

12.9 Planar Projection

We are finally ready to produce 2D images of 3D scenes. It is clear that if we were to seek a function from E^3 to E^2 that executes that mapping, the function will always leave something to be desired.

- Points at infinity project to vanishing points, yet E^3 does not include points at infinity.
- Points as distant from the view plane as the observer also project to points at infinity. But again the image plane E^2 does not include such points.

A mapping could be defined, but due to these weaknesses, the model would remain tenuous. As is clear from Chapter 11, we must instead pass through projective spaces. The mapping function, or the *graphics pipeline*, will consist of four functions:

$$\boxed{E^3} \xrightarrow{+1} \boxed{P^3} \xrightarrow{M} \boxed{P^3} \xrightarrow{V} \boxed{P^3} \xrightarrow{P} \boxed{P^3} \xrightarrow{D} \boxed{E^3} \xrightarrow{\backslash z} \boxed{E^2}$$

We consider each of the mapping functions:

1. Since P^3 is a superset of E^3, the first function (labeled "+1") simply consists of finding affine points in P^3 that represent the Euclidean points. Each Euclidean point (x, y, z) is mapped to the canonical representative of a projective point $(x, y, z, 1)$.
2. The functions M, V, and P effect the modeling, viewing, and perspective transformations.
3. D returns from projective space P^3 to Euclidean space E^3. We assume for now that the points mapped are affine.
4. $\backslash z$ maps from E^3 to E^2 by simply ignoring the z value. (In practice computing visibility does rely on the z value.)

We describe an implementation of the pipeline using a *function object*. Function objects, or *functors*, are described following the source code for the first example. We leave optimization to a distinct stage of development and are only concerned at this time about the correctness of the system. A main focus will be to identify the type of each object at the different stages of the graphics pipeline. For clarity the pipeline is presented as a concrete, not a generic, function object.

```
class Pipeline
{
  const Transformation_P3d T;
  const Perspective_divide_d my_perspective_divide;
  const Project_on_xy_d my_projection;
public:
  Pipeline( const Transformation_P3d& _T ) :
    T(_T), my_perspective_divide(), my_projection() {}

  Point_E2d operator()( const Point_E3d& P1 )
  {
    Point_P3d P2 = Point_P3d( P1 );
    Point_P3d P3 = T( P2 );
    Point_E3d P4 = my_perspective_divide( P3 );
    Point_E2d P5 = my_projection( P4 );
    return P5;
  }
};
```

An object of type Pipeline is initialized by passing it the combination of the modeling, viewing, and perspective transformations. The constructor also creates a function object for performing perspective divide (division by w):

```
template<typename NT>
struct Perspective_divide
{
  Point E3<NT> operator() (const Point_P3<NT>& p) const
  {
    return Point_E3<NT>(
                  p.hx()/p.hw(),
                  p.hy()/p.hw(),
                  p.hz()/p.hw());
  }
};
```

and another function object for projecting to E^2 (ignoring z):

```
template<typename NT>
struct Project_on_xy
{
  Point_E2<NT> operator() (const Point_E3<NT>& p) const
  {
    return Point_E2<NT>(p.x(), p.y());
  }
};
```

After an object of type Pipeline is created, it can be used as any other object—it can, for example, be sent to other functions as a parameter. One then has only to invoke the *object* as if it were a *function* and the **operator**() of the object will be invoked. The member function **operator**(), which takes the input to the pipeline and returns the final result, combines the four stages of the pipeline. We turn our attention to the line

```
Point_E3d P4 = my_perspective_divide( P3 );
```

which performs the perspective transformation proper.

Perspective Transformation

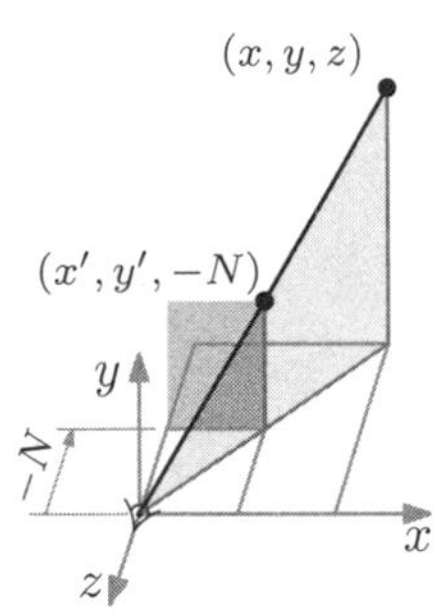

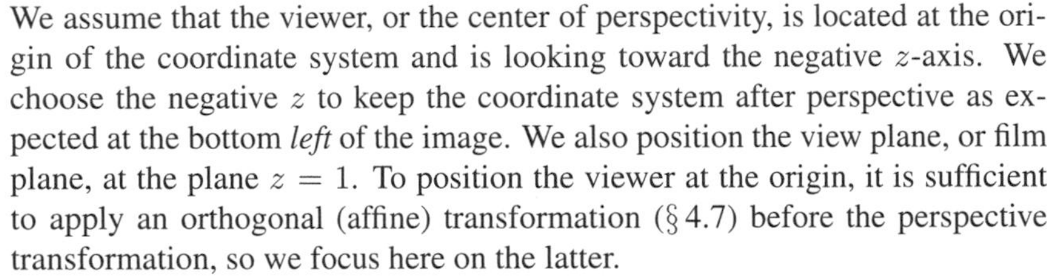

We assume that the viewer, or the center of perspectivity, is located at the origin of the coordinate system and is looking toward the negative z-axis. We choose the negative z to keep the coordinate system after perspective as expected at the bottom *left* of the image. We also position the view plane, or film plane, at the plane $z = 1$. To position the viewer at the origin, it is sufficient to apply an orthogonal (affine) transformation (§ 4.7) before the perspective transformation, so we focus here on the latter.

The perspective transformation is applied by projecting on an image plane. The image, or projection, plane is positioned such that it is orthogonal to the z-axis at $z = -N$, where N is some positive value. The x- and y-coordinates of the projected point $(x', y', -N)$ can be found by similar triangles:

$$x' = -N\frac{x}{z}, \qquad y' = -N\frac{y}{z}.$$

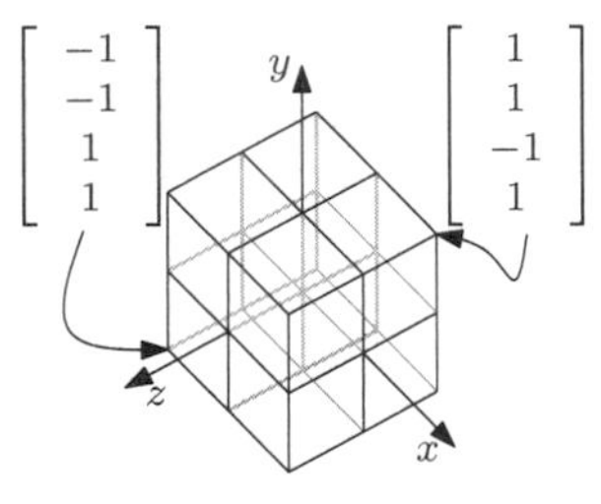

In the interface adopted by OpenGL, the view volume is passed as six scalar values describing two corners of the view volume on the image plane as well as the location of the distant, or *far*, clipping plane. The viewer is implicitly assumed to lie at the origin, and the view axis is implicitly assumed to be the negative z-axis. The function **glFrustum** has the following interface:

```
void glFrustum(
        GLdouble left, GLdouble right, GLdouble bottom,
        GLdouble top, GLdouble near_val, GLdouble far_val )
```

Both the near and far values are negative, but since it would be unwieldy to force users of the API to recall that the view axis is the negative z-axis, these two values are passed as positive z values and are negated internally by the **glFrustum** function.

The canonical view volume after the projection transformation is a cube with corners at $(-1, -1, 1)$ and $(1, 1, -1)$.

The desired matrix is [50]

$$M = \begin{bmatrix} \frac{2N}{R-L} & 0 & \frac{R+L}{R-L} & 0 \\ 0 & \frac{2N}{T-B} & \frac{T+B}{T-B} & 0 \\ 0 & 0 & \frac{-(F+N)}{F-N} & \frac{-2FN}{F-N} \\ 0 & 0 & -1 & 0 \end{bmatrix}.$$

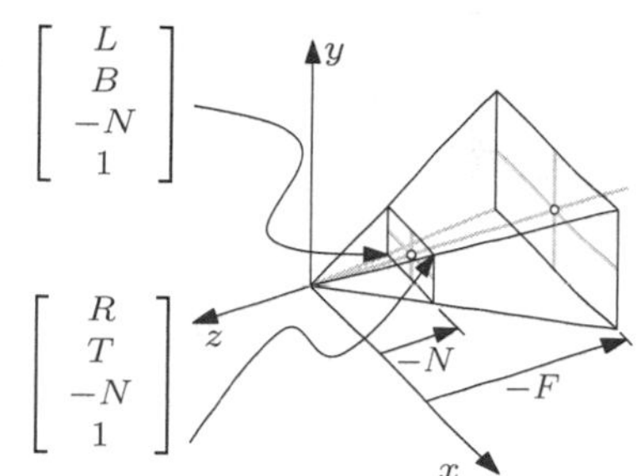

We confirm that M does indeed effect the desired mapping by finding the points to which the corners of the view volume map. The two corners

$$\begin{bmatrix} L & B & -N & 1 \end{bmatrix}^T \text{and} \begin{bmatrix} R & T & -N & 1 \end{bmatrix}^T$$

map to two corners of the canonical view volume. Also, the viewer at

$$\begin{bmatrix} 0 & 0 & 0 & 1 \end{bmatrix}^T$$

maps to an ideal point ($w = 0$):

$$M \begin{bmatrix} L \\ B \\ -N \\ 1 \end{bmatrix} = \begin{bmatrix} -N \\ -N \\ -N \\ N \end{bmatrix},$$

$$M \begin{bmatrix} R \\ T \\ -N \\ 1 \end{bmatrix} = \begin{bmatrix} N \\ N \\ -N \\ N \end{bmatrix},$$

$$M \begin{bmatrix} 0 \\ 0 \\ 0 \\ 1 \end{bmatrix} = \begin{bmatrix} 0 \\ 0 \\ \frac{-2FN}{F-N} \\ 0 \end{bmatrix}.$$

And so at the same time that the projection transformation takes the truncated pyramid to a cube, it also takes an affine point (the viewer at the origin) and maps it to an ideal point. This is not surprising; indeed all points lying at $z = 0$ map to ideal points and the plane $z = 0$ itself maps to the ideal plane.

Confusing an External for an Internal Segment

Now consider that we use the pipeline described above and generate a set of images as a viewer approaches a given cylinder. We argued earlier that segments are not a projective notion and that we would not define a class segment. Yet now we surely need some notion of segment to be able to render segments on the PostScript device. The (incorrect, as we will see) solution we adopt is to pass points (in the various geometries) through the pipeline and then render segments as specified by the connectivity of the object rendered. Notice that we are not taking any precautions when mapping back from P^3 to E^3, and so we could indeed end up dividing by zero. This is yet another problem we leave out for the moment; we simply ensure not to render images when any of the cylinder's vertices has a zero depth.

As long as the cylinder is in front of the viewer, all is well. Already starting in the second image, the viewer is so close that parts appear outside the viewport. But discarding these portions is simple since we would only need to clip in E^2.

Unusual diagonal lines start to appear in the last two images. These arise because some edges of the cylinder pierce the $z = 0$ plane. The points where

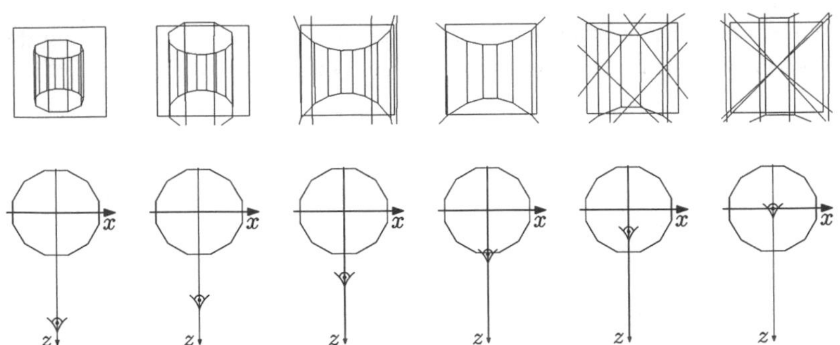

the edges and the plane intersect map to ideal points. We are in effect rendering external segments as if they were internal segments (§ 11.2). We have already encountered instances of internal segments that map to external ones on pages 111 and 115.

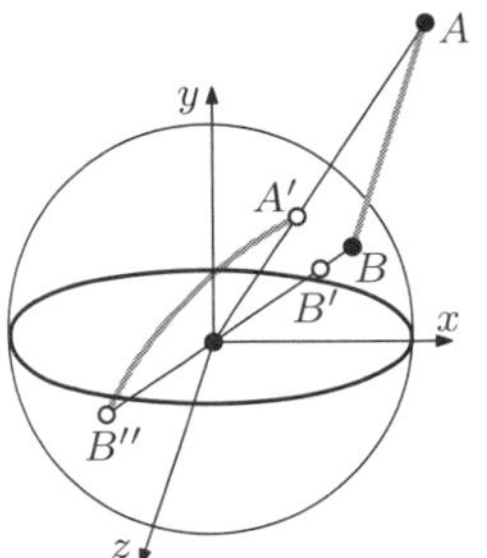

But why does an external segment cross the image plane diagonally? If an edge to the right and above the observer intersects the image plane, should it not remain to the right and above the observer? To see what is happening we appeal to the intuition we gained from spherical geometry in Chapter 9. If a segment AB intersects the $z = 0$ plane and if A is in front and B is behind the observer, then A will project to A' and B will project to B' on a small sphere centered at the observer. Yet the nonseparability of projective geometry makes it unable to distinguish between the front half and the back half of the sphere; projective geometry effectively collapses the two sides of the sphere centered at the observer! Points B' and B'' are confused and point B'' is the closest projective geometry can provide for a point on the Euclidean image plane as the projection of B. When we now render an internal instead of an external segment, it appears as a diagonal segment.

Inability to Clip in Projective Spaces

Clearly we need to clip. Discarding the portions of the scene behind the observer and those outside the view frustum before projecting would yield the planar projection we are seeking. Let us look once more at the graphics pipeline and attempt to choose the stage at which we would discard features behind the observer:

$$\boxed{E^3} \xrightarrow{+1} \boxed{P^3} \xrightarrow{M} \boxed{P^3} \xrightarrow{V} \boxed{P^3} \xrightarrow{P} \boxed{P^3} \xrightarrow{D} \boxed{E^3} \xrightarrow{\backslash z} \boxed{E^2}$$

Since the two stages $+1$ and $\backslash z$ take points in a separable space (E^3) as input whereas M, V, P, and D take points in a nonseparable space (P^3), we already know that clipping must be performed along with either $+1$ or $\backslash z$. Planes in P^3 have one side, so it would make no sense to attempt to decide whether the two endpoints of a segment lie on the same side of a plane in P^3.

One can indeed define solutions for this *w-wraparound problem* [47] while nominally staying in projective spaces, but distinguishing between signs of w [107, 17] signals that the geometry we are tackling is of a different nature—

the topic of Chapters 14 and 15. The best we can do if we are to stay within projective geometry is to generate images on a one-sided projection plane.

We contemplate whether it would be possible to perform clipping after returning from projective to Euclidean space.

$$E^3 \xrightarrow{+1} P^3 \xrightarrow{M} P^3 \xrightarrow{V} P^3 \xrightarrow{P} P^3 \xrightarrow{D} E^3 \xrightarrow{\textbf{clip}} E^3 \xrightarrow{\backslash z} E^2$$

But after D is performed it is too late; we would have already passed through projective space and the front and back sides of the image plane (as we shall shortly call them) would have collapsed. If we are to use projective geometry, our only hope is to perform clipping on the sides of the view volume in E^3 before we pass to P^3 through $+1$. But the modeling transformation M and the viewing transformation V are both affine, so there is no difficulty in applying them while remaining in Euclidean space E^3. Conceivably, we could clip after applying the modeling transformation.

$$E^3 \xrightarrow{M} E^3 \xrightarrow{V} E^3 \xrightarrow{\textbf{clip}} E^3 \xrightarrow{+1} P^3 \xrightarrow{P} P^3 \xrightarrow{D} E^3 \xrightarrow{\backslash z} E^2$$

Yet because the viewer would be at an arbitrary location, the viewing frustum would have to be defined anew for each image. We prefer instead to wait until the viewing transformation has been applied. The clipping planes then only depend on the projection used.

$$E^3 \xrightarrow{M} E^3 \xrightarrow{V} E^3 \xrightarrow{\textbf{clip}} E^3 \xrightarrow{+1} P^3 \xrightarrow{P} P^3 \xrightarrow{D} E^3 \xrightarrow{\backslash z} E^2$$

Using the six sides of the frustum needed in this case is still unwieldy. That four of them are not axis-parallel means that it is more expensive to determine whether two points lie on the same side of a plane: Rather than make do with two subtractions and sign tests, we must pay the price of dot products. The canonical view volume obtained after P would be perfect. Its six sides are parallel to the coordinate system. Is there a way to achieve efficiency while performing sane sidedness tests? We see how that is possible when we retake the theme of projective geometry by recovering separability [104] in Chapters 14 and 15, but we first take a short break and discuss barycentric coordinates, which exhibit a peculiar property of ideal points that will be illuminating.

12.10 Exercises

12.1 Provide your own "artistic rendering" of the projective plane and indicate on it each of the following points:

- (1,0,0)
- (0,0,1)
- (1,0,1)
- (1,1,1)
- (0,1,0)
- (1,1,0)
- (0,1,1)

12.2 What are the coordinates of the ideal point on the line $l(a, b, c)$ lying in the projective plane?

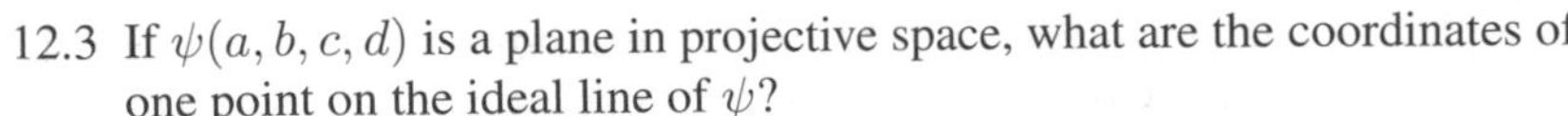

12.3 If $\psi(a, b, c, d)$ is a plane in projective space, what are the coordinates of one point on the ideal line of ψ?

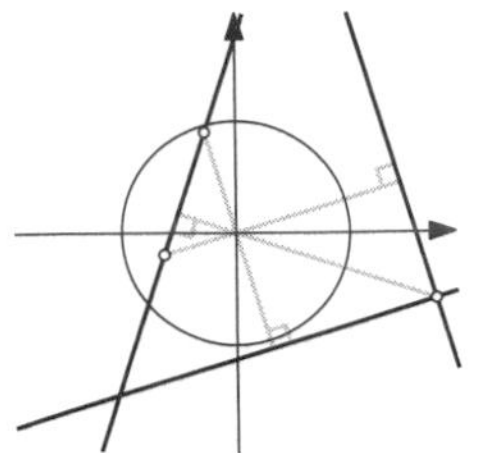

12.4 Compile and run the program labeled "duality" accompanying this text and then modify the code so that the underlying geometry classes are those for projective geometry rather than those for Euclidean geometry. Write a few lines commenting on your experience. Is the code more elegant after using projective objects? Why? Which of the two foundation layers (Euclidean or projective) is more fitting for this application?

12.5 Modify the start-up code labeled "duality" so that inserting a point snaps to another nearby point either when the point itself is near a line or when the line dual of the point passes by two points that have already been entered. After inserting n point-line duals, do you need to make $\mathcal{O}(n^2)$ checks before snapping?

12.6 *Fake shadowing* is a simple method for rendering the shadow of a 3D object. A plane in space is declared the "ground" and scene objects are projected on it. Each projected object is displayed as a set of gray polygons. Given a point P, a plane Π, and a point L, the position of the light source, determine the coordinates of the projection of P from L on Π. Your derivation must handle the case when L is an ideal point.

12.7 Solve Exercise 12.6 and then modify the start-up code labeled "tank" to generate shadows for the monoliths. What drawbacks do you observe in this shadow-generation scheme?

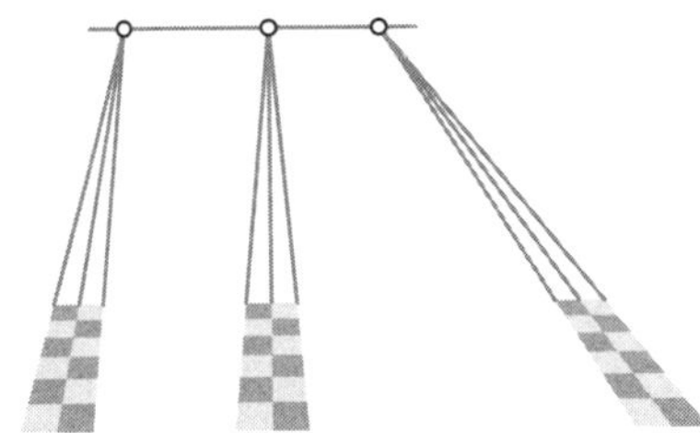

12.8 The objective of this exercise is to be convinced that perspective projection may map ideal points to affine points and an ideal line to an affine line.

Develop a system that generates a PostScript file showing the projection of three long strips in a checkerboard pattern as shown in the adjacent figure. Print the resulting file then confirm that the three vanishing points are indeed colinear. Start from the class Postscript and the set of classes in P^3 in the directory geometry_P3.

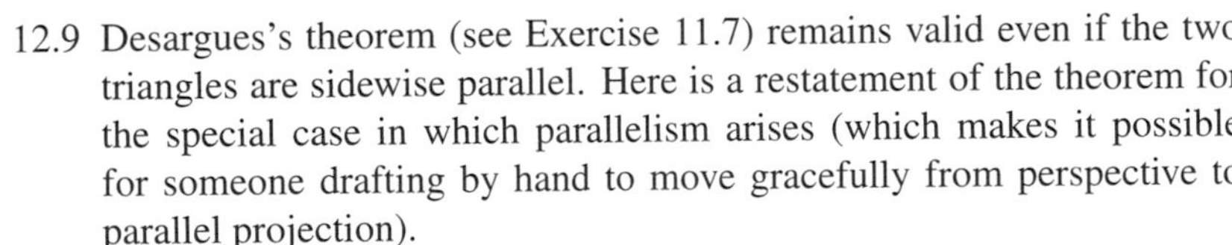

12.9 Desargues's theorem (see Exercise 11.7) remains valid even if the two triangles are sidewise parallel. Here is a restatement of the theorem for the special case in which parallelism arises (which makes it possible for someone drafting by hand to move gracefully from perspective to parallel projection).

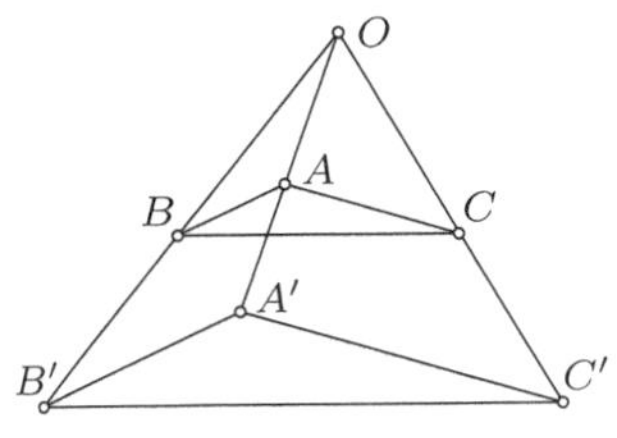

> *If ABC and A'B'C' are two triangles with distinct vertices, so placed that the line BC is parallel to $B'C'$, CA to $C'A'$, and AB to $A'B'$, then the three lines AA', BB', and CC' are either concurrent or parallel* [27].

Reimplement Exercise 11.7 using homogeneous coordinates. Was the logic of your new implementation significantly simpler than the one using Cartesian coordinates?

13 Barycentric Coordinates

Where is the center of gravity of a discrete set of masses? Möbius observed in 1827 that this elementary computation goes beyond its mere use for engineering, for the masses themselves can be used as coordinates that replace "parallel coordinates" [70]. His new coordinate system, which he dubbed *barycentric* or weight-centric, was, as mentioned at the start of Chapter 12, one instance of four near-simultaneous discoveries of homogeneous coordinates.

Readers who wish to pursue the original manuscripts will find pointers to translations of many fundamental works in Grattan-Guinness's recent volume [43] (though, strangely, Möbius' work is omitted; even more strangely, his *Barycentric calculus* appears to have never been translated). After encountering ideal points, or points at infinity, in Chapters 11 and 12 through projection, we encounter them once again in this chapter through algebraic manipulation.

13.1 Barycentric Coordinates in One Dimension

Consider putting two masses a and b at two distinct points A and B. The masses can have arbitrary values, just so long as both do not simultaneously vanish. This condition is necessary for otherwise it would make no sense to attempt to determine the center of mass of a massless system. Either or both masses may also be negative; the center of mass remains well-defined. We can avoid considering the notion of antimatter by thinking instead of (positive or negative) electric charges and determining the resulting location of the perceived charge.

We assume an ideal setting in which two infinitesimal objects can have arbitrary masses. Alternatively, we may consider that the objects are not infinitesimal, but that their centers of mass lie at the points A and B. The center of mass, also called the centroid or the barycenter, of the two point masses evidently lies on the line defined by the two points. If the mass at one of the two points vanishes, the barycenter lies at the other point. If both masses have the same sign, the barycenter lies at a point along the segment AB that divides the segment in the inverse proportion of the weights. This suggests that we can use the masses (a, b) for coordinates. The coordinates of A become $(1, 0)$, $(2, 0)$, or, indeed, $(\lambda, 0)$ for any $\lambda \neq 0$. Likewise the coordinates of B are $(0, \lambda)$. The midpoint has coordinates (λ, λ).

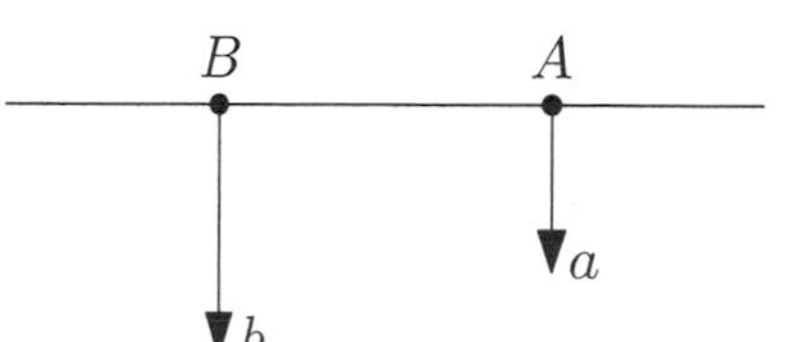

Suppose we start from two equal (positive) masses and wish to move the barycenter to A. We can proceed by increasing the mass a indefinitely until b is insignificant next to it. Or else we can simply set $b = 0$. It is easy to deduce

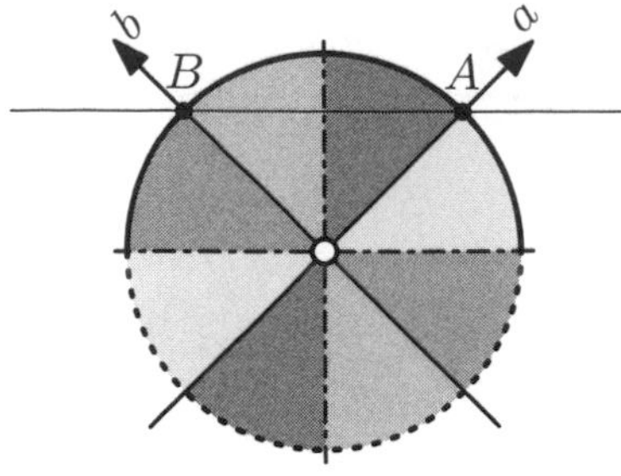

the coordinates of points in the segment AB, but what about the remaining point on the line AB? If one point, say B, has a negative charge b, then that charge acts as a repulsive force and the barycenter is closer to A than to B and outside the segment AB. And so the mapping between points in the (Euclidean) plane parameterized by (a, b) and the center of mass becomes clear as a relationship between the relative absolute magnitude of the two masses. If the plane is divided into eight octants, then the points in the first and fifth octants correspond to a barycenter in the interior of AB and closer to A than to B. Points in the fourth and eighth octants also correspond to points closer to A—those lying outside the segment AB on the carrying line, and so on.

If one of the two masses vanishes or if the two are equal, we are at three of the lines separating the octants; the location of the barycenter is then obvious. What if in the preceding experiment (reducing the value of b relative to a positive a), b approaches a in magnitude, but with opposite sign? We still ought to be able to determine the barycenter, if only in the limit. As b approaches $-a$ from above, the barycenter moves farther from A and approaches the ideal point, the point infinitely far on the line AB on the side of A (c.f. § 11.1). Even though it becomes awkward to attempt to determine the centroid of a massless system when $a + b = 0$, there is no difficulty in finding the limit. As long as $a + b \neq 0$, we can normalize by dividing by the sum, yielding relative values such that $a + b = 1$, but such normalization is no longer possible if the point intended is the ideal point, fully mirroring the canonical representatives for P^1 discussed in § 12.1.

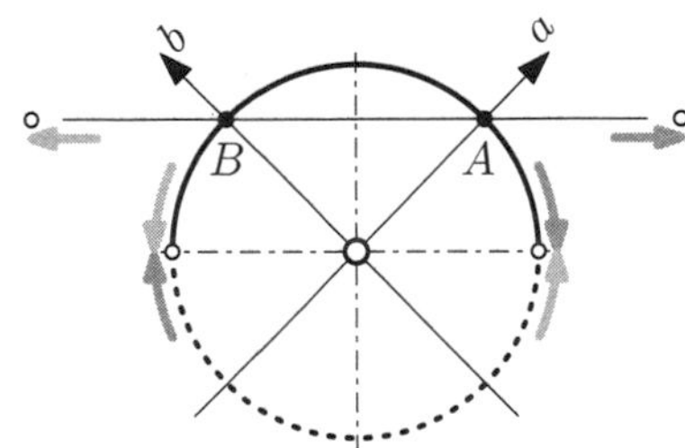

Because the case of $a = b = 0$ is disallowed, the center of the coordinate system does not map to a point on the line. Barycentric coordinates are homogeneous: Any two points that are colinear with the origin represent the same barycenter (c.f. § 12.1). This suggests that we can take *all* colinear points passing by the origin—lines passing by the origin—and map each line to a point on the line AB.

Continuing with updating b, what if b approaches $-a$ from below? The barycenter in that case is the ideal point on B's side. The total mass/charge of the system is now negative and we are in the dotted half of the circle in the figure. Because the limit line is identical regardless of whether we reach it using a clockwise or counterclockwise turn, there is only one ideal point. Each line passing by the origin maps to its intersection with AB.

Using Weights

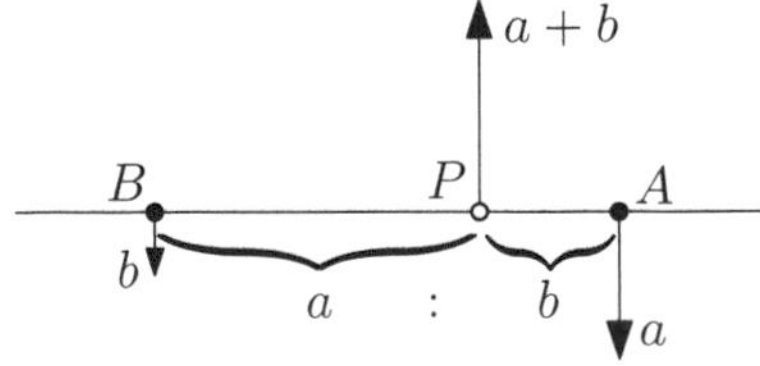

We have no need for some attractor to be able to calculate either the center of mass or the inertia of a system, but positioning some planet or star with gravity g somewhere in the plane (other than at A or B) makes it possible to calculate the resultant of the two forces. The center of gravity of ag and bg is then the location on the line where we can put the combined mass $(a + b)g$ to replace—or, in the figure, to counteract—the two masses. If the planet is positioned at the origin of the plane above, for instance, then the barycenter can be determined by intersecting the vector of the resultant with the line AB. If the planet is sufficiently far for the forces to be parallel, then the barycenter is likewise the location that acts as a counterweight.

Pure Geometry

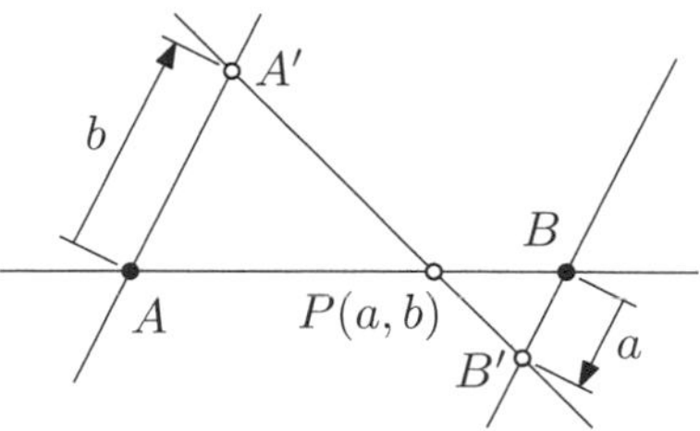

Yet a third way to determine the location of a point $P(a, b)$ is possible. We can discard any physical notion and remain purely geometric by erecting two parallel lines through A and B that are not themselves parallel to AB. If two points A' and B' that are b and a, respectively, from A and B and that lie on opposite sides of the line AB are defined, then P is the intersection of the two lines AB and $A'B'$.

13.2 Barycentric Coordinates in Two Dimensions

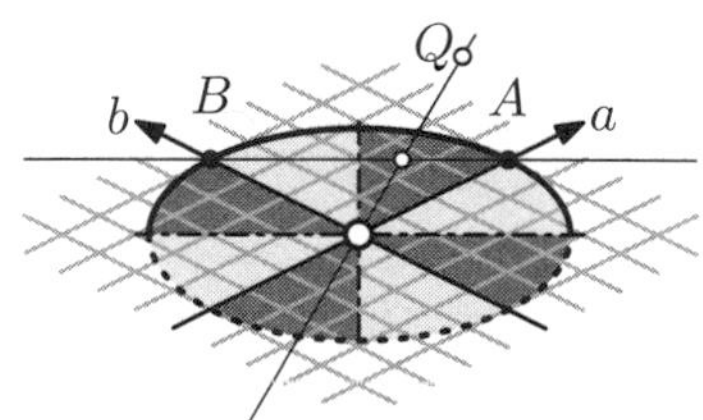

Before moving to two dimensions, let us revisit the mapping from a pair of coordinates to points on the line AB. We have been able to find an origin O in some plane carrying AB such that O is equidistant from the two *fundamental points* and such that the resulting coordinate frame is orthogonal. But is this last condition truly necessary? As the figure suggests, any coordinate system would indeed do. A point Q in the plane represents the point on AB that lies at the intersection of AB and OQ. Q can be replaced by any other point that is colinear with it and the origin O.

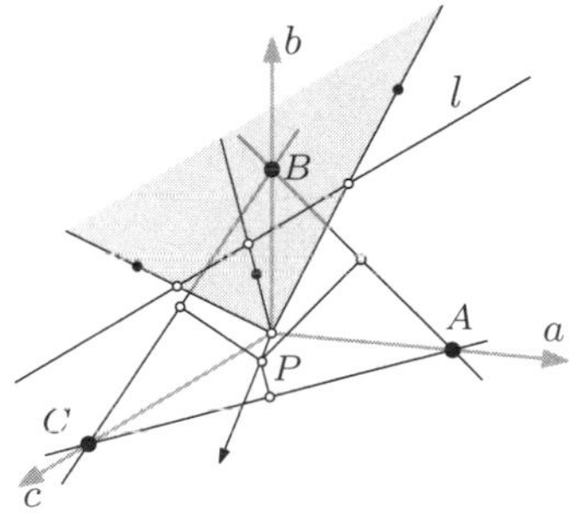

We are now given three noncolinear fundamental points and will study whether it is possible to position weights at the three points such that the center of mass will span the plane of the three points. We are looking for a point that will serve as the origin and that is equidistant from the three points, so we determine P, the intersection of the perpendicular bisectors of the segments AB, BC, and CA, and we erect a perpendicular to the ABC plane from P. Any point on the resulting line will do as origin O. If one of the three masses is zero, then the setup is identical to the one previously seen. The barycenter of three points can be determined in two stages; the barycenter of two of the three points is first determined.

Any point in the plane can be parameterized by the coordinates (a, b, c). As before, 3D points of the form $(\lambda a, \lambda b, \lambda c)$, $\lambda \neq 0$ capture the same 2D point—the intersection of the line in 3D passing by that point and the origin O with the plane ABC. The point P itself has coordinates $(\lambda, \lambda, \lambda)$.

We seek an equation for a line l in the 2D plane. Consider three points $P_i(a_i, b_i, c_i), i = 1 \ldots 3$, on the line l. Because the three 3D vectors $\overrightarrow{OP_i}$ lie in the same subspace, we have

$$\begin{bmatrix} a_1 & a_2 & a_3 \\ b_1 & b_2 & b_3 \\ c_1 & c_2 & c_3 \end{bmatrix} = 0,$$

the same form we saw in § 12.2 and nearly the same as the one in § 2.2. If two points are known and the third $P(a, b, c)$ is unknown, we can compute the minors in the matrix above to yield the equation for a line as

$$\alpha a + \beta b + \gamma c = 0,$$

also as in § 12.2. Evidently there is a mapping between 3D planes and 2D lines. This is convenient. The intersection of two 2D lines is the common subspace between two 3D planes passing by the origin. The solution is reached by

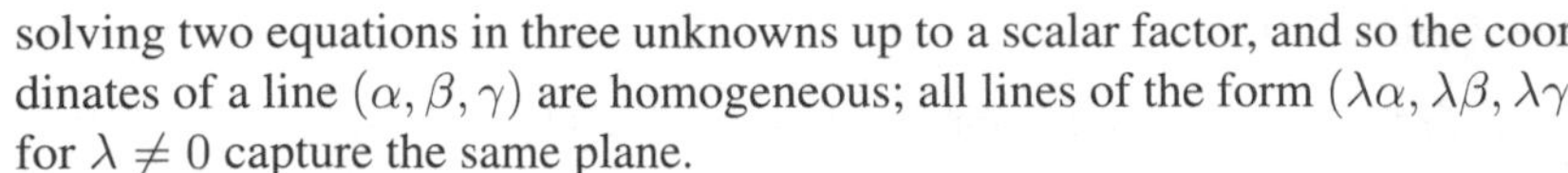

solving two equations in three unknowns up to a scalar factor, and so the coordinates of a line (α, β, γ) are homogeneous; all lines of the form $(\lambda\alpha, \lambda\beta, \lambda\gamma)$ for $\lambda \neq 0$ capture the same plane.

Now suppose that we partition one unit of mass in three (positive) parts. The barycenter must then lie inside the triangle ABC. This parameterization is useful for color interpolation in a triangular mesh or, conversely, for finding the relative weights at the three vertices of a triangle after ray casting [96]. Directions (§ 3.2) are usually only known at the vertices of a solid object described by its boundary (Chapter 26). And so shading (Chapter 21) is determined only at vertices. Barycentric parameterization makes it possible to generate shading values in the interior of triangles.

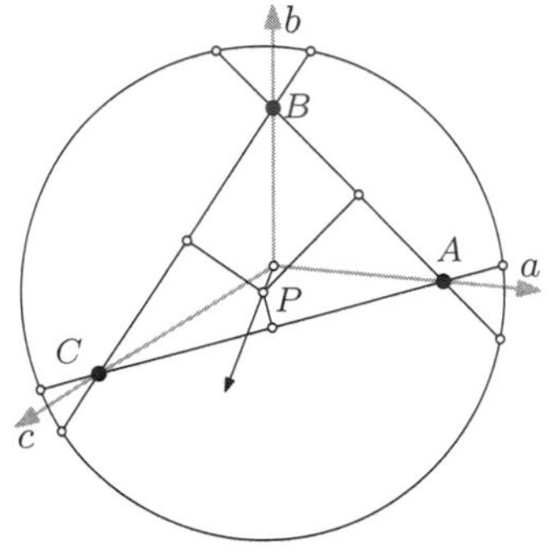

No discussion of homogeneity is complete without considering ideal points. As with projective homogeneous coordinates, barycentric coordinates capture ideal points on the plane ABC, not those of the embedding 3D space. The three ideal points of the three lines AB, BC, and CA are straightforward generalizations of the case in 1D. Given that the three ideal points have $\alpha + \beta + \gamma = 0$, we know that that must be the characterization of the ideal line.

Converting from Euclidean to Barycentric Coordinates

One occasionally knows the Euclidean coordinates of a point and wishes to determine its barycentric coordinates. Given a point $P(x, y)$ amidst the fundamental points A, B, and C, we seek a, b, and c of P.

The answer, which we will need in Exercise 20.1, reduces to a simple area computation [27]. Since

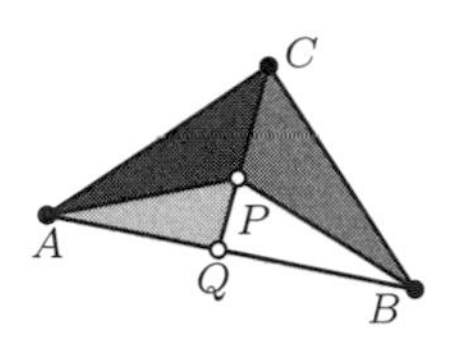

$$\frac{a}{b} = \frac{QBC}{AQC} = \frac{PQC}{AQP} = \frac{QBC - PQC}{AQC - AQP} = \frac{BPC}{CPA}$$

and likewise for the ratios b/c and c/a, the ratio $a : b : c$ is that of $BPC : CPA : APB$. Determining the areas of the three triangles (Exercise 7.6) thus also provides the barycentric coordinates.

13.3 Exercises

13.1 Draw two fundamental points A and B on a line and then determine the location of the points with barycentric coordinates $(2, 1)$, $(2, -1)$, $(2, 4)$, and $(6, -3)$.

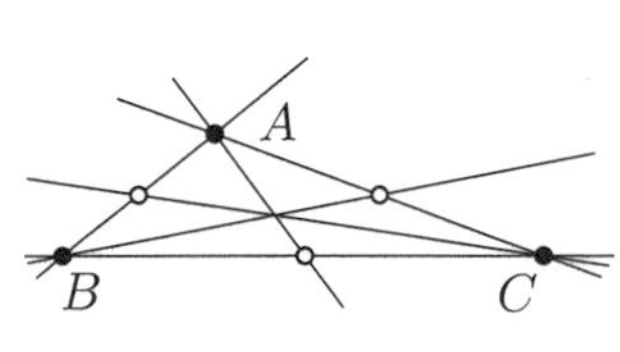

13.2 Determine the coordinates of the midpoints shown in the adjacent figure as well as those of the median of the triangle ABC.

13.3 Determine the equation of the three median lines in the adjacent figure.

13.4 Determine the coordinates of the ideal point on each of the three median lines.

13.5 Write a characterization for the coordinates of the ideal point on a line AB with a mass a at A and another b at B.

13.6 Devise two methods, one based on a circle and the other based on parallel lines, to find the barycentric coordinates of a point P lying on a line AB.

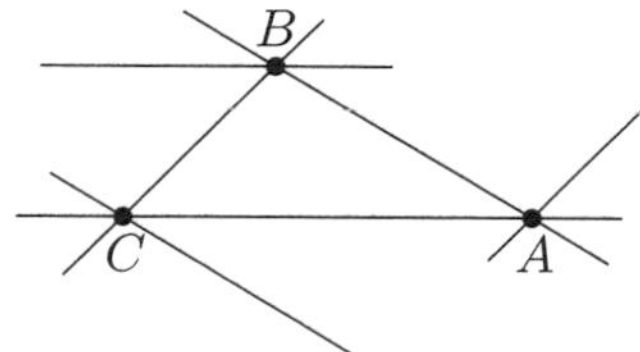

13.7 Find the equation of the line l that passes by two points P and Q. P lies on the line AC and has coordinates $(p_a, 0, p_c)$ and Q lies on the line BC and has coordinates $(0, q_b, q_c)$.

13.8 Determine the equation of each of the six lines shown in the figure as a function of weights at the three fundamental points A, B, and C.

14 Oriented Projective Geometry

Something disconcerting lies at the heart of computer graphics. Our main objective is to simulate a camera using a computer. Because the camera we simulate has a planar film, we naturally appeal to projective geometry. We declare the film plane to be the projection plane and the eye to be the center of perspectivity. We create computer models of the scene and position the eye. But unless the scene model is incomplete or very modest, the eye will be surrounded by the scene—parts of the scene will lie behind the camera. Because projective geometry is nonseparable, all points on a line passing by the eye will project to the same point on the image plane—regardless of whether the point being projected lies in front of or behind the eye. This is of course not how a real camera works. It probably is also not what we want to render. We would prefer objects behind the eye not to be included in the image.

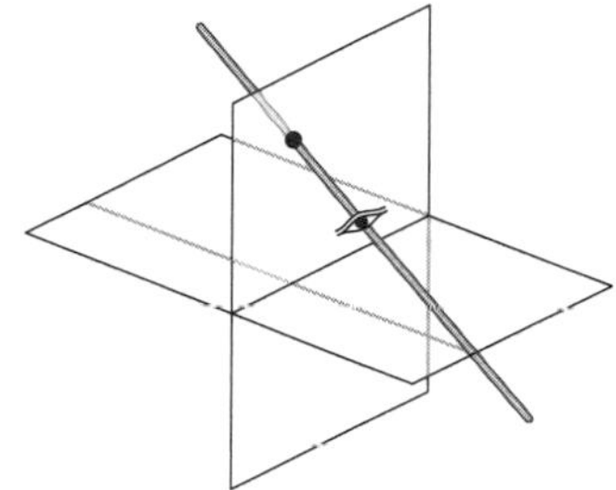

14.1 Double-Sided Projection

So what do we do? One option is to discard objects behind the eye before projecting them on the image plane. In any case, we want to do that for efficiency reasons. But that also smacks of hacking. We still do not have a theory for arguing about what would have happened to the objects behind the eye had we not performed clipping. Besides, if the computer makes it possible for us to do something that is impossible with a physical camera, why would we want to throw out that power and tailor our model to the limitations of a physical device? Perhaps we wish to implement an all-viewing system, where an omni-eye is capable of seeing everywhere at the same time. We will see in Part VII that we can do just that by projecting on a sphere. This is convenient, but we still want to produce planar projections. After projecting on a sphere the problem has been distilled to its essence, but it is still there. If we declare a center of interest, the location where the eye is most interested, and position the view plane such that it is orthogonal to the view direction, then *two* points will map to the same point on the image plane.

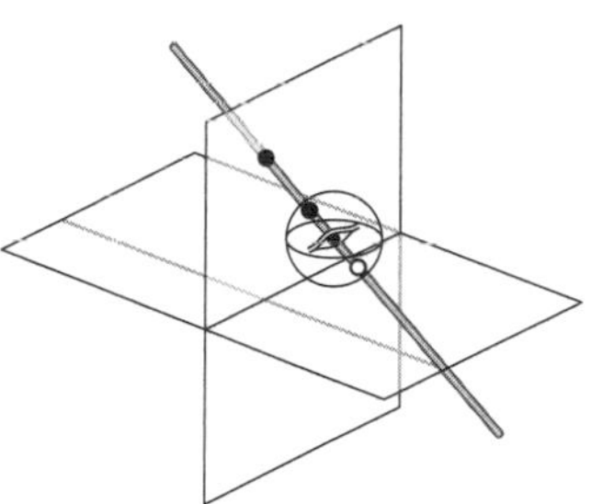

Stolfi's oriented projective geometry has the answer [104]. Instead of one, he assumes there are two projection planes—infinitesimally separated from each other. The dual projection planes can be positioned anywhere but cannot pass by the eye. Their distance from the eye is a matter of focal distance and relates in a physical camera with using one lens or another. The perspective is not affected by positioning the double projection plane, only by positioning the camera. Regardless of where the double projection plane lies, imagine that

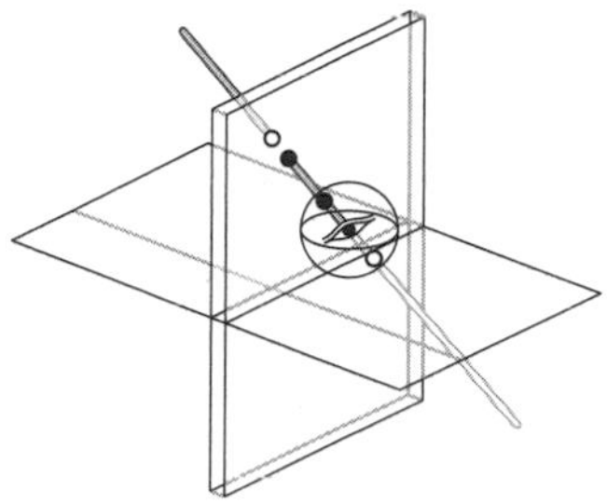

there is one plane parallel to it and passing by the eye. When we talk about objects in front of or behind the camera, we refer to that plane, not the double projection plane. Now declare the projection plane on the side of the eye to be the front projection plane and declare the plane on the opposite side of the eye the back side. (In a film camera the front side of the projection plane is the side of the film photons impinge on and the side of the film touching a plate—were it to be possible to use it to capture pictures—would be the back side.)

Two types of projection take place. Objects in front of the eye (including those between the eye and the projection plane) project on the front projection plane and objects behind the eye project on the back projection plane. Oriented projective geometry makes it possible to model flipping the orientation of the camera since pointing the camera in the opposite direction should also have an algebraic effect. Elements of the transformation matrix will change sign in addition to the front and the back image planes swapping roles [62]. In classical projective geometry the sign of a transformation matrix does not affect the projection and two opposite orientations of the camera cannot be distinguished algebraically.

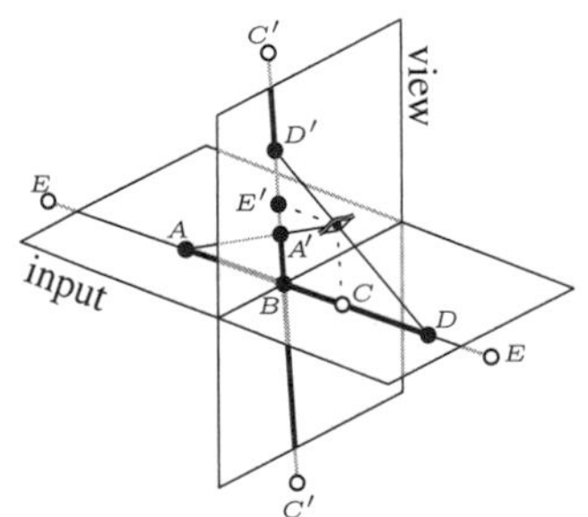

Recall the setup in Chapter 11. A line segment AD lies in an input plane and is projected through a center of perspectivity onto a segment $A'D'$ on a view plane. Yet turning the view projection plane into a double-sided plane will make it easy to distinguish between the internal and the external segments $A'D'$.

Points as distant from the double projection plane as the eye are special. They project on the ideal line of the double projection plane. But this ideal line is also of a more exotic variety than a spherical circle. Like a spherical circle (§ 9.3), antipodal points are *not* identified. The surface on which we are projecting has the same topology as a sphere, though geometrically it is an unusual sphere: The two antipodal ideal points are shared between the front and the back projection planes in a criss-cross fashion. To see this, consider projecting AD where A lies in front of the eye eye and D lies behind the eye. The interval $[A, C)$ projects on the front plane and the interval $(C, D]$ projects on the back plane. The point C', the projection of C, is shared between one side of the front plane and the *opposite* side of the back plane. Just as with barycentric coordinates it was rather awkward to talk about the center of mass when the total mass in the system was zero, and so we talked instead about two distinct limits for approaching a massless state, so here also we talk about two distinct limits. If we approach C from A's side, the limit falls on the front plane. If we approach it from D's side, the limit falls on the back plane. But ultimately the projection is indeed continuous because the two limits are identical.

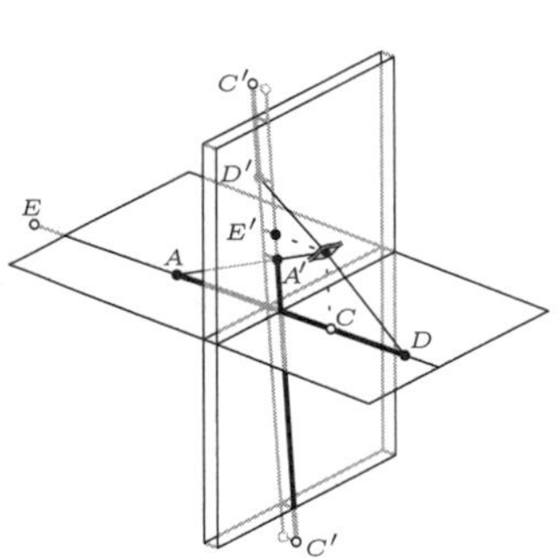

Another indication that the double projection plane, oriented projective plane, or T^2 for simplicity, is not simply a flattened sphere can be seen by considering the orientation of the two segments, $A'C'$ on the front plane and $C'D'$ on the back plane. A spherical circle in a flattened sphere would have opposing orientations.

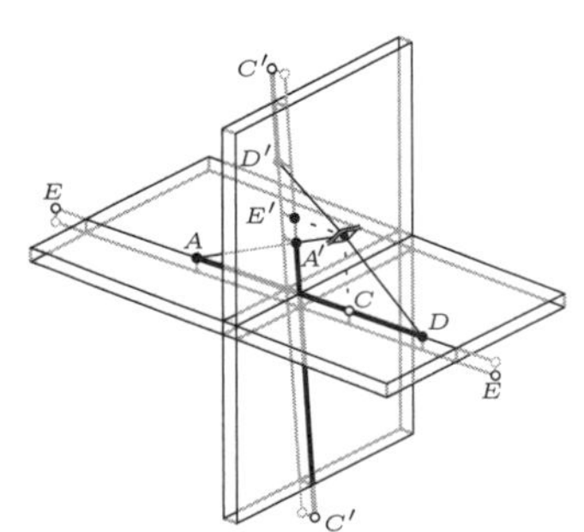

One detail remains awkward. If the image plane is made double-sided, should the object (or input—in computer graphics) plane not also be made double-sided? If the input is devised in Euclidean space and if we map the input points to oriented projective space by appending $+1$ to the coordinates, the full richness of the double-sidedness of the input plane will not be observed. But just as completing the mapping between two projective lines required appending ideal points on both lines (§ 11.1), so here also must we make both planes double-sided to complete the mapping.

14.2 The Oriented Projective Line and Plane

The Oriented Projective Line

We are now ready to construct a new kind of projective line, the oriented projective line T^1. As with the (classical) projective line, P^1, the new line can be studied by considering either the points on the unit circle—the spherical model—or the points on two lines, a front and a back side—the straight model [104].

An axiomatic development [59] does not bring us any closer to designing computing machinery, so we continue by studying double-sided linear sets embedded in Euclidean spaces. Points on the front side of a line have $w > 0$, and points on the back side have $w < 0$. The coordinates are homogeneous: A point represented by (x, w) can equally well be represented by $(\lambda x, \lambda w)$. The difference is that we now insist, as with S^1 (§ 8.1), that $\lambda > 0$. Notice that x and w have the reflected order with respect to the one previously considered. This is convenient to ensure that the counterclockwise rotation (positive rotation) in the spherical model agrees with the positive sense of the x-axis.

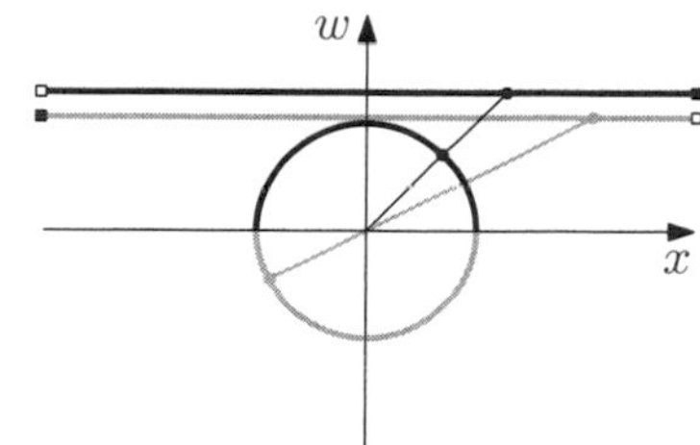

T^1 has two ideal points. For simplicity let us call them $+\infty$ and $-\infty$. These two points occupy their expected locations in the front side of T^1, but they exchange places in the back side.

To justify the unusual setting of the two ideal points, we appeal to projection. But there is no need to have a terribly elaborate setup of embedding in T^2 if we are convinced that central projection through the origin is general enough. (This view is taken further in Part VII.)

We consider two triangles ABC and DEF. In 2D scenes we posit that the appropriate sense for triangles (or any polygon) is to have a clockwise orientation. The interior of a triangle then lies in the negative halfspace defined by the boundary. This is a widely adopted convention for 3D models (see Chapter 26) and consistency will make it possible to implement a single algorithm once for more than one geometry and for more than one dimension in that geometry (see Chapters 28 and 29).

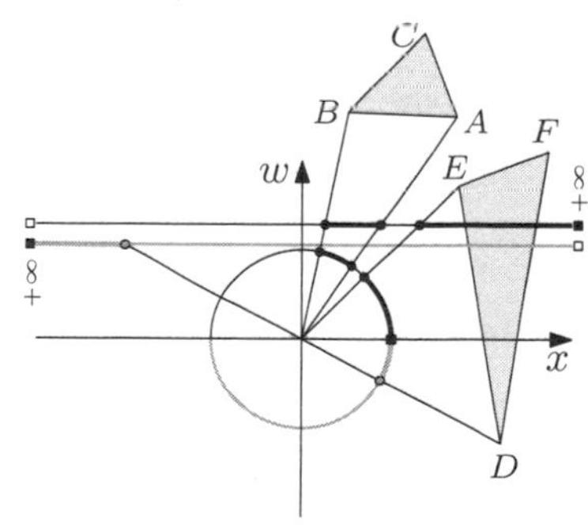

Of the six edges in the two triangles, only two, AB and DE, are seen by (front-facing to) the projection point. AB projects wholly on the front side of T^1, whereas DE has one point that projects on one ideal point. One part of the projection of DE lies on the front side and another part lies on the back side. The ideal point labeled $+\infty$ is shared between the two parts.

Something may appear to be amiss concerning ideal points. T^1 has two ideal points, but because each ideal point lies on opposite sides of the front and back sides, it may appear that we are not gaining much concerning projection. If we apply a transformation matrix and find that the resulting point

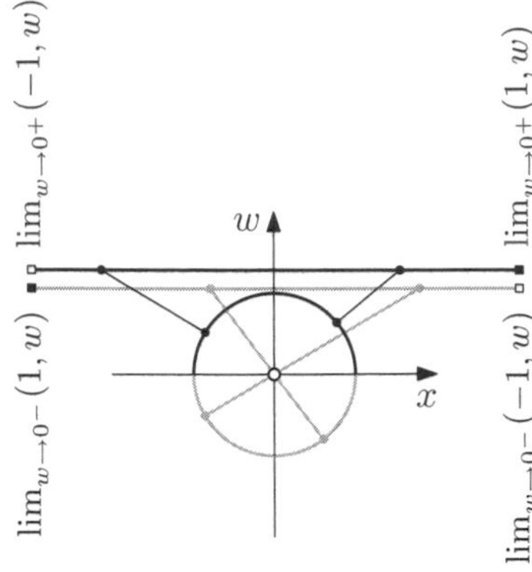

has coordinates $(1, 0)$, it would not be evident whether we are on the front side or on the back side. The answer is not to take the coordinates of ideal points by themselves, but to consider how we got there. To reach the $+\infty$ point on the front side, we take $\lim_{w\to 0^+}(1, w)$. To reach the same point on the back side, we take instead $\lim_{w\to 0^-}(1, w)$. Likewise the $-\infty$ point can be reached on the front side by taking $\lim_{w\to 0^+}(-1, w)$ or on the back side by taking $\lim_{w\to 0^-}(-1, w)$.

That it is not possible to determine on which side of infinity an ideal point lies by observing the final coordinates alone may seem to suggest that moving to oriented projective geometry is an exercise in pedantry in the first place. Not so. Oriented projective geometry adds coherence and makes it possible to determine whether we are subsequently "seeing" the front side or the back side of a film plane by choosing to render either points with $w > 0$ or those with $w < 0$. Crucially, oriented projective geometry also makes it possible to perform clipping without the belief suspension inherent in explanations of clipping using a plane that does not separate space into two parts.

The details of an implementation would be too similar to those so far discussed for there to be a need to go through an exposition here. Classes and predicates are needed for Point_T1 and Segment_T1 that parallel those seen in Chapter 8 and a class Transformation_T1 is needed that parallels that seen in Chapter 12.

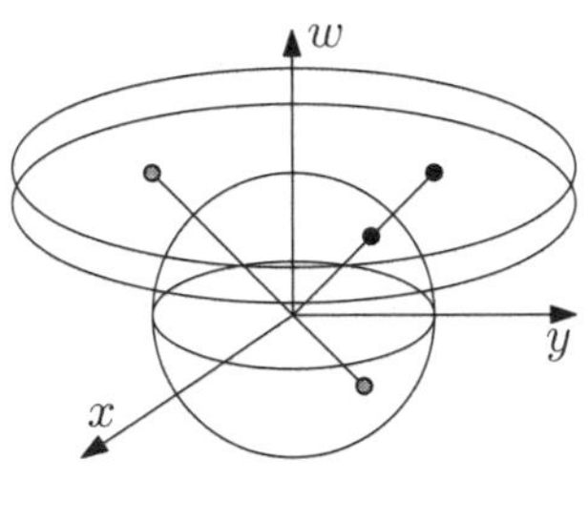

The Oriented Projective Plane

A point can be on either the front side of the oriented projective plane T^2 (if $w > 0$) or on the back side (if $w < 0$). A point is represented using a 3D vector. Two points are equal iff one vector is a positive multiple of the other.

A line in T^2 is an instance of T^1. If T^2 is considered to be embedded in 3D, then lines in T^2 are in one-to-one correspondence with oriented planes in 3D and a line is represented by the 3D vector normal to its corresponding plane.

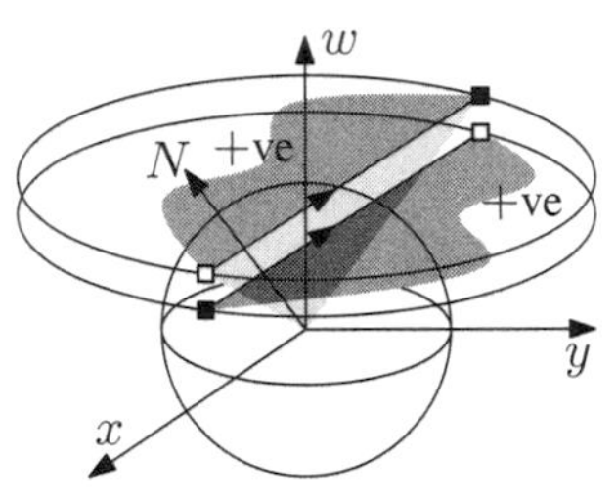

A line splits T^2 into two parts. A point P lies in the positive side of a line l iff $l \cdot P > 0$.

14.3 Oriented Projective Transformations

If oriented projective geometry is a superset of projective geometry (identifying antipodal points yields classical projective geometry), one would guess that oriented projective transformations are also more powerful than projective transformations. The additional power is that oriented projective transformations can map points from the front side to the back side and vice versa, a mapping that would be meaningless in projective spaces.

A 2×2 real-valued matrix can be interpreted in four distinct settings, leading to four distinct one-dimensional geometries. If we interpret a matrix

$$M = \begin{bmatrix} a & b \\ c & d \end{bmatrix}$$

in the 1D spherical geometry of S^1, we constrain the matrix to be orthogonal (§ 10.1). If we interpret the same matrix in the 1D Euclidean geometry represented using homogeneous coordinates, then, as we will see in § 16.5, we must impose the constraint $c = 0$. Projective transformations are more general because an arbitrary matrix can be used in P^1, with matrices connected by a scalar multiple representing an identical transformation. Oriented projective transformations are yet more general because an arbitrary matrix can be used in T^1, where matrices connected by a *positive* scalar multiple represent the same transformation.

We contemplate the effect of various 2×2 matrices on the projective line and on the oriented projective line. We start with the the projective line P^1.

Projective Transformations

As discussed in Chapter 12, projective space P^n is the space of lines passing by the origin in E^{n+1}. The projective line, for instance, is the space of lines passing by the origin in the plane (§ 12.1).

Consider choosing a set of sample points in P^1 and applying different transformations on them. The points in P^1 are uniformly sampled lines defined by antipodal points on the canonical circle. The effect of applying the matrix

$$\begin{bmatrix} i & 0 \\ 0 & 1 \end{bmatrix}$$

for $i = 1 \ldots 6$ is illustrated in Figure 14.1. Negating the matrix or, in fact, multiplying by an arbitrary (positive or negative) scalar would result in the same transformation.

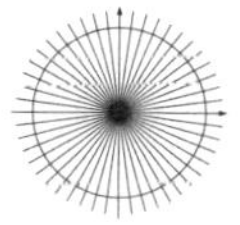
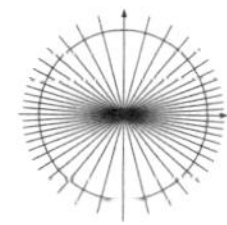
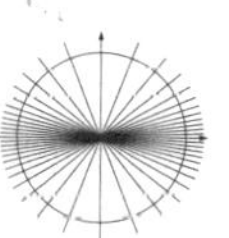
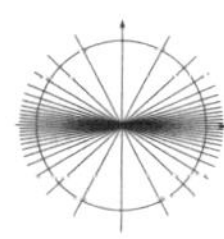
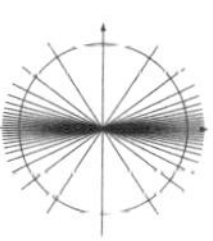
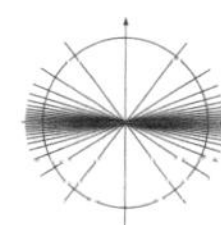

Figure 14.1
Effect of x-scaling on P^1

What is perceived when projected on the canonical line representation of P^1 as a translation results from the transformation

$$\begin{bmatrix} 1 & i \\ 0 & 1 \end{bmatrix} \begin{bmatrix} x \\ w \end{bmatrix} = \begin{bmatrix} 1 \cdot x + i \cdot w \\ w \end{bmatrix},$$

also for $i = 1 \ldots 6$. The result is illustrated in Figure 14.2.

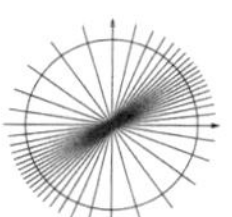
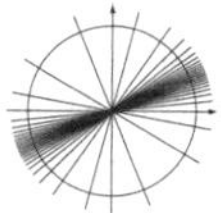
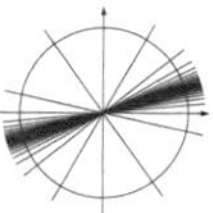
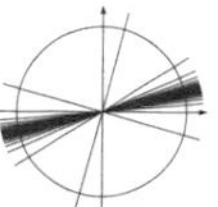
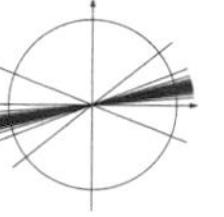
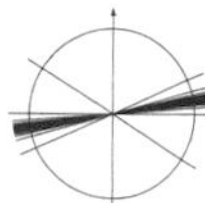

Figure 14.2
Effect of translation on P^1

The previous two transformations are equivalent to determining a projection between two parallel lines. In the first case the center of perspectivity is affine. In the second it is ideal. To effect a transformation equivalent to projecting between two nonparallel lines, a transformation of the form

$$\begin{bmatrix} 1 & 0 \\ i & 1 \end{bmatrix} \begin{bmatrix} x \\ w \end{bmatrix} = \begin{bmatrix} x \\ i \cdot x + 1 \cdot w \end{bmatrix}$$

is applied. Its effect is illustrated in Figure 14.3.

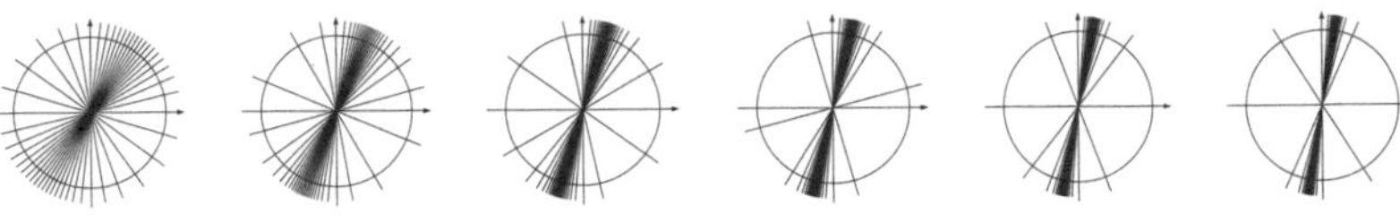

Figure 14.3
Effect of perspective on P^1

Finally, w-scaling can be applied using the matrix

$$\begin{bmatrix} 1 & 0 \\ 0 & i \end{bmatrix}$$

for $i = 1 \ldots 6$. The result is illustrated in Figure 14.4.

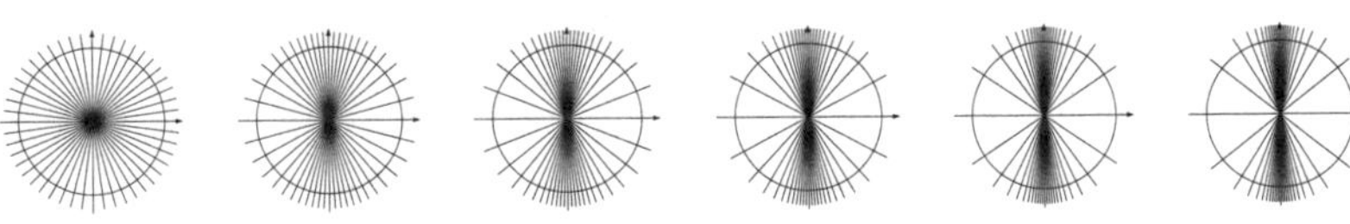

Figure 14.4
Effect of w-scaling on P^1

Oriented Projective Transformations

We choose a set of sample points in T^1 on the front and on the back side. The two sets are shown in the spherical model of T^1 as points on a unit circle. The scale transformation

$$\begin{bmatrix} i & 0 \\ 0 & 1 \end{bmatrix}$$

effects the mapping illustrated in Figure 14.5. The added power is the ability to map between points on the front and the back sides; multiplying a transformation by -1 no longer yields an identical transformation.

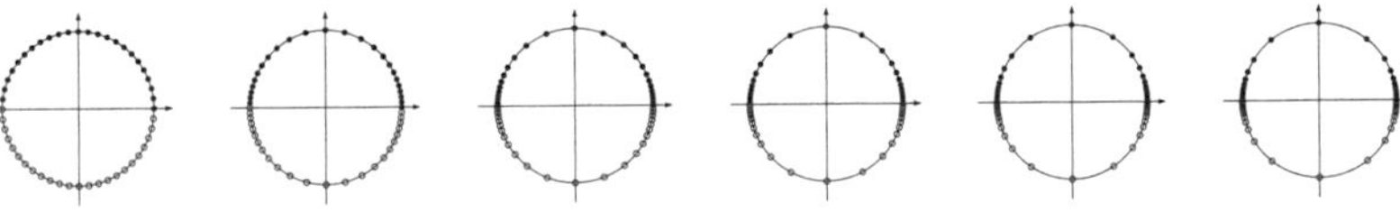

Figure 14.5
Effect of x-scaling on T^1

The effect of a translation

$$\begin{bmatrix} 1 & i \\ 0 & 1 \end{bmatrix}$$

on the points on the front and the back sides is illustrated in Figure 14.6.

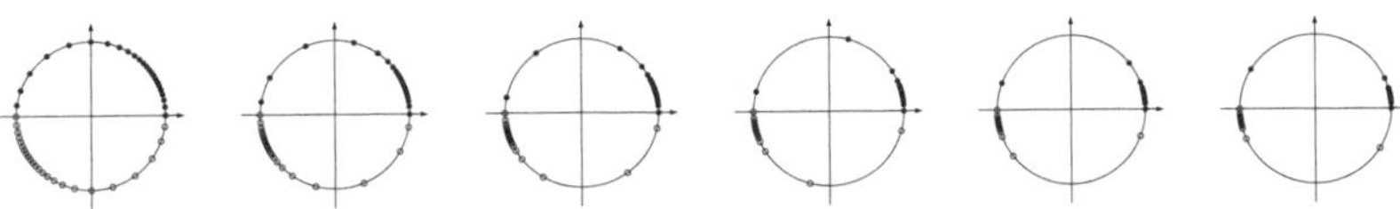

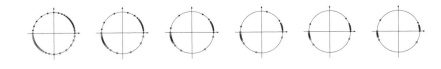

Figure 14.6
Effect of translation on T^1

The effect the perspective transformation

$$\begin{bmatrix} 1 & 0 \\ i & 1 \end{bmatrix}$$

is shown in Figure 14.7. It is now possible to distinguish between two colocated viewers looking in opposite directions [41]. To model a projection that simulates a viewer looking in an antipodal direction, one simply negates the matrix. The effect is to swap the front and the back projection planes.

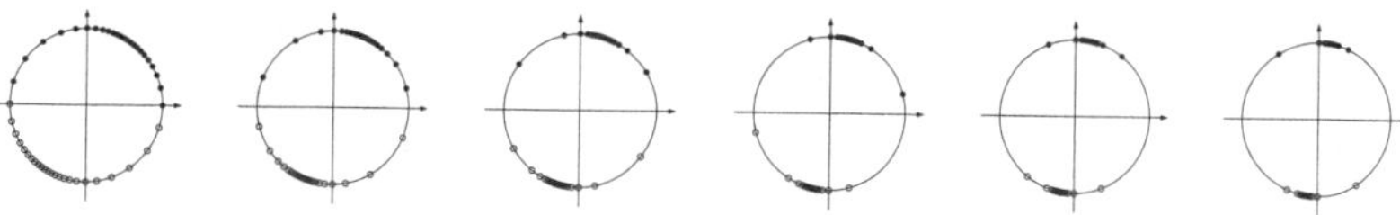

Figure 14.7
Effect of perspective on T^1

Finally, the effect of scaling by w is shown in Figure 14.8.

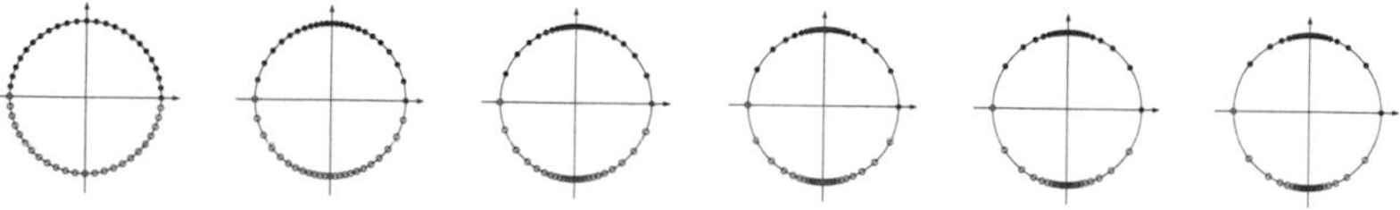

Figure 14.8
Effect of w-scaling on T^1

14.4 Should One Collapse Geometric Libraries?

Seeing how similar the implementation of a set of classes for the four geometries presented leads one to ask the obvious question: Is there a way to implement a single library that can be subsequently specialized to capture all or most classes for E^n, S^n, P^n, and T^n? A related question, whether the "right" approach is to provide a set of less type-strict, and hence fewer, classes and leave application programmers worry about the geometry in which they operate, is left to § 18.6.

The answer to the specialization question, which would maintain type strictness, appears easy if we would be content with a library that does impose an efficiency hit. If we are willing to pay a time and space penalty, it is clear that a library for oriented projective geometry alone could be implemented and then specialized by overriding equality operators and sidedness predicates, as well as throwing exceptions if $w = 0$ when it should not, and so on.

A more significant objective would be to seek a single library for any geometry and for any dimension that incurs no run-time penalties. It is not clear whether such a library can exist, but this question, as well as building a single coherent body of knowledge that blends geometric algebra [33] with the set of libraries described here, are topics for future work.

14.5 Which Model Is Right?

An important question the reader will no doubt ask is: Was Blinn working in oriented projective spaces all along? In other words, was homogeneous clipping [17, 15], which distinguishes between positive and negative w, an oriented projective geometry algorithm? The answer likely depends on whom you ask.

Since this chapter suggests that that is the case, it is worthwhile to conclude with the opposite view. One may argue that the projection plane not only need not be two-sided, but ought not either. After all it is impossible to construct two images on a physical film; light reaching the film from either side will have the same result on its particles. The physical film is, effectively, one-sided.

Two details are necessary for this view to be valid. The first is that one must distinguish between internal and external segments; the latter includes an ideal point whereas the former does not. This works well. The second detail does not work so well, for for this view to be valid one must put up with the notion that it is possible to do clipping, an operation that requires testing sidedness, in (classical) projective geometry, where a cutting plane has only one side. If one is willing nevertheless to perform intersection in a nonseparable space and then use the sign of w to determine whether an interior or an exterior segment is intended, then classical projective geometry is a suitable, if awkward, model. So, ultimately, oriented projective geometry may perhaps continue to suffer from being an instance of the "what can you do with it that you could not do without it" syndrome [89, p. 48]. Looking back at Blinn and Newell's manuscript from 1978 [17], it is easy in retrospect to determine the space they are in. Their writing $x+w=0$ does not suggest an answer one way or the other, but writing $x+w>0$ indicates that they were indeed in oriented projective space all along. Once one is used to the idea of discarding projective geometry and adopting oriented projective geometry instead, the naturalness of it all will become evident—so it is perhaps fitting to conclude by quoting what Sutherland and Hodgman wrote in an appendix [107]:

> *If we choose w_1 and w_2 to have the same sign, we represent the internal line segment. If we choose w_1 and w_2 to have opposite signs, we are representing the external line segment.*

14.6 Exercises

14.1 Modify the start-up code labeled "p1-transformation" such that the user is provided with a starting segment and is able to manipulate using sliders the transformation matrix. Show the effect of the transformation on the initial segment using both the straight-line as well as the spherical model.

14.2 Solve Exercise 14.1, distinguishing between the front and the back side of T^1 by displaying them on two adjacent, but distinct, lines.

14.3 Solve Exercise 14.2, but track instead the motion of a set of points that are initially equally spaced on the front side of T^1.

15 Oriented Projective Intersections

We are given a scene, a viewer, and a projection plane in Euclidean space E^3. Projection requires that we move to projective space P^3. We are unable to perform clipping in classical projective geometry because a necessary predicate that reports whether two points lie on the same side of a cutting plane cannot be implemented. So we move instead from E^3 to oriented projective space T^3.

We do not wish to project objects that lie behind the viewer. To discard such parts of the scene, we could define a viewing pyramid in E^3 and perform clipping operations in Euclidean space, but that would entail too many floating point operations just to determine whether a given segment straddles each of the six clipping planes bounding the viewing pyramid.

So we choose instead to move to T^3. A transformation matrix will map the six corners of the viewing pyramid to the corners of a unit cube. Clipping is performed using the planes carrying the sides of the unit cube. This chapter discusses a revision of the graphics pipeline (§ 12.9), adding oriented projective clipping.

15.1 The Need for Clipping

Projective geometry satisfies all our imaging needs—with the exception of clipping. We already saw the issues involved in § 12.9, but let us take another look at projection once again in oriented projective space.

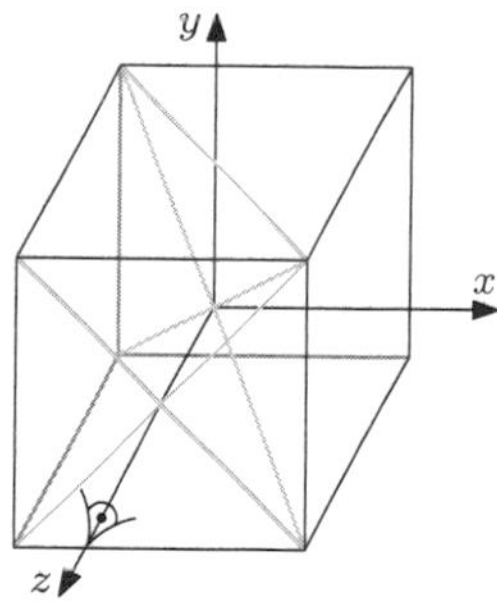

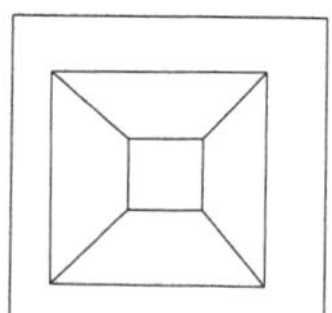

Analysis: Approaching a Cube

Consider a viewer located near the positive z-axis, looking in the negative z-direction, and moving along that direction toward a cube centered at the origin. Since two faces of the cube are always parallel to the image plane, their projections will evidently remain two squares. But what is the projection of the four edges parallel to the z-axis?

What the viewer sees when the image plane does not intersect the cube is simple. The farther of the two faces parallel to the image plane will project, due to perspective, to a smaller square than the one closer to the image plane. The remaining edges will connect these two faces. To break the symmetry the viewer is situated on a line parallel to the z-axis and near, but not coincident to, that axis.

But we are most interested in what happens when the viewer continues moving forward and crosses the near face of the cube. As the viewer becomes

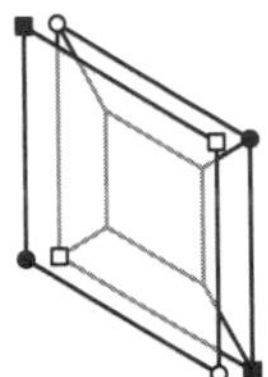

infinitesimally close to that face, its four vertices will project to four points that approach four (not two) ideal points on the front projection plane. At the moment of crossing, the four vertices will project to ideal points. These ideal points are located simultaneously on the front and the back image planes. If we are convinced that the projective plane P^2 is no more than a sphere S^2 with antipodal points identified, it is not hard to believe that each point on just one line (the ideal line) consists of identifying two points—one from the front and the other from the back image plane.

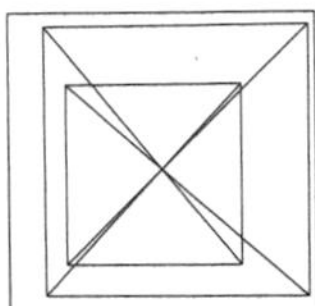

After crossing, because the image of a point lying *behind* the viewer is reversed on the image plane (not unlike the reversal of photographs on the film plane in physical cameras), the projection of the four edges parallel to the view direction will appear diagonally across the image plane. The elementary perspective rule still applies: Features closer to the viewer (either in front or behind the viewer) will appear larger than farther features. But even though our model has a front and a back image plane, the paper we print on has only one side—and so two images appear superimposed on paper.

The image thus obtained has two flaws. The less harmful of the two flaws is that it may not necessarily make sense to assume that the viewer is capable of seeing simultaneously in two opposite directions, but perhaps we wish to consider that that is a feature. The more serious problem is that, as encountered in Chapter 11, a segment AB that crosses the image plane projects to a segment $A'B'$ on the image plane, but the resulting oriented projective segment is not the affine combination of the projections of the two endpoints A' and B'. The result must by necessity contain an ideal point, the projection of the intersection of AB with a plane passing by the viewer and parallel to the image plane. To see that that is the case, glide a point C from A to B and determine its projection C'. The set of points C' is the desired projection. When we adopt oriented projective geometry as our model, the external segment no longer consists of a single segment. There are now two segments, one on the front and the other on the back image plane.

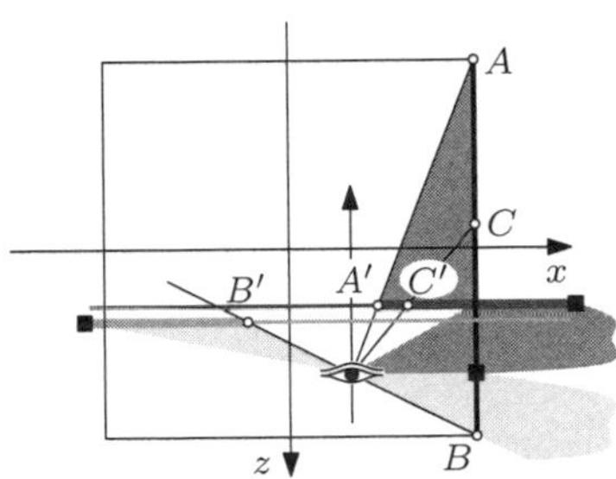

One may clip to ensure that neither of these two problems arises. And so the objective of clipping is to discard the portions of the scene that lie behind the viewer, but points lying arbitrarily close to the plane passing by the viewer and parallel to the image plane would also project to points arbitrarily far (arbitrarily "close to" an ideal point) and would also need to be discarded. In practice one sets a *near clipping plane* in front of, but not too near to, the viewer.

15.2 Clipping in Oriented Projective Spaces

As always, to clip is to determine the intersection of an object of interest with a halfspace or a combination of halfspaces—a rectangle or a parallelepiped. A halfspace in oriented projective geometry is defined by a hyperplane. The hyperplane has two sides: a positive side and a negative side—and so also we can talk about either the positive halfspace or the negative halfspace. To clip using a halfspace is to discard those portions of an object that do not lie in it.

We first consider clipping in the spherical model of the oriented projective line T^1. Consider that the "object" we are clipping is no other than the entire line T^1, shown as the circle in Figure 15.1(a). We define two points P_1 and P_2 on that line and clip twice. Were we to clip using the positive halfspace defined by P_1, we would obtain the set of points highlighted in Figure 15.1(b). Clipping using P_2's negative halfspace would give us those points highlighted in Figure 15.1(c). The intersection of the two sets is shown in Figure 15.1(d).

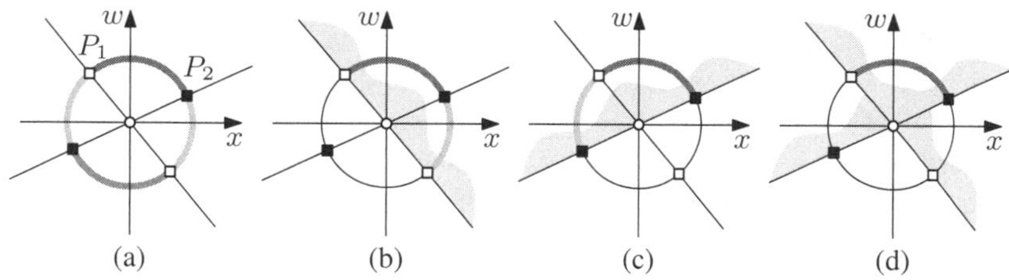

Figure 15.1
Clipping illustrated in the spherical model of T^1

In the linear model of T^1, the two hyperplanes also define four segments—even though we are tempted to perceive them as six [Figure 15.2(a)]. Clipping using one hyperplane or another is illustrated in Figures 15.2(b,c). Clipping with both is illustrated in Figure 15.2(d).

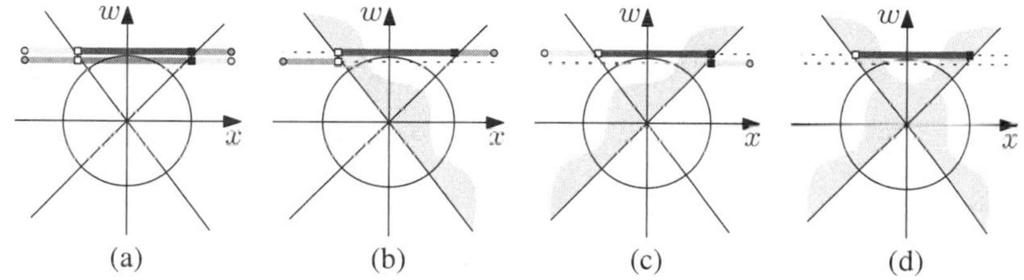

Figure 15.2
Clipping illustrated in the linear model of T^1

Clipping proceeds in much the same way in T^2. Now we have a line l that we wish to clip using the positive side of a line (hyperplane in T^2) m and the negative side of another line n [Figure 15.3(a)]. Clipping with the former is shown in Figure 15.3(b) and clipping with the latter in Figure 15.3(c). The result of performing the two clipping operations successively is illustrated in Figure 15.3(d). Naturally, the line l is itself oriented, and the ending fragment has the same orientation as l.

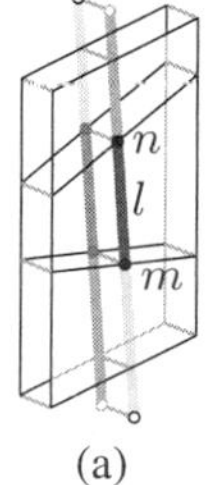

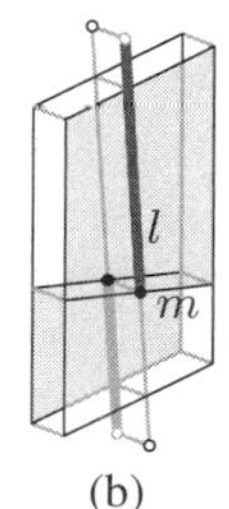

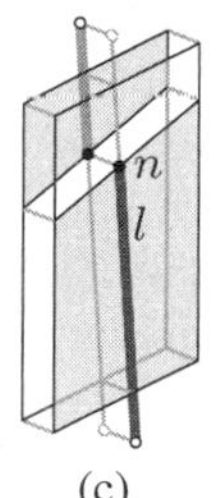

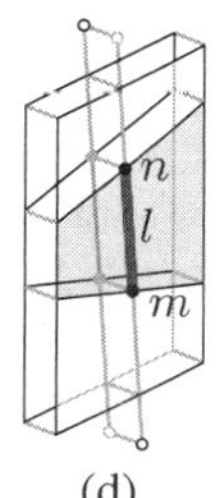

Figure 15.3
Clipping illustrated in the linear model of T^2

15.3 The Algebra of Intersections

Negative w

Chapter 12 discussed the difficulty one encounters when attempting to render a segment in the projective plane by mapping the segment to the Euclidean plane. Simply dividing x by w for each endpoint and assuming that the segment consists of the affine set of points colinear and *between* the two points (as in the left of Figure 15.4) is inadequate. Doing so presumes that the segment does not contain an ideal point. Yet *two* segments are plausible in a projective space, one containing an ideal point and the other fully affine.

If one then renders the segment and ensures that a segment containing an ideal point is indeed rendered as the *external* segment between two points, the segment would still appear incorrectly. As shown in the middle of Figure 15.4, the two film planes should not be superimposed.

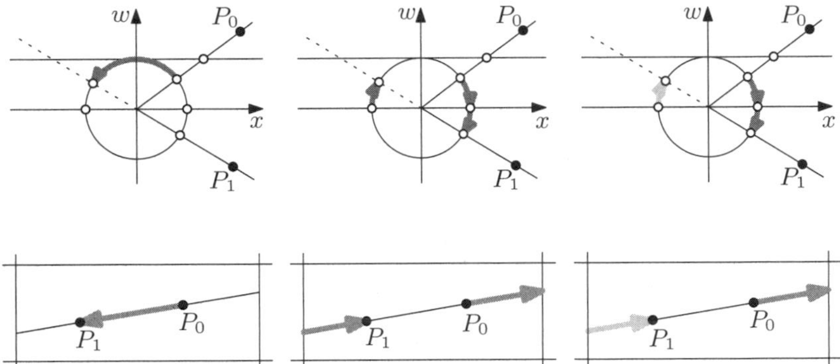

Figure 15.4
Three interpretations of a segment in T^1 are possible.

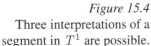

The only correct interpretation (shown on the right of Figure 15.4) of a segment with one endpoint having a negative w is to treat the segment as consisting of two portions, one with positive w and the other with negative w. The portion with (strictly) positive w maps easily to the Euclidean plane — though some cutoff is needed so that the ideal point is not approached, in practice leading to multiple clipping operations. The portion with negative w lies on the back image plane and is not rendered.

When a projective segment is rendered in a viewport (three bottom images in Figure 15.4), the two false interpretations lead to either the left or the middle images: The segment is either wholly affine or the sign of the homogenizing variable w is ignored. In the desired image on the right of the figure only a subset of the segment appears to simulate correctly a camera aimed at one direction.

Point Combination

We can interpolate between two nonantipodal endpoints of a segment in T^3 defined by (x_1, y_1, z_1, w_1) and (x_2, y_2, z_2, w_2) using the equations

$$x = x_1 + (x_2 - x_1)t, \qquad y = y_1 + (y_2 - y_1)t,$$
$$z = z_1 + (z_2 - z_1)t, \qquad w = w_1 + (w_2 - w_1)t.$$

Plane-Segment Intersection

To find the intersection of an oriented plane in T^3 defined by the vector

$$(X, Y, Z, W)$$

and a segment in T^3 defined by the two points

$$(x_1, y_1, z_1, w_1) - (x_2, y_2, z_2, w_2),$$

we substitute in the equation of the plane

$$Xx + Yy + Zz + Ww = 0$$

by the combination of the two points [14]

$$\begin{aligned}
& X[x_1 + (x_2 - x_1)t] \\
& +Y[y_1 + (y_2 - y_1)t] \\
& +Z[z_1 + (z_2 - z_1)t] \\
& +W[w_1 + (w_2 - w_1)t] = 0 \\
\Longrightarrow\ & Xx_1 + X(x_2 - x_1)t \\
& +Yy_1 + Y(y_2 - y_1)t \\
& +Zz_1 + Z(z_2 - z_1)t \\
& +Ww_1 + W(w_2 - w_1)t = 0,
\end{aligned}$$

and so

$$\begin{aligned}
Xx_1 + Yy_1 + Zz_1 + Ww_1 = & X(x_1 - x_2)t \\
& + Y(y_1 - y_2)t \\
& + Z(z_1 - z_2)t \\
& + W(w_1 - w_2)t,
\end{aligned}$$

yielding the expression for t:

$$t = \frac{Xx_1 + Yy_1 + Zz_1 + Ww_1}{X(x_1 - x_2) + Y(y_1 - y_2) + Z(z_1 - z_2) + W(w_1 - w_2)}.$$

15.4 The Revised Graphics Pipeline

Using projective geometry as a foundation software layer for computing projectivities had several inconsistent and therefore undesirable features:

- In the projection transformation the observer was moved to a point at infinity along the view axis. But which infinity? In projective geometry only one ideal point exists on a line. This would appear to say that the observer is moved to the ideal point on *either* side of the view axis.

- Because the projection plane is nonorientable, writing a function that determines whether the "front" face of a polygon is visible makes no sense. The back face of the polygon is equally visible. The notion of a polygon itself is awkward; one may rigorously talk about the endpoints of a polygon and of the lines carrying the edges of a polygon, but not of the segments defining a polygon or of the polygon itself.

The input scene is described in $E^3(\mathbb{R})$, which consists of the classes for 3D Euclidean geometry instantiated for a data type simulating the set of real numbers. We can for simplicity think of $E^3(\mathbb{R})$ as equivalent to $E^3($**float**$)$ or $E^3($**double**$)$, subject to keeping in mind the traps discussed in Chapter 7.

The output image is described in a subset of $E^2(\mathbb{R})$ if we are interested in a vector image, or in a subset of $E^2(\mathbb{N})$ if we are interested in a raster image.

If we choose to operate purely through sampling, then remaining in Euclidean spaces is possible. Perspective discrete images can indeed be performed using solely Euclidean objects through ray casting (Chapter 23). But if we wish to remain, even partially, in a continuous domain, then some move into non-Euclidean spaces becomes necessary.

The two operations that we wish to perform are

1. Clipping: We need to discard part of the image that will not fall within the viewing window. If the two corners of the image in $E^2(\mathbb{N})$ are declared to be $(0, 0)$ and (width, height), we would not wish any features to appear outside that rectangle.

2. Projection: We need to find a function that maps points from $E^3(\mathbb{R})$ to $E^2(\mathbb{R})$ (leaving rasterization as a distinct question).

We retake the graphics pipeline discussed in § 12.9 and present a revised version that passes through oriented projective space rather than through projective space:

$$E^3 \xrightarrow{+1} T^3 \xrightarrow{M} T^3 \xrightarrow{V} T^3 \xrightarrow{P} T^3 \xrightarrow{D} E^3 \xrightarrow{\backslash z} E^2$$

When we map the input primitives from Euclidean to oriented projective space, we choose to append $w = 1$. In other words, we choose the "front side of the world." The modeling transformation M must now maintain the primitives in that side. Viewing the scene continues to move the viewer to the origin and is performed by an orthogonal transformation V. The perspective transformation P is performed next, followed by perspective divide, and finally z is discarded (after computing visibility).

Just as we were able to clip in Euclidean space when using a projective geometry engine, we can still do so after the modeling transformation.

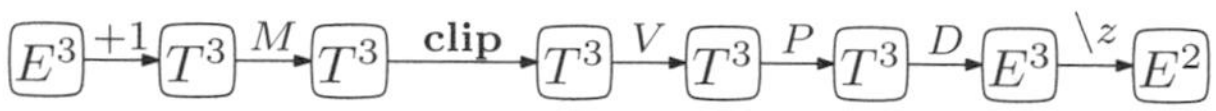

$$E^3 \xrightarrow{+1} T^3 \xrightarrow{M} T^3 \xrightarrow{\text{clip}} T^3 \xrightarrow{V} T^3 \xrightarrow{P} T^3 \xrightarrow{D} E^3 \xrightarrow{\backslash z} E^2$$

But that is unappealing since it is expensive. We contemplate clipping after the viewing transformation. Because the viewer is then at the origin, four of the six clipping planes would pass by the origin.

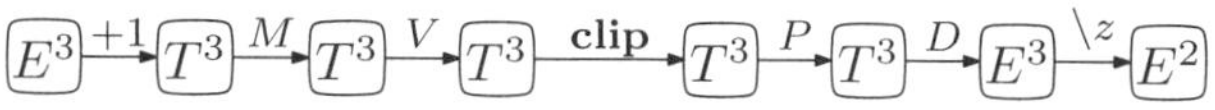

$$E^3 \xrightarrow{+1} T^3 \xrightarrow{M} T^3 \xrightarrow{V} T^3 \xrightarrow{\text{clip}} T^3 \xrightarrow{P} T^3 \xrightarrow{D} E^3 \xrightarrow{\backslash z} E^2$$

Yet it is best to perform clipping after the perspective transformation. As we see in the next section, the sides of the unit cube make computation in oriented projective space particularly easy.

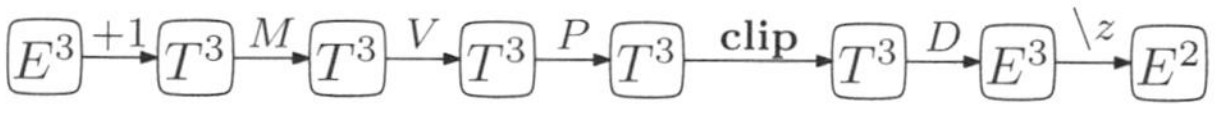

$$E^3 \xrightarrow{+1} T^3 \xrightarrow{M} T^3 \xrightarrow{V} T^3 \xrightarrow{P} T^3 \xrightarrow{\text{clip}} T^3 \xrightarrow{D} E^3 \xrightarrow{\backslash z} E^2$$

15.5 Clipping or Splitting a Segment

The projective transformation discussed in § 12.9 maps points to the cube with diagonal $(-1, -1, -1, 1)$–$(1, 1, 1, 1)$. After this transformation clipping can be performed using a predetermined clipping plane with known coefficients.

Plane	T^3 Plane	E^3 Plane	T^3 Halfspace	E^3 Halfspace
Xmax	$-x + w = 0$	$x = 1$	$-x + w > 0$	$x < 1$
Xmin	$x + w = 0$	$x = -1$	$x + w > 0$	$x > -1$
Ymax	$-y + w = 0$	$y = 1$	$-y + w > 0$	$y < 1$
Ymin	$y + w = 0$	$y = -1$	$y + w > 0$	$y > -1$
Zmax	$-z + w = 0$	$z = 1$	$-z + w > 0$	$z < 1$
Zmin	$z + w = 0$	$z = -1$	$z + w > 0$	$z > -1$

Since many of the coefficients of the set of canonical clipping planes vanished, it would be advantageous not to reduce unit-cube clipping to six instances of general plane clipping, but to write six individual routines that cater to the canonical clips. Doing so discards all four multiplications and replaces them with the simpler multiplications by 0 or by ± 1.

In an implementation the six clipping planes map to six oriented projective planes in T^3.

```
const Plane_T3d Xmax = Plane_T3d(−1, 0, 0, 1);
const Plane_T3d Xmin = Plane_T3d( 1, 0, 0, 1);
const Plane_T3d Ymax = Plane_T3d( 0, −1, 0, 1);
const Plane_T3d Ymin = Plane_T3d( 0, 1, 0, 1);
const Plane_T3d Zmax = Plane_T3d( 0, 0, −1, 1);
const Plane_T3d Zmin = Plane_T3d( 0, 0, 1, 1);
```

Plane	Xmax	Xmin	Ymax	Ymin	Zmax	Zmin
$\begin{bmatrix} X \\ Y \\ Z \\ W \end{bmatrix}$	$\begin{bmatrix} -1 \\ 0 \\ 0 \\ 1 \end{bmatrix}$	$\begin{bmatrix} 1 \\ 0 \\ 0 \\ 1 \end{bmatrix}$	$\begin{bmatrix} 0 \\ -1 \\ 0 \\ 1 \end{bmatrix}$	$\begin{bmatrix} 0 \\ 1 \\ 0 \\ 1 \end{bmatrix}$	$\begin{bmatrix} 0 \\ 0 \\ -1 \\ 1 \end{bmatrix}$	$\begin{bmatrix} 0 \\ 0 \\ 1 \\ 1 \end{bmatrix}$

To determine whether a point $P(x, y, z, w)$ lies in the positive halfspace defined by a plane $\pi(X, Y, Z, W)$, determine the sign of the inner product $Xx + Yy + Zz + Ww$.

A point on a line segment $A(x_a, y_a, z_a, w_a)$–$B(x_b, y_b, z_b, w_b)$ passes by the plane $x = 1$ or, homogeneously, the plane $x = w$, when

$$x_a + (x_b - x_a)t = w_a + (w_b - w_a)t,$$

and so

$$t = \frac{w_a - x_a}{(w_a - x_a) - (w_b - x_b)}.$$

An intersection with the plane $x = -1$, homogeneously the plane $x = -w$, is characterized by

$$x_a + (x_b - x_a)t = -w_a - (w_b - w_a)t,$$

and so

$$t = \frac{w_a + x_a}{(w_a + x_a) - (w_b - x_b)},$$

and likewise when clipping at $y = \pm 1$ and at $z = \pm 1$.

If a point on the segment AB is parameterized by t such that $t = 0$ at A and that $t = 1$ at B, the point can be described by

$$x = x_a + (x_b - x_a)t, \qquad y = y_a + (y_b - y_a)t,$$
$$z = z_a + (z_b - z_a)t, \qquad w = w_a + (w_b - w_a)t.$$

As discussed in § 15.3, replacing in the equation of the splitting plane

$$t = \frac{Xx_a + Yy_a + Zz_a + Ww_a}{X(x_a - x_b) + Y(y_a - y_b) + Z(z_a - z_b) + W(w_a - w_b)},$$

and then in the equation of the segment yields the desired coordinates of the intersection point.

Polygon clipping using homogeneous coordinates is otherwise identical to clipping a polygon in Euclidean space (§ 5.4).

15.6 The Graphics Pipeline as a Function Object

The similarity between projective space P^3 and oriented projective space T^3 is such that we can write a function object for the oriented graphics pipeline that closely mimics that written for the (classical) graphics pipeline (c.f. page 137).

```
class T3_pipeline_without_clipping
{
  const Transformation_T3d T;
  const Perspective_divide_d my_perspective_divide;
  const Project_on_xy_d my_projection;
public:
  T3_pipeline_without_clipping( const Transformation_T3d& _T ) :
    T(_T), my_perspective_divide(), my_projection() {}

  Point_E2d operator()( const Point_E3d& P1 )
  {
    Point_T3d P2 = Point_T3d( P1 );
    Point_T3d P3 = T( P2 );
    Point_E3d P4 = my_perspective_divide( P3 );
    Point_E2d P5 = my_projection( P4 );
    return P5;
  }
};
```

But in T^3 we also gain the ability to perform clipping.

```
class T3_pipeline_with_clipping
{
  const Transformation_T3d T;
  const Perspective_divide_d my_perspective_divide;
  const Project_on_xy_d my_projection;
public:
  T3_pipeline_with_clipping( const Transformation_T3d& _T )
    : T(_T), my_perspective_divide(), my_projection() {}

  Point_T3d pre_clip( const Point_E3d& P1 )
  {
    Point_T3d P2 = Point_T3d( P1 );
    Point_T3d P3 = T( P2 );
    return P3;
  }

  Segment_T3d pre_clip( const Segment_E3d& S1 )
  {
    Point_T3d source = pre_clip( S1.source() );
    Point_T3d target = pre_clip( S1.target() );
    return Segment_T3d( source, target );
  }

  Point_E2d post_clip( const Point_T3d& P3 )
  {
    Point_E3d P4 = my_perspective_divide( P3 );
    Point_E2d P5 = my_projection( P4 );
    return P5;
  }

  Segment_E2d post_clip( const Segment_T3d& S3 )
  {
    Point_E2d source = post_clip( S3.source() );
    Point_E2d target = post_clip( S3.target() );
    return Segment_E2d( source, target );
  }

  std::pair<bool, Point_E2d>
  operator()( const Point_E3d& P1 )
  {
    Point_T3d P3 = pre_clip( P1 );

    if(
      oriented_side(Xmax, P3) == ON_POSITIVE_SIDE &&
      oriented_side(Xmin, P3) == ON_POSITIVE_SIDE &&
      oriented_side(Ymax, P3) == ON_POSITIVE_SIDE &&
      oriented_side(Ymin, P3) == ON_POSITIVE_SIDE &&
      oriented_side(Zmax, P3) == ON_POSITIVE_SIDE &&
      oriented_side(Zmin, P3) == ON_POSITIVE_SIDE )
    {
      Point_E2d P5 = post_clip( P3 );
      return std::make_pair( true, P5 );
    }
    else
      return std::make_pair( false, Point_E2d() );
  }

  std::pair<bool, Segment_E2d>
  operator()( const Segment_E3d& S1 )
  {
    Segment_T3d S3 = pre_clip( S1 );

    if( positive_half_space_clip( Xmax, S3 ) &&
      positive_half_space_clip( Xmin, S3 ) &&
      positive_half_space_clip( Ymax, S3 ) &&
      positive_half_space_clip( Ymin, S3 ) &&
      positive_half_space_clip( Zmax, S3 ) &&
      positive_half_space_clip( Zmin, S3 )
      )
    {
      Segment_E2d S5 = post_clip( S3 );
      return std::make_pair( true, S5 );
    }
    else
      return std::make_pair( false, Segment_E2d() );
  }
```

The above code treats all six clipping planes as ordinary planes, which results in many more floating point operations than are necessary. Production code would introduce the (simple) refactoring by writing specialized functions for each of the six clipping operations. Both are functionally equivalent and so we leave this code as is.

Were we to perform clipping in x only, we would obtain the images in Figure 15.5. After x and y clipping, we obtain the images in Figure 15.6, and after x, y, and z clipping, we obtain those in Figure 15.7.

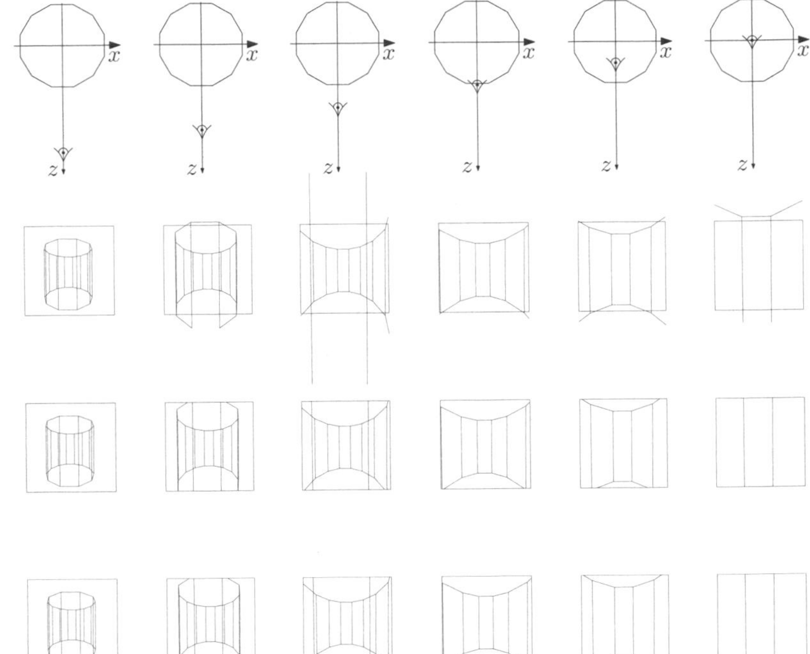

Figure 15.5
Clipping is performed in x only.

Figure 15.6
Clipping is performed in x and y only.

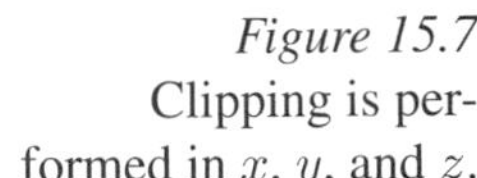

Figure 15.7
Clipping is performed in x, y, and z.

15.7 Characterizing the Four Geometries

The distinction between Euclidean, spherical, projective, and oriented projective geometries can be made clearer by observing (as Klein did in 1872 [58]) that the characteristic distinction is that of the transformations that can be applied. The set of possible transformations in Euclidean geometry of n dimensions is the set of of nonsingular matrices of order n—and an order-n translation vector. The set of possible transformations in projective geometry of n dimensions is the set of nonsingular matrices of order $n+1$ *up to scalar multiples*. The topological connection between oriented projective geometry and spherical geometry is discussed extensively by Stolfi [104], but it is again easy to characterize a difference between the two geometries. Whereas the set of transformations for oriented projective geometry of order n is the set of nonsingular matrices of order $n+1$ up to a *signed* scalar multiple, only orthogonal transformations are allowed for spherical geometry.

In other words, as illustrated in Figure 15.8, take S^2 and identify each two antipodal points, but do not give any special status to the ideal line—or treat it homogeneously—and you obtain P^2. The three lines defined by a triangle partition P^2 into only four regions. Continue to identify antipodal points, but declare a special line to be the ideal line and disallow transformations that map points on the ideal line to affine points (or vice versa), and you obtain the affine plane. We can also restrict the transformations to disallow affine transformations that do not preserve congruence (rigid-body and uniform scale remain, but shear and nonuniform scale are excluded) to obtain the Euclidean plane. In computing it does not seem worthwhile to make the distinction so we can continue to apply an arbitrary affine transformation in Euclidean geometry knowing that only some preserve angles and lengths ratios. Interpret the points in S^2 as the two sides of an oriented projective plane and you obtain T^2.

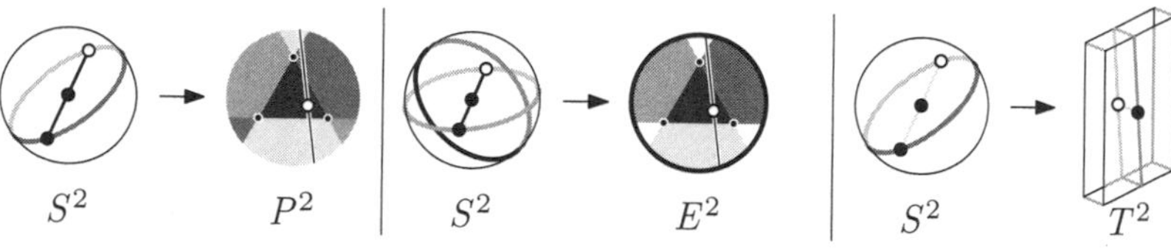

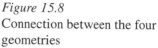

Figure 15.8
Connection between the four geometries

15.8 Exercises

15.1 Draw one circle on S^2 and define a region as a connected set of points such that it is not possible to move continuously from points in one region to another without crossing the circle.

a. Label the regions separated by the circle on S^2. How many regions are there?

b. Label the regions that result when antipodal points are identified, resulting in P^2. How many regions are there?

c. Label the regions that result from interpreting S^2 as an oriented projective plane T^2. How many regions are there?

15.2 Repeat Exercise 15.1, but start with two distinct circles on S^2. Into how many regions is each plane partitioned?

15.3 Repeat Exercise 15.1, but partition S^2 using three circles. How many regions result in each geometry?

15.4 The adjacent figure is a redrawing of the (incorrect) image of a cube that we encountered earlier in this chapter. Only this time the front and the back faces of the cube are shown in a heavier line. Draw by hand a correct image by replacing each of the remaining lines. Justify your drawing.

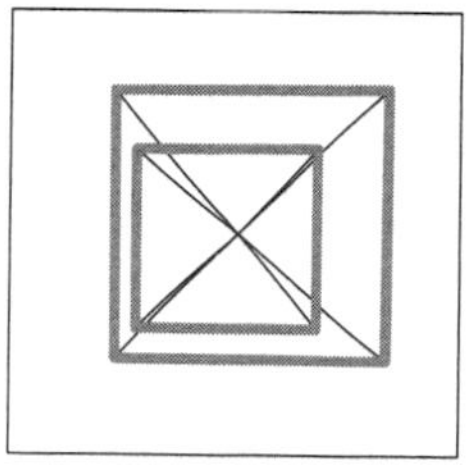

15.5 Once again draw by hand the adjacent figure, but this time use two colors (or shades). Use one color for the projection on the front side of the image plane and another color for the projection on the back side.

15.6 We continue with the same cube, but now we focus on the four faces that intersect the plane passing by the viewer and parallel to the image plane. The projection of these four faces is shown in Figure 15.9. Draw the four faces on the spherical model of the oriented projective image plane.

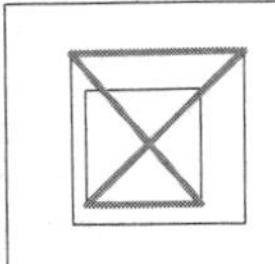
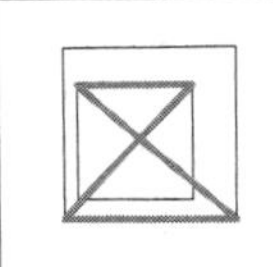
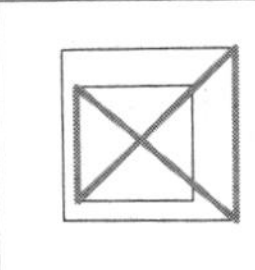
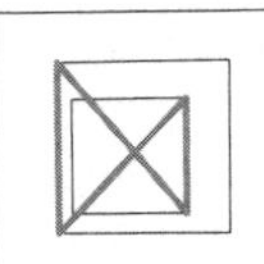

Figure 15.9
Four faces of the cube

Part III

Coordinate-Free Geometry

16 Homogeneous Coordinates for Euclidean Geometry

Homogeneous coordinates were introduced as a representation for projective geometry. We consider in this chapter how homogeneous coordinates can also be used in Euclidean geometry. When discussing homogeneous coordinates as an analytical tool for projective geometry, the case of the homogenizing variable being set or acquiring a zero value turned out to be especially useful since it made it possible to capture ideal points. This chapter discusses why it may be useful to use homogeneous coordinates even when the case of $w = 0$ is disallowed by discarding the transformations that map between affine and ideal points.

16.1 Homogeneous Coordinates for Efficiency

It is possible to design a library for Euclidean geometry using homogeneous coordinates. A point P has coordinates (x, y, w) and a line L has coefficients (X, Y, W). P is incident to L iff $Xx + Yy + Ww = 0$ and is on the positive side of L if $Xx + Yy + Ww > 0$. But the two operations of line and point construction become simpler when one chooses to use homogeneous coordinates. Using Cartesian coordinates (Chapter 1), the intersection of a line with coefficients (a_1, b_1, c_1) capturing the line $a_1x + b_1y + c_1 = 0$ with the line (a_2, b_2, c_2) is found by computing

$$x = \frac{\begin{vmatrix} b_1 & c_1 \\ b_2 & c_2 \end{vmatrix}}{\begin{vmatrix} a_1 & b_1 \\ a_2 & b_2 \end{vmatrix}}, \qquad y = \frac{\begin{vmatrix} c_1 & a_1 \\ c_2 & a_2 \end{vmatrix}}{\begin{vmatrix} a_1 & b_1 \\ a_2 & b_2 \end{vmatrix}}.$$

Finding the Cartesian coordinates (x, y) of the point of intersection requires six multiplications, three subtractions, and two divisions.

The alternative is to use homogeneous coordinates. Intersecting a line described by the coefficients (X_1, Y_1, W_1) that captures the points satisfying $X_1x + Y_1y + W_1w = 0$ with the line (X_2, Y_2, W_2) is found by

$$x = \begin{vmatrix} b_1 & c_1 \\ b_2 & c_2 \end{vmatrix}, \qquad y = \begin{vmatrix} c_1 & a_1 \\ c_2 & a_2 \end{vmatrix}, \qquad w = \begin{vmatrix} a_1 & b_1 \\ a_2 & b_2 \end{vmatrix}.$$

Computing the point of intersection in homogeneous coordinates requires six multiplications and three subtractions, but no divisions.

The treatment for parallelism under homogeneous coordinates need not be any different than under Cartesian coordinates. Whichever action the library designer chooses to adopt when $a_1b_2 - a_2b_1 = 0$ under Cartesian coordinates (exception, assertion, warning, and annulment—in the case of an interactive system) can also be taken under homogeneous coordinates. The flag is then simply that $w = 0$.

16.2 Homogeneous Coordinates for Exactness

Because division operations are no longer necessary under homogeneous coordinates, one may use integer instead of floating point numbers. A system basing its predicates exclusively on integers (see § 7.6) will always make correct branching decisions and will have no chance of crashing for making contradictory branching decisions.

But there is a price to pay. If two integer coordinates of n bits each are used in, say, an intersection computation, their product will in general require $2n$ bits. Using a pair of such numbers in another intersection operation will require $4n$ bits. This suggests that machine built-in integers would suffice for algorithms that can be expressed with a very modest algebraic degree (§ 6.4) and, even then, only if the input also can be described with limited bit size.

16.3 Point Representation

If homogeneous coordinates are used to capture a point in E^2, then a sketch for a class Point_E2 may look as follows:

```
template<typename T>
class Point_E2
{
private:
    T _x, _y, _w;
public:
    Point_E2() : _x(0), _y(0), _w(1) {}
    Point_E2(const T& x, const T& y, const T& w=1) : _x(x), _y(y), _w(w) {}
    ...
};
```

Even though the class fully mirrors the class Point_P2 for a point in the projective plane P^2 (§ 12.2), the above sketch assumes that the two implementations would be kept entirely separate. The issues surrounding the possibility of merging libraries for distinct geometries is discussed in § 18.6.

16.4 Line Representation

We saw in § 1.5 that a line was more naturally captured by the coefficients of $ax+by+c=0$ rather than through the coefficients of the normalized equation $Ax + By + 1 = 0$. Using the coefficient c was at that time seen as not only convenient to avoid two division operations when constructing the line, but also as necessary to capture lines passing by the origin.

```
template<typename T>
class Line_E2 {
private:
    T _X, _Y, _W;
public:
    Line_E2(const Point_E2& P1, const Point_E2& P1);
    ...
};
```

Perhaps the most elegant formulation for a line that passes by $P_1(x_1, y_1, w_1)$ and $P_2(x_2, y_2, w_2)$ is similar to an expression we have already encountered in § 2.2. If $P(x, y, w)$ is a point on the line, then P, P_1, and P_2 must be colinear: hence,

$$\begin{bmatrix} x & x_1 & x_2 \\ y & y_1 & y_2 \\ w & w_1 & w_2 \end{bmatrix} = 0.$$

This form makes it easy to recall that one can determine the values of X, Y, and W as minors of the matrix.

16.5 Affine Transformation Objects

We were content in § 4.4 to represent an affine transformation while masquerading its six elements for Cartesian coordinates as a 2×2 matrix in addition to a vector. Effectively, the following sketch could have been used:

```
template<typename T>
class Transformation_E2
{
protected:
    T m00, m01, m02;
    T m10, m11, m12;
    ...
};
```

But following the discussion in Chapter 12 we now notice that the set of affine transformations can be considered a subset of projective transformations. A sketch for a homogeneous transformation may look as follows:

```
template<typename T>
class Transformation_E2
{
protected:
    T m00, m01, m02;
    T m10, m11, m12;
    T  m22;
    ...
};
```

The two perspective elements m_{20} and m_{21} are not stored since they are both implicitly 0. To ensure that the homogenizing variable w is not set to 0, the value of m_{22} should also be constrained such that it does not equal 0.

16.6 Sorting Points on the Euclidean Line E^1

It remains possible to sort points on the Euclidean line even when using homogeneous rather than Cartesian coordinates. We defined in § 2.5 a predicate oriented_side that returns the relative order of two points in E^1 by calculating the difference between their coordinates. If the two points P_1 and P_2 are given as

$$P_1 = \begin{bmatrix} x_1 \\ w_1 \end{bmatrix}, \qquad P_2 = \begin{bmatrix} x_2 \\ w_2 \end{bmatrix},$$

then $P_1 < P_2$ if and only if

$$\begin{vmatrix} x_1 & x_2 \\ w_1 & w_2 \end{vmatrix} \text{sign}(w_1 w_2) < 0.$$

This form is equivalent to normalizing the two homogeneous x-coordinates by dividing by w and then computing the subtraction, but it is preferable since it avoids two division operations and uses instead two multiplications in addition to sign tests.

16.7 Exercises

16.1 Since homogeneous coordinates require the storage of one number more than Cartesian coordinates in any dimension, they may seem not to be too appealing. Yet for any application one must weigh the extra storage with the gain in the number of (floating point) operations needed.

Compare the number of floating point operations needed when one uses homogeneous rather than Cartesian coordinates to find the intersection of three planes in Euclidean space. Also, compare the two coordinate types for finding the intersection of a line and a plane.

16.2 Repeat Exercise 16.1 using CGAL. How does the efficiency gained from homogeneous coordinates compare to that obtained with Cartesian coordinates?

16.3 Consider the following design option. A distinct class is declared for an orthogonal transformation. This would make it possible to know through only the type of the transformation object that the transformation therein is in fact an orthogonal transformation.

Sketch a design option for that approach and debate how it compares to a less stringent approach that confuses transformation objects with various properties.

17 Coordinate-Free Geometric Computing

Programmers are tempted to extract the coordinates of a geometric object in intermediate stages. They frequently do so because they need to implement some function that is not available in the geometry library they use. Coordinates are necessary for input and for output, but extracting coordinates in intermediate stages is dangerous since it easily becomes the source of many bugs that are hard to find.

This chapter discusses *coordinate-free geometry*, or the premise that one does not need (nor wants) to manipulate coordinates in the intermediate steps of a geometric system. An input to a geometric system must, by necessity, include coordinates since these would ordinarily be read from a file. The output also must include coordinates since they are the only way to interface with the graphical subsystem. But it is possible and desirable to ensure that all intermediate computations are performed without coordinate manipulation.

The irony of calling for coordinate freedom is that before Fermat and Descartes all geometry was coordinate-free. Here is how E.T. Bell chose to describe their impact in his inimitable style:

> *...The idea behind it all is childishly simple, yet the method of analytic geometry is so powerful that very ordinary boys of seventeen can use it to prove results which would have baffled the greatest of the Greek geometers...* [8]

This chapter tries to convince the reader that we may wish after all to do geometry as ancient geometers would have—even in geometric computing. Taking this purist approach means ignoring the existence of coordinates and stepping on higher ground while only calling underlying "oracles." That underlying predicates would be resolved via coordinates should be left as a detail and ignored during development of noncore code.

17.1 Stages of a Geometric System

Programming quickly makes us study geometry through coordinates. In doing so we can forget that geometry was long practiced with no appeal to coordinates. The simple idea of assigning a pair of numbers to a point in the plane did not occur to the ancients. Establishing geometric facts using algebraic manipulation is a relatively recent event. Even though the previous chapters took the notion of coordinates for granted, this chapter suggests instead that the careless use of coordinates in intermediate stages of computation is a bad idea.

Indeed, the source of very many geometric programming bugs is the extraction of coordinates halfway through a system. *Coordinates are necessary for input and for output, but all computation should otherwise be coordinate-free.* By following this advice, geometric programmers could indeed save a great deal of anguish.

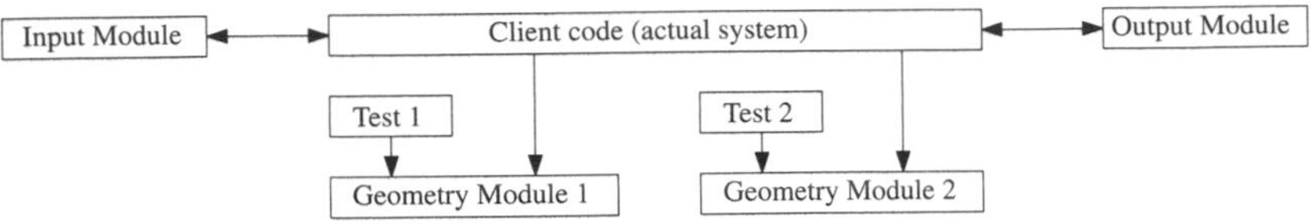

Figure 17.1
Layout of a geometric system

The designer of a geometric library consisting of the modules shown in Figure 17.1 may wish to shield beginning programmers from writing coordinate-dependent code. One option that could be considered is to provide no accessor functions for reporting the x-, y-, and w-coordinates of a point represented using homogeneous coordinates in the plane. Only reader modules, which interface with the filesystem, and writer modules, which interface with both the filesystem and with the graphics library, need and are provided access to coordinates. Client programmers could thus be effectively (and aggressively) shielded from writing poor-quality and error-prone code. But doing so is usually not an option for the simple reason that developers frequently rely on coordinates to debug geometric systems. A better approach is to train developers using a given API to write coordinate-free code—the topic of this chapter.

The chapter is also a prelude to introducing the notion of a geometric kernel. Suppose that when starting to implement a geometric system one remains undecided about whether to use Cartesian or homogeneous coordinates. Could one delay this decision as long as possible? In other words, could a set of classes for points, lines, segments, etc. be written such that they capture either Cartesian or homogeneous coordinates by using a minimal modification of the source code? CGAL, which is based on the design of such a *geometric kernel* or simply a *kernel*, is discussed in Chapter 18.

17.2 Dangers of the Lack of Coordinate Freedom

While looking in Chapter 1 at the simple task of writing constructors for a segment in the plane, we decided not to write a constructor that takes four coordinates.

```
class Segment_E2d {
private:
    Point_E2d source, target;
public:
    ...
    Segment_E2d(double x1, double y1, double x2, double y2)
        : source(x1, y1), target(x2, y2)
    {}
    ...
};
```

One of the strongest arguments in favor of coordinate freedom is also the simplest. If we design our geometric library with a constructor such as the one above, we can be sure that some day, some client programmer is going to construct a segment using Segment_E2d(x1, x2, y1, y2). Experience also shows that no one is beyond falling for this trap—a trivial bug that is particularly insidious because it is so easy to omit even looking for it.

Those considering that it would be ludicrous to misinterpret the order of the parameters in the constructor above only need to look at the following constructor for a bounding box in the plane—one that was not included in the set of constructors discussed for a Bbox_E2d in Chapter 4.

```
class Bbox_E2d
{
public:
    ...
    Bbox_E2d(double xmin, double xmax, double ymin, double ymax)
        : LL(xmin, ymin), UR(xmax, ymax)
    {}
    ...
};
```

Even if listing the x-bounds before listing the y-bounds appears quite reasonable, introducing a bug in a program by writing the parameters in an incorrect order is also a subtle bug. It is easiest simply to omit such a constructor.

17.3 Coordinate Freedom as Bug Deterrent

We see in Chapter 18 how CGAL delays both the choice of the floating point type and the choice between the use of Cartesian or homogeneous coordinates. Now consider the unfortunate scenario of a library designer providing functions P.x(), P.y(), and P.w() to extract the homogeneous coordinates of a point P in the Euclidean plane. Consider also that communication failure among members of the design team led one person to use the first two access functions as if Cartesian coordinates were used when in fact homogeneous coordinates are. The team would be on the road to hunting down a bug that has been introduced entirely unnecessarily and that could have been guarded against using coordinate freedom.

A simple answer is to use different names for accessing the homogeneous coordinates of a point, perhaps P.hx(), P.hy(), and P.hw(). Doing so is wise but is an orthogonal issue. What is stressed here is that coordinate freedom leads to the design of elegant and easily maintainable systems that have a lower risk of succumbing to trivial programming bugs. Even if the system architect has no intention of switching from Cartesian to homogeneous coordinates or vice versa, the use of different names for accessor functions makes it possible to reveal the segments of code that are not coordinate-free—simply by switching temporarily from one representation to the other and recompiling.

Giving access to coordinates provides a different venue for introducing errors. Writing high-level geometric code is challenging and rewarding, but writing coordinate-dependent code is often tedious since virtually identical computations often need to be performed for several coordinates. The bored programmer is unlikely to type a statement a second and a third time for, say, the y- and the z-coordinates, but will prefer to cut and paste the line or lines for the x-coordinate and then modify the coordinates individually. Programmers who have spent time hunting for a bug they perceived as deep only to discover that the error lay in the simplest part of their code (and hence the least likely place for them to search) will have eventually decided that it is wise to isolate coordinate-dependent code.

17.4 Coordinate Freedom as a Force to a Cleaner Interface

Another argument in favor of coordinate freedom is the memorability of the resulting interface. Consider the simple example of viewport mapping (§ 4.5). One may design a transformation function that takes the coordinates of the lower left and the upper right points in both viewports.

```
Point_E2d viewport_map(
            const Point_E2d& p,
            const Point_E2d& from_LL, const Point_E2d_E2d& from_UR,
            const Point_E2d& to_LL, const Point_E2d_E2d& to_UR );
```

One may go as far as choosing an interface consisting of the individual coordinates of the two points at the corners of the two viewports.

```
Point_E2d viewport_map(
            const Point_E2d& p,
            double from_LL_x, double from_LL_y,
            ... );
```

Neither of these solutions is as elegant nor as safe for the application programmer as an interface that requires passing only the two bounding boxes.

```
Point_E2d viewport_map(
            const Point_E2d& p,
            const Bbox_E2d& from,
            const Bbox_E2d& to );
```

Coordinate Freedom Determines the Design of a Function

Coordinate freedom also acts as a force that determines how the components of a geometric API are to be organized. For example, the function dominant was introduced in § 2.4 as a nonmember function. Since that function includes the statements

```
double dx = segment.target().x() − segment.source().x();
double dy = segment.target().y() − segment.source().y();
```

it becomes clear that dominant must be either a member function of the class Segment_E2 or a nonmember function in a directory capturing all those functions that are allowed to access coordinates. The function dominant must be classified as part of the core or kernel of a geometric library.

17.5 Coordinate Freedom in Practice

The discussion so far perhaps risks suggesting that coordinate freedom means performing geometric computation without coordinates, and so defining what could be termed the *principle of coordinate freedom* is warranted.

> A coordinate-free geometric system is one that isolates the set of functions that access (read or write) coordinates to a small core (or kernel). The input and output stages can read and write coordinates, but the processing stages operate by delegating requests to the kernel while never directly accessing coordinates.

The reader would be rightly suspicious to wonder how far coordinate freedom can be pushed in a practical application. This section outlines how coordinate freedom can be adopted and suggests the limits to its adoption.

Enforcing Coordinate Freedom

From the perspective of a geometry library designer, one possible scheme for enforcing coordinate freedom is to provide no coordinate accessor functions. Constructors would take coordinates as input and the output subsystems would be allowed to extract coordinates. A graphics subsystem, for example, would extract coordinates for visualization.

```
template<typename K>
struct GLdraw {
  static void draw(
            const Point_E2<K>& p,
        const Color& pointColor = Color::yellow,
        float pointSize = 5.0 )
  {
  glPointSize(pointSize);
  glColor3f(pointColor.r(), pointColor.g(), pointColor.b());
  glBegin(GL_POINTS); {
    glVertex2f(p.x(), p.y());
  } glEnd();
  }
  ...
};
```

In the example above the class GLdraw is a wrapper for all visualization functions. GLdraw would be granted friendship liberally so that the *private* functions `Point_E2<K>::x()` and `Point_E2<K>::y()` may be accessed. These functions in turn isolate whether Cartesian or homogeneous coordinates are used.

There is another advantage to keeping output, and especially visualization, routines strictly apart from the main library. A frequently adopted design constraint is to write a system such that it can be compiled for two target architectures, perhaps one built on OpenGL and the other on Direct3d. By isolating the output routines, the code base shared between the two targets would be maximized, and the developers responsible for each would not "pollute" the logic of the main code. Even if the developer has no intention of producing an executable for a different architecture, it is still desirable to keep the output routines separate. Experience suggests that output functions that share only a little similarity can go unnoticed and that code duplication can occur, wasting programmer time and bloating the size of the program. Naturally, the decision to migrate to a different output library is significantly less costly if the output modules were isolated in the first place.

As discussed next, and despite all its attractiveness, it is seldom the case in the real world that heavy-handed enforcement of coordinate freedom would be tolerated.

Compromise Between Elegance and Efficiency

Standard computing dogma suggests that only the relative complexity of algorithms and implementations should be compared. Yet a geometry library meant to be used outside the confines of academia could hardly justify preventing clients from accessing individual coordinates. For example, orthogonal views are widely used in engineering and architecture drawing since measurements can easily be taken in the field from diagrams. A system that generates orthogonal views of a scene needs to decide whether a polygon faces the camera. Not insisting on a coordinate-free formulation would forego the computation of an inner product and would simply compare x, y, or z with zero.

Likewise, a viewport clipping routine (§ 5.4) designed for a rectangle of arbitrary orientation could be made significantly faster if the clipping rectangle is known to be axis-parallel. Both determining whether a point is inside a region of interest and finding the intersection of two lines can be made more efficient if the bounding line is known to be axis-parallel.

The above example for computing orthographic projections effectively depends on comparing one coordinate of a vector with zero. Suppose one wishes to sort a set of points in 3D along one of the three main axes. The solution CGAL takes is to define a function:

```
Comparison_result compare_x(Point_E2 P, Point_E2 Q);
```

Incorporating compare_x promotes coordinate freedom by making it possible to compare the x-coordinates of two points while only indirectly accessing those coordinates. Such a predicate may then be used to sort a set of points without exposing the coordinates.

The option to design the code such that all functions and classes that do access coordinates are isolated under one directory is particularly attractive. The code accompanying this text uses this convention: The code in directories whose name starts with geometry may and does access coordinates, whereas the code outside these directories does not. The consensus among developers would be that even though coordinates *may* be read throughout the system, they *should* only be read in an agreed-upon set of directories.

The Importance of Training

Possibly the strongest argument for adopting coordinate-free geometry arises from the collective experience of seasoned programmers who eventually observe that a considerable fraction of bugs was found in coordinate-specific code. Aside from the efficiency-based need for access to coordinates outlined above, beginning programmers will frequently have the need to access coordinates to implement functionality that they perceive is missing from the geometry library they use. The answer is to ensure that programmers joining a team are aware of the abilities of the libraries available and to stress that, on the chance they require new functionality, such functionality should be added to the kernel rather than to their own code.

Chapter 1 and this chapter outlined general guidelines for designing geometric systems. These guidelines are frequently collectively called "coordinate-free geometry" [40, 64, 69]. The convention used here is instead to assume that suggestions, such as the separation of the types for points and vectors, are merely geometrically sensible. That they also make it harder for a client programmer to introduce bugs by using absurd constructions (allowing the scaling of vectors, but not of points) does not make them fall under the guidelines for writing coordinate-free code. The latter is used here to refer to code that does not extract coordinates until the final (output) stages, in addition to the inevitable use of coordinates to read input geometry.

17.6 Exercises

17.1 Write two programs that contain one function for performing viewport mapping (refer to Chapter 4). The first function uses a viewport and point objects while the second uses neither but only declares and directly passes floating point variables as parameters. Compare the assembly language programs your compiler produces for the two functions and deduce what efficiency hit, if any, results from using a well-designed program compared to the direct, but inelegant and unmaintainable, approach.

17.2 If you have used a version control system during the development of a geometric program and recall that there was a programming error that took significantly longer than average to track down, look for the change that was eventually needed and determine whether or not the error was in coordinate-dependent code.

17.3 An extreme way of writing coordinate-dependent code is not to define a class for a point and simply to use individual variables for a geometric computation. Does one lose anything in efficiency (time and space) when writing coordinate-free code? To find the answer, write a program for intersecting two segments both using individual variables and by relying on the libraries accompanying this text. Do the two executables that your compiler generated have the same size? When running each program a large number of times, is there a time penalty to coordinate freedom?

18 Introduction to CGAL

Computational geometry is the art of designing efficient algorithms for answering geometric questions. Traditionally, the field has focused on the efficiency of algorithms, but 1996 represented a departure for the field from a purely theoretical to also a practical one [75]. The reasoning was that geometric algorithms are intricate, and so for its many results to see their way into practice, the algorithms' designers themselves should also implement accompanying systems.

But most algorithms need the same software foundation layer, benefit from the same optimizations, and use Euclidean geometry. This suggested that a single *kernel* [36, 48] could act as a collection of geometric classes on which all algorithms would be built. At the time of this writing, the *Computational Geometry Algorithms Library* (CGAL) remains an active, and widely adopted, software project.

18.1 Typename Aliasing

With an implementation of two classes Point_E2c_f and Point_E2h_f for a point in the Euclidean plane, one using Cartesian and the other homogeneous coordinates (Chapters 1 and 16), one can use typename aliasing to switch from one representation to another.

```
class Point_E2c_f
{
private:
  float _x, _y;
public:
  Point_E2c_f() : _x(0), _y(0) {}
  Point_E2c_f( float x, float y )
    : _x(x), _y(y) {}

  bool operator==(const Point_E2c_f& P)
    const
  {
    return (this == &P) ||
      ( _x == P._x ) &&
      ( _y == P._y );
  }
};
```

```
class Point_E2h_f
{
private:
  float _hx, _hy, _hw;
public:
  Point_E2h_f() : _hx(0), _hy(0), _hw(1) {}
  Point_E2h_f( float hx, float hy, float hw = 1 )
    : _hx(hx), _hy(hy), _hw(hw) {}

  bool operator==(const Point_E2h_f& P)
    const
  {
    return (this == &P) ||
      are_dependent(
              _hx, _hy, _hw,
              P._hx, P._hy, P._hw);
  }
};
```

This can be done by commenting out one of the following two **typedef** lines:

```
// typedef Point_E2c_f Point_E2f;
typedef Point_E2h_f Point_E2f;

int main()
{
  Point_E2f Pc1(2,3);
  Point_E2f Pc2(2,3);

  assert( Pc1 == Pc2 );
}
```

In this approach, shown to be viable and adopted by the Library for Efficient Data structures and Algorithms (LEDA) in its geometric components [67], the application programmer typically creates an alias for each geometric type used (point, line, segment, etc.). The aliases resolve to class implementations. LEDA provides a set of classes based on Cartesian floating point (point, ...) and another set based on homogeneous arbitrary-sized integers (rat_point, ...).

To change from one set of classes to another, a group of typename aliasing lines is replaced by another. That the modification must be done for each class used is a small inconvenience. A more serious limitation is that the design of a given set of classes, or *kernel*, is final. Each class is aware only of those classes that are in the same kernel. Making a modification to any class means that this tight coupling is broken and that one must then reimplement all accompanying classes.

Is there a way to design a geometric library such that one

1. may easily switch between Cartesian and homogeneous representations,
2. can extend a given set of classes with localized modifications, and
3. does not pay any run-time penalty?

The last constraint is crucial. The innermost loops of geometric algorithms consist of manipulations on geometric objects. If the functions on these objects are not known at compile time, a penalty will be paid at run time in accessing virtual function tables. Worse, not knowing which function will be called prevents optimizers from inlining those functions that are reasonably concise. This suggests that inheritance is a poor choice for an extension mechanism of geometric kernels.

18.2 Traits

A particular variation on *traits* is the answer. Traits [71, 112] is a mechanism for defining functions on types. Both the argument as well as the return type of a function are normally objects, but C++ templates make it also possible to define a function that takes a type and that returns another type. Using the class of classes in the Java language, Class (note the capital C), it is possible to implement functions that take and return types. C++ goes further. The evaluation is done at compile time.

```
#include <cassert>
#include <typeinfo>

struct First { typedef int Parameter_type; };
struct Second { typedef float Parameter_type; };

template<typename T>
struct F
{
  typedef typename T::Parameter_type Return_type;
};

int main()
{
  typedef F<First>::Return_type Return1;
  typedef F<Second>::Return_type Return2;

  assert( typeid( Return1 ) == typeid( int ) );
  assert( typeid( Return2 ) == typeid( float ) );
}
```

The class F acts as a function between types. It takes the argument T and returns Return_type. F is invoked on a given type such as First by instantiating the concrete type F<First>. Once instantiated, Return_type becomes an alias to whichever type T (i.e., First here) declares as its Parameter_type. Since the type First::Parameter_type resolves to **int**, that is also the type to which Return_type resolves.

The use of the keyword **typename** in

```
typedef typename T::Parameter_type Return_type;
```

is necessary. It signals to the compiler that the identifier Parameter_type will not resolve to an object but to a type (which it must for type aliasing to be legal).

18.3 Nested Typedefs for Name Commonality in CGAL

Consider the two implementations Point_E2c and Point_E2h.

```
template<typename NT>
class Point_E2c
{
protected:
  NT _x, _y;
public:
  Point_E2c() : _x(0), _y(0) {}
  Point_E2c(
        const NT& x,
        const NT& y,
        const NT& w=1)
    : _x(x/w), _y(y/w) {}

  bool operator==(const Point_E2c<NT>& p)
    const
  {
    return (this == &p) ||
      (_x == p._x) && (_y == p._y);
  }
};
```

```
template<typename NT>
class Point_E2h
{
protected:
  NT _x, _y, _w;
public:
  Point_E2h() : _x(0), _y(0), _w(1) {}
  Point_E2h(
        const NT& x,
        const NT& y,
        const NT& w=1)
    : _x(x), _y(y), _w(w) {}

  bool operator==(const Point_E2h<NT>& p)
    const
  {
    return (this == &p) ||
      are_dependent(_x, _y, _w, p._x, p._y, p._w);
  }
};
```

Rather than expose the two names Point_E2c and Point_E2h directly, the two classes Cartesian and Homogeneous each contain a **typedef** statement from the common name Point_E2 to the corresponding Cartesian or homogeneous point class.

```
template<typename NT>
class Cartesian
{
public:
  typedef NT Number_type;
  typedef Point_E2c<NT> Point_E2;
};
```

```
template<typename NT>
class Homogeneous
{
public:
  typedef NT Number_type;
  typedef Point_E2h<NT> Point_E2;
};
```

The common name Point_E2 will in turn be used for a class that is parameterized by the kernel: either Cartesian or homogeneous. Because Point_E2 is derived from whichever class Kernel::Point_E2 maps to, Point_E2 effectively becomes an alias for either class [6].

```
template<typename Kernel>
class Point_E2 : public Kernel::Point_E2
{
  typedef typename Kernel::Number_type NT;
  typedef typename Kernel::Point_E2 Base;
public:
  Point_E2(const NT& x, const NT& y) : Base(x, y) {}
  Point_E2(const NT& x, const NT& y, const NT& w) : Base(x, y, w) {}

  bool operator==(const Base& p) const {
    return this->Base::operator==(p);
  }
};
```

A program can then switch from one kernel to the other by changing the instantiation of the kernel.

```
// typedef Cartesian<double> Kernel;
typedef Homogeneous<double> Kernel;
typedef Point_E2<Kernel> Point_E2f;

int main()
{
  Point_E2f Pc1(2.0f,3.0f);
  Point_E2f Pc2(2.0f,3.0f);

  assert( Pc1 == Pc2 );
}
```

18.4 Defining a New CGAL Kernel

By way of example, let us make a small modification to the semantics of point equality. That operator was defined in the Cartesian kernel using a Boolean shortcut. If one object is being compared to itself (via two references), **true** is returned without incurring the more expensive cost of evaluating comparisons on coordinates. Suppose, merely as a minimal example, that we wish two objects to be equal only if an object is compared to itself. This would lead to a class such as Point_E2c_new.

```
template<typename NT>
class Point_E2c_new
{
protected:
  NT _x, _y;
public:
  Point_E2c_new(const NT& x, const NT& y, const NT& w=1)
    : _x(x/w), _y(y/w) {}

  bool operator==(const Point_E2c_new<NT>& p) const {
    return (this == &p);
  }
};
```

Point_E2c_new is then aliased to Point_E2 within New_Cartesian; the new kernel class is itself derived from Cartesian.

```
template<typename NT>
class New_Cartesian : public Cartesian<NT>
{
public:
  typedef NT Number_type;
  typedef Point_E2c_new<NT> Point_E2;
};
```

The new kernel can be instantiated just as before, but now two distinct point objects with identical coordinates are no longer equal.

```
typedef New_Cartesian<double> Kernel;

typedef Point_E2<Kernel> Point_E2f;

int main()
{
  Point_E2f Pc1(2.0f,3.0f);
  Point_E2f Pc2(2.0f,3.0f);

  assert( (Pc1 == Pc1) && !(Pc1 == Pc2) );
}
```

Coupling

Observe that the coupling between the Cartesian and the homogeneous representations of a point remains visible in the implementation of Point_E2. Because Point_E2 resolves to either Point_E2c or to Point_E2h: the actual class Base must provide constructors for two and three arguments. Since either

```
template<typename Kernel>
class Point_E2 : public Kernel::Point_E2
{
  typedef typename Kernel::Number_type NT;
  typedef typename Kernel::Point_E2 Base;
public:
  Point_E2(const NT& x, const NT& y) : Base(x, y) {}
  Point_E2(const NT& x, const NT& y, const NT& w) : Base(x, y, w) {}

  bool operator==(const Base& p) const {
  return this->Base::operator==(p);
  }
};
```

Point_E2c or Point_E2h is the base class for Point_E2, both classes must provide Cartesian and homogeneous constructors. In practice in both the case of Point_E2c

```
Point_E2c::Point_E2c(const NT& x, const NT& y, const NT& w=1)
  : _x(x/w), _y(y/w) {}
```

as well as in that of Point_E2h,

```
Point_E2h::Point_E2h(const NT& x, const NT& y, const NT& w=1)
: _x(x), _y(y), _w(w) {}
```

only one constructor will do, but it is worthwhile noticing that the isolation is not complete. Not that it is desirable for the isolation to be complete; if coordinates are necessary to input data to a set of classes, but not needed for execution (Chapter 17), and if it is easily isolated in the output stage, then it is of value to make it possible for application programmers to switch between Cartesian and homogeneous coordinates by changing one keyword, which makes it necessary for both point classes to support two and three parameters.

18.5 Using Pointers in a Geometric Kernel

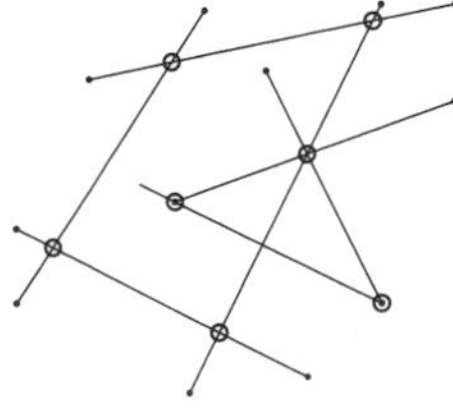

Consider a classical problem in computational geometry, *segment intersection*. The input is a set of segments in the Euclidean plane and the output is the set of segments that result from partitioning the input segments at intersection points. Since an arbitrary number of segments may meet at a point, a reasonable way to implement a segment intersection algorithm is to avoid representing the output as a set of segments. Many of the output points may be shared, and so it makes sense to compute the intersection points and then to create a special segment class that stores pointers to a pair of points among the input and the intersection points.

Using pointers simplifies subsequent equality tests since it then suffices to compare the pointers rather than the coordinates. If the coordinate representation chosen is not of constant size (Chapter 7), pointers can potentially save significant storage.

Because this potential saving is true not just for segments, but for any geometric object, it makes sense to design a geometric library based on such a sharing principle. Such a design means that the onus for the details (reference counting and destruction) would be encapsulated in the geometric library rather than be a task for application programmers. A point would be a wrapper for a pointer to a given point representation, and assignment operators and copy constructors would copy the pointers rather than the representations.

18.6 Future Libraries

So far we have operated on the assumption that the *type* of the geometry in which we perform computation needs to be specified at the outset. By contrast, the graphics pipeline is notorious for being type-free. Such a view is perfect for a hardware implementation, where values are treated in quadruples and the interpretation of each quadruple depends on what one will finally do with it.

Is it too fussy to insist that in a simulation that precedes a hardware implementation, or in development entirely detached from such an implementation, we would classify each stage of the graphics pipeline in the geometry in which it operates? The notion is tempting, for it means that development is significantly safer. We can then always interpret the validity and the significance of the results.

One option for a sequel to CGAL [23] is to follow the outline given here. Four distinct libraries are defined for Euclidean, spherical, projective, and oriented projective (Stolfian?) geometries. Loose coupling is provided via a set of functions or function objects.

But even though the four sets of geometries are distinct, their implementations share considerable similarities, which suggests that another option is possible. Only classes and operations for oriented projective geometry are defined. Which geometry is intended is left to the application programmer. Distinct predicates and other functions are provided in the library and the choice of which predicates to use implicitly defines the geometry that the application programmer has in mind. If one has in mind only prototyping a hardware implementation, then T^3 would suffice; the dimension intended could also be left to the application programmer, who would operate in a linear subspace ($z = 0$, or $y = 0$ and $z = 0$).

Still, starting from a library for oriented projective geometry that then gets specialized may be too type-constrained. Blinn debates whether an *algebraic layer* that leaves the *geometric interpretation implicit* in the code may replace an *explicit geometric layer* [16]. There are several levels to which such distinctions can be made. At the limit, duality in non-Euclidean geometries makes it possible for us to confuse such basic constructs as a point and a line. In P^2 one may, for example, replace point and line classes with a single class. There is then no need to define two predicates, one testing for the colinearity of three points and the other for the concurrency of three lines. Since the two tests are identical algebraically [26], it is possible to define a more malleable type that gets used by clients in whichever way they wish to interpret it. Is facilitating duality at the level of a geometry API desirable?

The answer may in any case be a personal one. Many programmers flock to type-rigorous languages such as C++ because no matter their focus most of the time they would rather be spared having to search for bugs that could have been caught by the type system. Others may prefer a looser system that allows them to interpret results differently by changing a minimal set of, say, predicate calls. If one chooses to deploy a library for use in a large system by a variety of programmers, then accommodating those who prefer type safety is likely the better option—even if they are not the majority, though they in any case most probably will be.

To recapitulate, notice that the development of libraries in a computing medium closely mirrors the abstract or mathematical view. To mirror the discussion at the conclusion of § 15.7, we note that if one takes two distinct points and declares them the fundamental points on a line while permitting arbitrary affine transformations, one gets E^1. If they are on a circle, but one allows only rotations, one gets S^1. That the antipodal points of the two fundamental points are distinct is also a characterization of spherical geometry. If one next identifies each pair of antipodal points and declares them the same "point" (an observation first made, according to Coxeter [28, p. 13], by Klein), but allows an arbitrary homogeneous linear transformation, one gets the projective line P^1. Declare the antipodal points distinct, and continue to allow an arbitrary homogeneous linear transformation, and one gets the oriented projective line T^1.

Part IV

Raster Graphics

19 Segment Scan Conversion

Given a segment in the plane and a set of grid points, or pixels, we wish to flag the set of pixels that would be an appropriate discrete rendition of the segment. Finding the set is straightforward if we are content with using floating point arithmetic, but since *segment rasterization* or *segment scan conversion*, the alternative names for the problem, is performed many times in interactive and real-time rendering, an algorithm that uses only integer arithmetic provides significant gains. This chapter discusses Sproull's derivation [102] of Bresenham's algorithm [21] using *refactoring*, or incrementally modifying a program while verifying its correctness to enhance either its efficiency or its structure.

19.1 What Is a Pixel?

A raster image is displayed on a computer monitor as an array of pixels. A pixel may be displayed on a cathode ray tube using three monochromatic phosphor discs and on a liquid crystal display as a set of rectangular, also monochromatic, regions. Neither is adequate for defining a pixel.

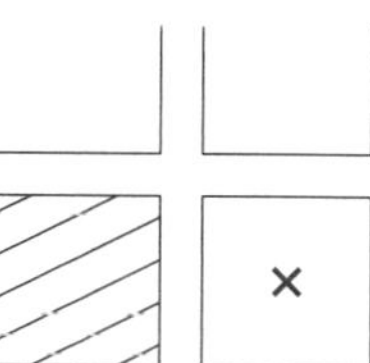

The smallest addressable fragment of an image may be a suitable definition, but one must still remember that a pixel is not some minuscule square [101]. A pixel is indeed *displayed* as such a square only to interface with humans, but the only correct abstraction for a pixel is a point in the (Euclidean) plane.

19.2 Raster Images

Before discussing how geometric objects can be converted from a continuous to a discrete space—a process called *rasterization* or *scan conversion*—we discuss a data structure that can capture the output. A *raster image* or, more briefly, an image is a 2D array of grayscale or color values. Such a grayscale or a color image is a useful abstraction of a raster display device. An image consists of a 2D array of *pixels*, or picture elements.

Raster images of sufficiently high resolution can convey significant information. Raster images can also be generated at enough speed that a sequence of images can be faithfully perceived by an observer as an animation. The information conveyed by a single pixel will arise from sampling some geometric object at a point. Such information appears on a display device as a small square, but this is merely so that it becomes perceptible. It is important to remember that a raster image remains no more than a set of *samples*.

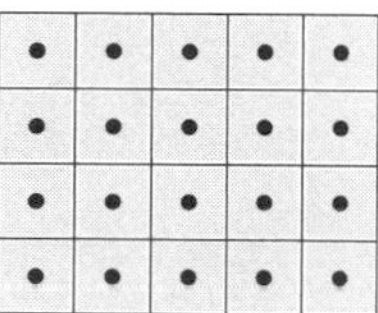

A Class for a 2D Array

Since an image will be stored in a 2D array, we start by looking at a generic class Array2.

```
template<class Type>
class Array2 {
    typedef Color_4<unsigned char> Color_4uc;
    friend class Image<Color_4uc>;
    int xsize, ysize;
    Type* table;
public:
    void* getTable() const
    { return (void*) table; }

    void init(const Type& t)
    {
        table = new Type[xsize * ysize];
        for(int i=0; i< xsize * ysize; ++i)
            table[i] = t;
    }

    Array2() : xsize(2), ysize(2)
    { init(Type()); }

    Array2(
            int xsizein,
            int ysizein,
            const Type& t = Type())
        : xsize(xsizein), ysize(ysizein)
    { init(t); }

    virtual ~Array2()
    {
        delete[] table;
    }

    Array2(const Array2& rhs)
        : xsize(rhs.xsize), ysize(rhs.ysize)
    {
        table = new Type[xsize * ysize];
        for(int i=0; i<xsize*ysize; ++i)
            table[i] = rhs.table[i];
    }

    Array2& operator=(const Array2& rhs)
    {
        if(this != &rhs) {
            delete[] table;
            table = new Type[xsize * ysize];
            xsize = rhs.xsize;
            ysize = rhs.ysize;
            for(int i=0; i<xsize*ysize; ++i)
                table[i] = rhs.table[i];
        }
        return *this;
    }

    int getXsize() const { return xsize; }
    int getYsize() const { return ysize; }
    bool withinBounds(int x, int y) const
    {
        return
            x>=0 && x<xsize &&
            y>=0 && y<ysize;
    }
    // matrix is stored in row-order
    const Type&
    operator()(int x, int y) const
    {
        assert( withinBounds(x,y) );
        return table[x + y * xsize];
    }
    Type&
    operator()(int x, int y)
    {
        assert( withinBounds(x,y) );
        return table[x + y * xsize];
    }
};
```

The class needs no commenting, with the exception perhaps of the two **operator**() functions. Recall that C++ permits overloading on **const**-ness. Using two functions makes it possible to return a const object when the array object is const and to return a non-const object when the array object is not.

Defining grayscale or color images is discussed next, but notice that we choose to break the encapsulation and declare one concrete color image type a friend.

```
typedef Color_4<unsigned char> Color_4uc;
friend class Image<Color_4uc>;
```

Texture mapping requires access to an image. Since the two dimensional-array used here for an image happens to coincide with the one expected when the type is **unsigned char**, we avoid the penalty in efficiency that would arise from proper encapsulation and make it possible instead to access the color image directly.

A Generic Image Class

We can likewise define a class for an image, which acts largely as a wrapper for an Array2 object. The generic implementation will make it convenient to generate multiple types of images.

```
template<typename PixelType>
class Image
{
protected:
    int _xres, _yres;
    Array2<PixelType> H;
public:
    Image() : _xres(0), _yres(0) {}
    Image(int xres, int yres) : _xres(xres), _yres(yres), H(xres, yres) {}

    int xres() const { return _xres; }
    int yres() const { return _yres; }
    const PixelType& operator()(int x, int y) const
    {
        return H(x,y);
    }
    PixelType& operator()(int x, int y)
    {
        return H(x,y);
    }
};
```

A Color_image class as well as a GrayscaleImage class can now be instantiated.

```
template<typename ColorUnit>
struct Color_4 {
    ColorUnit r,g,b,a;
};

typedef Image<Color_4<float> > Color_image; // floats range from 0 to 1
typedef Image<char> GrayscaleImage;
```

Since at this point we are merely interested in testing the correctness of segment rasterization and comparing the efficiency of more than one approach, we will be content with the following rather rudimentary function for displaying Boolean images:

```
ostream& operator<< (ostream& os, const Image<bool>& I)
{
    for(int y=I.yres()−1; y>=0; −−y) {
        for(int x=0; x<I.xres(); ++x)
            os << (I(x,y) ? 'x' : '.');
        os << endl;
    }
    return os;
}
```

19.3 Segment Rasterization

The function B8 below implements a segment rasterization algorithm due to Bresenham [21], but rather than show the algorithm directly, we present a derivation of the algorithm due to Sproull [102]. The derivation is particularly delightful, not least because it shows how a discovery can be reduced to mere labor.

We assume that pixels lie at integer grid points and that an object Point_E2i captures the integer coordinates of such points.

```
struct Point_E2i {
  int x, y;
  Point_E2i(int x, int y) : x(x), y(y) {}
};
```

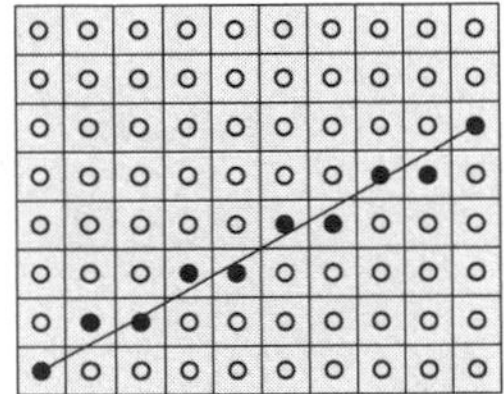

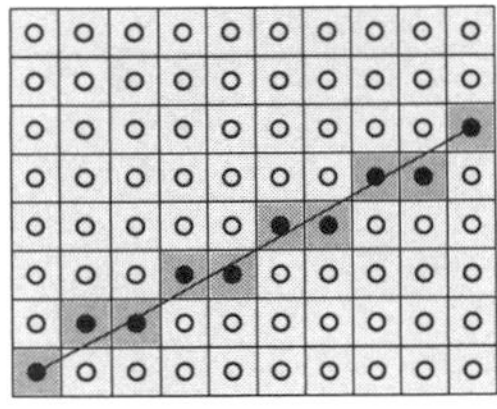

We also assume that the segment has integer coordinates, that one of its endpoints lies at the origin and the other at $P(x, y)$, and that $x >= y >= 0$. The line thickness is informally assumed to be a single pixel. This suggests that a suitable representation for segments lying in the first octant is to iterate over the x values and generate one pixel. Segment rasterization reduces to finding the integer y-coordinates.

Notice that it is unnecessary to specify that pixel *centers* are meant. A pixel is a Boolean sample, and the array of Boolean samples constituting the Boolean image will eventually be mapped to two tones for suitable display. The segment is *informally* of unit width because, as can be easily confirmed, the area covered by the line decreases as its slope approaches that of $x = y$. This is not ideal, since a rotating segment will appear to change tone, but we worry here only of generating the Boolean samples. This is in any case as well as we can do when visualizing a Boolean array (using either squares or circles).

The first function for rasterizing a segment simply iterates over the integer x values, which are multiplied by the slope of the segment. The resulting **float** yt is then rounded to the nearest integer.

```
void B1(Image<bool>& I, const Point_E2i& P)
{
  float yt;
  int xi = 0, yi;

  float Py_by_Px = float(P.y) / float(P.x);
  while(xi <= P.x) {
    yt = Py_by_Px * float(xi);
    yi = int(std::floor(yt + 0.5)); // floor returns float
    I(xi, yi) = true;
    ++xi;
  }
}
```

But rather than incur the cost of a floating point multiplication at each iteration by writing yt = Py_by_Px * **float**(xi), a floating point addition can be used instead.

```
void B2(Image<bool>& I, const Point_E2i& P)
{
    float yt = 0;
    int xi = 0, yi;

    float Py_by_Px = float(P.y) / float(P.x);
    while(xi <= P.x) {
        yi = int(std::floor(yt + 0.5));
        I(xi, yi) = true;
        yt += Py_by_Px;
        ++xi;
    }
}
```

A second program transformation initializes the floating point variable to 0.5 and thus avoids one floating point addition in the body of the loop.

```
void B3(Image<bool>& I, const Point_E2i& P)
{
    float ys = 0.5;
    int xi = 0, yi;

    float Py_by_Px = float(P.y) / float(P.x);
    while(xi <= P.x) {
        yi = int(std::floor(ys));
        I(xi, yi) = true;
        ys += Py_by_Px;
        ++xi;
    }
}
```

We can now see what happens at each iteration. The increment in x may or may not cause an increment by 1 in y. This can be explicitly simulated by storing a *residue*. The residue is incremented by the slope at each iteration. It is only if the new residue exceeds unity that y is incremented (and the residue decremented accordingly). The value of y is now stored as the sum of an integer component ysi and a floating point component ysf.

```
void B4(Image<bool>& I, const Point_E2i& P)
{
    float ysf = 0.5;
    int xi = 0, ysi = 0;

    float Py_by_Px = float(P.y) / float(P.x);
    while(xi <= P.x) {
        I(xi, ysi) = true;
        if(ysf + Py_by_Px >= 1) {
            ++ysi;
            ysf += Py_by_Px - 1;
        }
        else
            ysf += Py_by_Px;
        ++xi;
    }
}
```

Our final objective will be to make the computation use only integers; thus, we now avoid the floating point division by multiplying by $2P_x$. One of the two values v1 and v2 is now added to the residue r.

```
void B5(Image<bool>& I, const Point_E2i& P)
{
  float r = float(P.x);
  int xi = 0, ysi = 0;

  float v1 = 2.0 * float(P.x) − 2 * float(P.y);
  float v2 = 2.0 * float(P.y);
  while(xi <= P.x) {
    I(xi, ysi) = true;
    if(r + v2 >= 2 * P.x) {
      ++ysi;
      r −= v1;
    }
    else
      r += v2;
    ++xi;
  }
}
```

It is now unnecessary to use floating point variables. They can be replaced by integers.

```
void B6(Image<bool>& I, const Point_E2i& P)
{
  int r = P.x;
  int xi = 0, ysi = 0;

  int v1 = 2 * P.x − 2 * P.y;
  int v2 = 2 * P.y;
  while(xi <= P.x) {
    I(xi, ysi) = true;
    if(r + v2 >= 2 * P.x) {
      ++ysi;
      r −= v1;
    }
    else
      r += v2;
    ++xi;
  }
}
```

Because a comparison with zero is faster since only one bit is tested, an incrementally minor modification discards the comparison with $2P_x$.

```
void B7(Image<bool>& I, const Point_E2i& P)
{
  int r = − P.x;
  int xi = 0, ysi = 0;

  int v1 = 2 * P.x − 2 * P.y;
```

```
    int v2 = 2 * P.y;
    while(xi <= P.x) {
        I(xi, ysi) = true;
        if(r + v2 >= 0) {
            ++ysi;
            r -= v1;
        }
        else
            r += v2;
        ++xi;
    }
}
```

Finally, we avoid adding v2 inside the loop, resulting in Bresenham's segment rasterization algorithm.

```
void B8(Image<bool>& I, const Point_E2i& P)
{
    int r = 2 * P.y - P.x;
    int xi = 0, ysi = 0;

    int v1 = 2 * P.x - 2 * P.y;
    int v2 = 2 * P.y;
    while(xi <= P.x) {
        I(xi, ysi) = true;
        If(r >= 0) {
            ++ysi;
            r -= v1;
        }
        else
            r += v2;
        ++xi;
    }
}
```

Running any of the functions above via a code snippet such as

```
Point_E2i P(11,7);
Image<bool> I1(12,12); B1(I1, P); cout << I1 << endl;
```

yields the following ASCII art.

```
............
............
............
............
...........x
.........xx.
........x...
......xx....
....xx......
...x........
.xx.........
x...........
```

Since compiler-performed optimizations may perhaps render unnecessary one or more of the optimizations performed explicitly above, it is interesting to compare the running time of each under a different compiler optimization level. Running the eight preceding routines on a modern CPU and a modern

Optimization Level	B1	B2	B3	B4	B5	B6	B7	B8
none	4.35	3.92	4.38	2.64	2.60	2.50	2.38	2.35
1	2.07	1.71	1.31	0.67	0.64	0.53	0.51	0.50
2	2.07	1.88	1.76	1.03	0.53	0.52	0.47	0.44
3	1.97	1.75	1.25	0.74	0.83	0.63	0.64	0.42

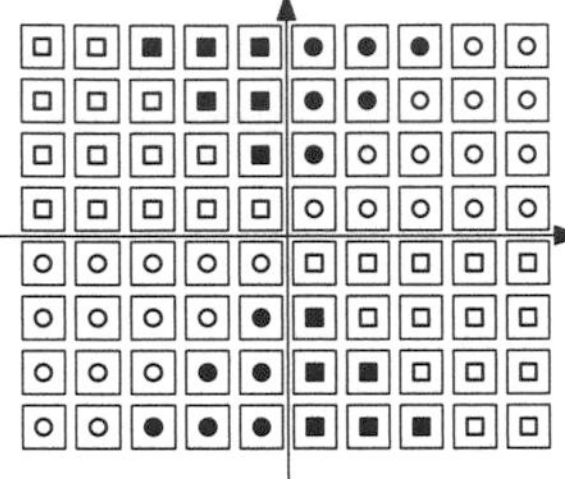

compiler (at the time of this writing) yielded the relative running times shown in the table.

Particularly intriguing is the increasingly more aggressive optimization possible as the algorithm is incrementally optimized by hand.

More than one possible partition of the target pixels into octants is possible, and a few symmetric cases round up the implementation and testing.

19.4 Exercises

19.1 Modify the start-up code labeled "segment-rasterization" to use only integer arithmetic while still handling all eight octants correctly.

19.2 This question connects this chapter with Chapters 7 and 23.

Often when generating ray-traced images, one uses the following optimization. The smallest sphere bounding each set of objects is determined. Before intersecting the objects within, the bounding sphere is tested. If a ray does not intersect it, there is no need to intersect the objects within.

This strategy is useful even if the ray-traced object is also a sphere. Develop a system for ray tracing one sphere by shooting parallel rays. To find the nearest intersection of a ray and a sphere, substitute in the equation of the sphere with $P + \alpha\overrightarrow{v}$ for a ray $(P, \overrightarrow{v})$. Now implement the following optimization. Before finding the intersection, determine whether there is one by calculating the distance from the ray to the center of the sphere.

Using one ray for each pixel in a raster image, determine the intersection using both the optimization and the actual test (then discard the value of α). Generate a raster image that highlights the pixels in which the two methods disagreed.

20 Polygon-Point Containment

This chapter discusses several strategies to determine whether a point is contained in a polygon. The batch versions of the problem, by having either multiple polygons, multiple points, or both, arise sufficiently frequently to warrant special attention.

20.1 Rays

One fundamental object in geometric computing in the Euclidean plane, yet one that was not introduced in Chapter 1, is that of a ray. A ray consists of the set of points $P + \alpha D$ for $\alpha \geq 0$, a point P, and a direction D. At times it will be necessary to capture the notion of a ray such that $\alpha > 0$. The set of points is still unbounded but is now also open at the point P. Yet because the two possible notions of a ray share the same attributes and most properties, it would be undesirable to capture them separately. Recall, for instance, that the distinction between an open and a closed halfplane in E^2 was not made explicit, but that both abstractions relied on that of a directed line; the client program is asked to partition the three possible orientations (left, right, colinear). Here also only one ray is defined. Functions seeking an open or a closed abstraction will need to cater for that need individually. It is in general clear from the context whether an open or a closed ray is intended. In particular, the ray is by default closed in discussions involving polygon scan conversion (this chapter) and open in ones involving ray tracing (Chapter 23).

```
typedef<typename T>
class Ray_E2
{
    Point_E2<T>       _source;
    Direction_E2<T> _direction;
public:
    ...
};
```

20.2 Convex Polygon-Point Containment

We are given one polygon and one point, and we wish to determine whether the point is inside the polygon. The batch version of the problem, discussed next, handles the case when containment needs to be determined for many points.

Convex Polygon-Point Containment by Halfplane Testing

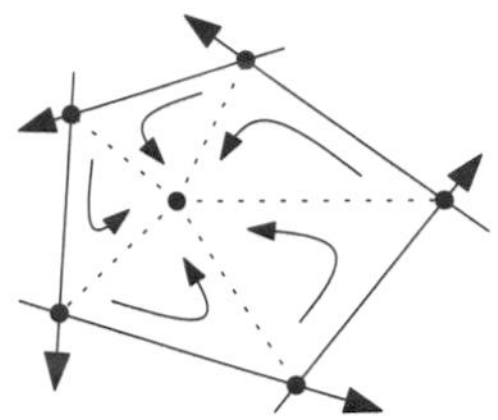

If only one polygon is considered, it is customary to assume that the polygon is a closed set—i.e., it includes the points on its boundary. If the polygon is convex, orient its n vertices counterclockwise and form n oriented lines along consecutive segments on the boundary. The set of points contained in the polygon is the set of points that simultaneously lies to the left or on each line.

Determining n-gon-point containment then reduces to the evaluation of n 3×3 determinants. If one determinant is less than zero, the point is outside the polygon. The test involves $9n$ multiplications and $5n$ additions/subtractions.

20.3 Concave Polygon-Point Containment

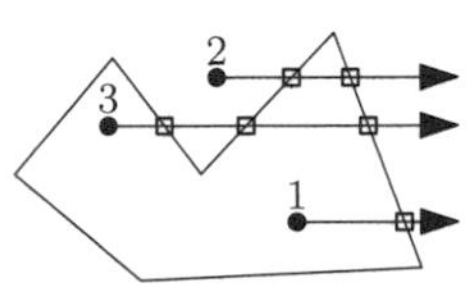

An alternative strategy, suitable for concave as well as convex polygons, is to "shoot" a ray from the query point in a given direction (say the positive x-direction). The general idea is to count the number of intersections. If the number is odd, the point is inside the polygon. If it is even, the point is outside. Before mentioning a strategy that does work, it is instructive to examine a few that fail.

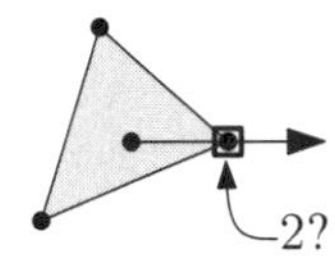

A special strategy must be taken to handle cases when the ray intersects the boundary of the polygon at one or more vertices. It is clear that each vertex cannot be counted twice—as belonging to both adjacent segments, for otherwise a point inside the polygon could have an even count, leading to the conclusion that it is outside the polygon (and vice versa). That would be the case for points with coordinates (x, y) if a polygon vertex is at (x', y), $x < x'$.

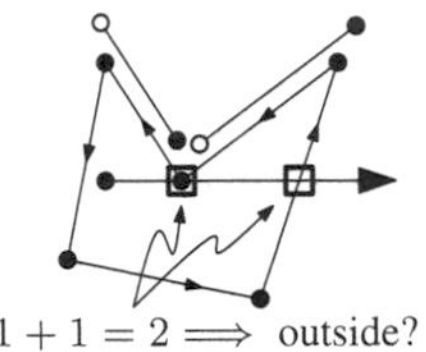

A tempting answer is to associate each boundary point with only one of the two adjacent segments in counterclockwise order around the boundary. But considering that the segments on the boundary of the polygon are open at one end, say the target side, and closed on the side of the source can also lead to concluding incorrectly that a point inside the polygon is outside (and vice versa).

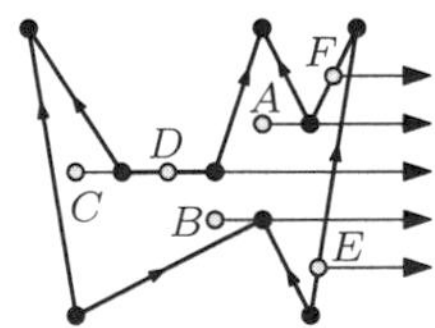

And so the segments need to be oriented not with respect to their order around the polygon, but in the orientation of their projection on a direction orthogonal to the direction of the ray. Count, say, an intersection with a segment at its endpoint with the smaller y-coordinate, but not at the endpoint with the larger y. This strategy computes the correct result at points A, B, and C. Some strategy still needs to be adopted for points such as D, which are on horizontal segments. The segment may bound the polygon either from above or from below, and any of the above strategies will be seen to lead to inconsistent classifications.

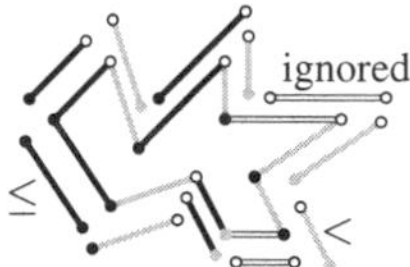

At this stage it is useful to diverge from the method adopted for handling points on the boundary of a convex polygon. If a query point coincided with the boundary of a polygon, the point was classified as contained in the polygon. What if the concave polygon was considered instead to be one of many polygons sharing boundaries? A consistent strategy is to assume that horizontal segments are contained in the polygon above the segment, but not the polygon below the segment.

20.4 Mesh Pixel Partition

In one batch version of the polygon-point containment problem we wish to partition a set of points into those lying inside (or on the boundary) and those lying outside the polygon. In the second we are given a *mesh*, or a set of polygons sharing boundaries, and are asked to partition the points into the polygon to which each belongs, if any. The second problem captures the first, and so we discuss it next.

The strategy outlined for a concave polygon is exactly the one that needs to be adopted for meshes—for both convex as well as concave polygons. Points in polygon interiors are easily classified. The difficulty lies in establishing rules for points lying on the boundaries between polygons.

Two rules summarize how to classify points in the interior of edges:

- If the neighborhood $(x + \epsilon, y)$ on the right of the boundary point (x, y) is inside some polygon, then the point is classified in that polygon.
- Points (x, y) on boundaries parallel to the x-axis are classified in the polygon at the neighborhood $(x, y + \epsilon)$.

Two rules summarize the classification of points coincident with vertices:

- A vertex of the mesh with coordinates (x, y) is classified in the polygon containing $(x + \epsilon, y)$.
- If the first rule is ambiguous (the adjacent edge is parallel to the x-axis), the vertex is classified in the polygon containing $(x + \epsilon, y + \epsilon)$.

Figure 20.1 illustrates the set of rules for a given mesh. We can now ignore that the problem is to be solved on a regular grid since we have just seen that the correctness of a solution for a regular grid relies on unambiguously classifying each point in $E^2(\mathbb{R})$, the set of points in the Euclidean plane with real coordinates. In the figure, vertices and edges are highlighted (darker marks/bolder edges), to indicate the polygon into which they are classified.

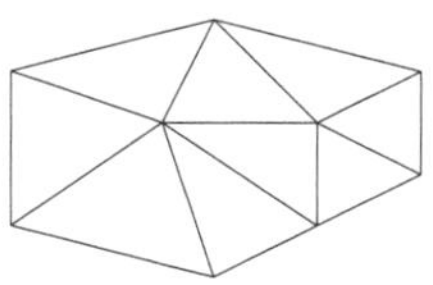

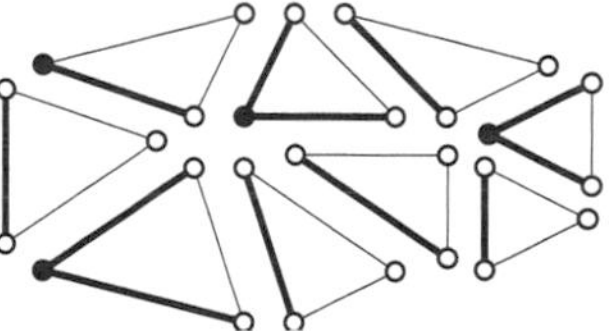

Figure 20.1
Partitioning points in the plane to triangles in a mesh

20.5 Exercises

20.1 Determine the barycentric coordinates of the points $P_1(5, 5)$ and $P_2(3, 5)$ in the triangle $(0, 0), (10, 0), (0, 10)$ in the Euclidean plane. Conclude with the characterization for a point inside a triangle.

20.2 The start-up code labeled "rasterize-polygon" lets the user draw a polygon by rubber-banding. The polygon is then rasterized by repeatedly performing polygon-point containment. Modify the code to scan-convert an arbitrary simple polygon efficiently.

20.3 Modify the start-up code labeled "rasterize-polygon" to display a mesh of a few triangles whose vertices move along a small circular trajectory. The code provided accumulates the effect of the rasterization; pay particular attention to the cases when mesh vertices or edges overlap pixels.

21 Illumination and Shading

Several functions—some physical, some heuristic—are used to generate smoothly varying colors. These *shading functions* are the topic of this chapter.

21.1 Diffuse Shading

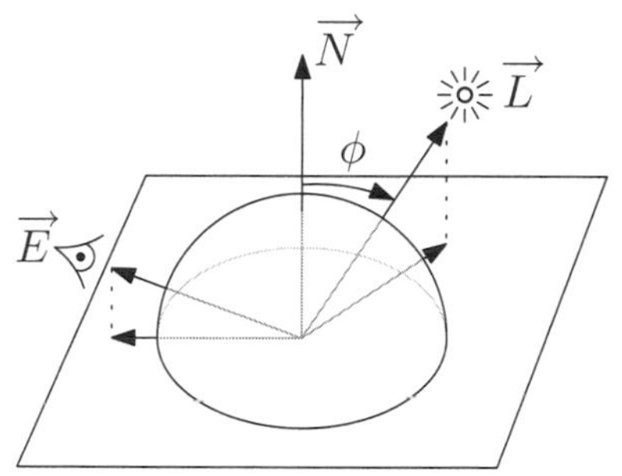

To *shade* a *point* on a surface is to compute a color that captures the lighting at that point. Shading is a function that takes light parameters, material parameters, and the surface normal vector and returns an estimate of the color seen at the (infinitesimal) point on the surface shaded.

To *shade* a *pixel* is to fill the square that represents the pixel's sample point using the shading of the object visible through a ray passing by the pixel's sample point.

The diffuse shading resulting from each light source is given by

$$I_d = \rho_d(\overrightarrow{N} \cdot \overrightarrow{L}),$$

where ρ_d is the diffuse color of the surface, $\overrightarrow{N}$ is the surface normal direction, and $\overrightarrow{L}$ is the light direction. (A direction is a distinct type for a normalized vector; see § 1.3 and § 3.2.) Since $\overrightarrow{N}$ and $\overrightarrow{L}$ are directions, their dot product $\overrightarrow{N} \cdot \overrightarrow{L}$ will range from -1 to 1. If $\overrightarrow{N} \cdot \overrightarrow{L} < 0$, no lighting from that light source will impinge the surface, and its contribution to the total illumination should be discarded.

If a colored light source is used, the equation is modified to

$$I_d = \rho_d \mathcal{L}_d(\overrightarrow{N} \cdot \overrightarrow{L}),$$

where $\mathcal{L}_d$ is the diffuse color of the light source.

Recall that if the angle separating $\overrightarrow{N}$ and $\overrightarrow{L}$ is θ and the two vectors are normalized (directions), then $\overrightarrow{N} \cdot \overrightarrow{L} = \cos\theta$.

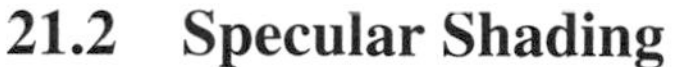

21.2 Specular Shading

A mirror or mirror-like surface would reflect incoming light in exactly the reflection direction. A larger class of surfaces, *glossy surfaces*, would reflect the light in some distribution that rapidly falls off from the reflection direction [80].

The input to a specular shading function consists of the eye direction $\overrightarrow{E}$, the normal direction $\overrightarrow{N}$, and the light direction $\overrightarrow{L}$. The output is a scalar

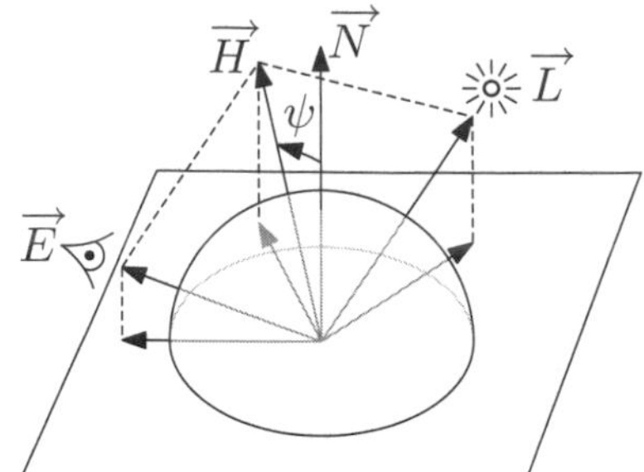

multiplier for the inner product of the specular material and the specular light coefficients. To determine how far $\vec{E}$ and $\vec{L}$ are to be the reflection direction of each other, the *halfway direction* $\vec{H}$ is defined as their sum: $\vec{H} = \vec{E} + \vec{L}$ [12]. The dot product of $\vec{H}$ and $\vec{N}$ determines the specular coefficient. To make it possible to tailor the size of the highlight to a more or less specular surface, an additional term, the *Phong coefficient* c, is used as the power of $\vec{N} \cdot \vec{H}$. The expression for specular lighting becomes

$$I_s = \rho_s \mathcal{L}_s (\vec{N} \cdot \vec{H})^c.$$

21.3 Ambient Shading

The combined contribution of diffuse and specular lighting generates synthetic pictures with unusually large variations in light to dark areas, which do not reconcile with our experience of scenes lit by even a small number of (point) light sources. The discrepancy is due to *light interreflection*: A fraction of the light impinging a surface is reflected according to a surface-specific distribution, called the *bidirectional reflectance distribution function*, or BRDF. The stable state reached by the interreflection in natural environments leads to all surfaces acting as secondary light sources. The field of *global illumination* studies methods for generating synthetic images, taking into account light interreflection. A simple heuristic that is frequently used in real-time systems is to add a constant to account for *ambient* light. A general scene contribution A as well as a light-specific factor, $\mathcal{L}_a$, and a surface-specific factor ρ_a, make it possible to fine-tune the heuristic to the desired lighting effect:

$$I_a = A + \rho_a \mathcal{L}_a.$$

To shade a single point, the sum of the three above terms is computed for each light source not hidden from the point:

$$I = I_a + I_d + I_s.$$

The three cross-sectional distributions shown in Figure 21.1 illustrate the relative contributions of ambient, diffuse, and specular lighting.

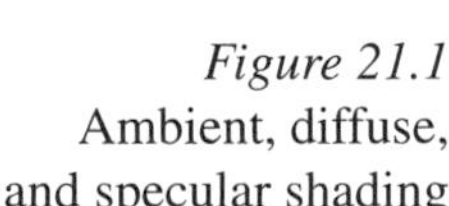

Figure 21.1
Ambient, diffuse, and specular shading

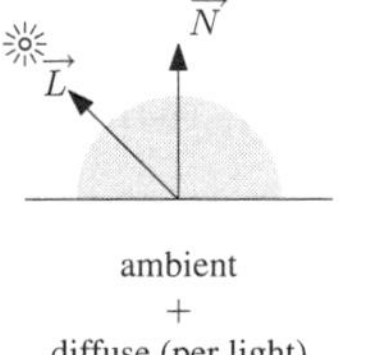

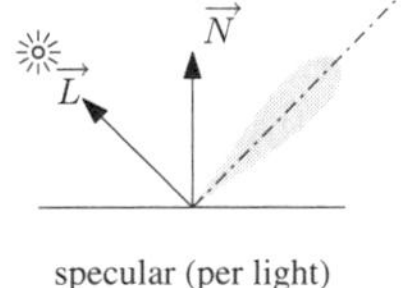

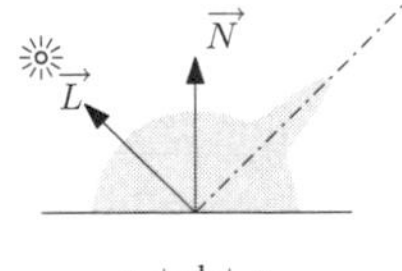

21.4 Gouraud and Phong Shading

The previous discussion assumes that normal vectors are defined for each face in a scene. Even though the eye and light directions would be distinct for each pixel rendered, the variation in illumination is in practice minuscule for the number of operations performed. Two options are possible. In the first, the number of operations will be significantly reduced, while images of comparable quality are obtained. In the second, the number of operations will be kept almost the same, but the quality of the image will be significantly increased.

Gouraud Shading

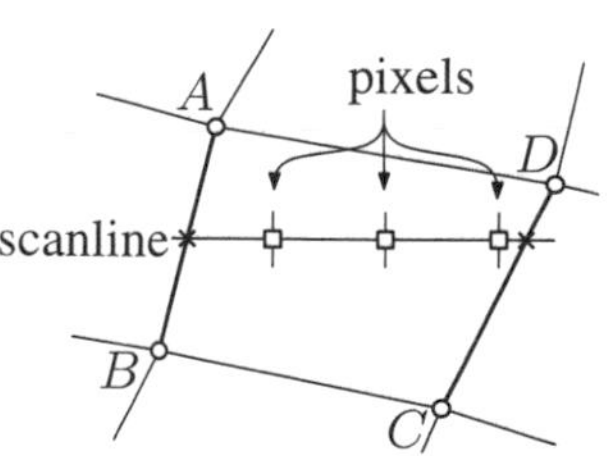

If the model defines normal directions for each vertex rather than for each face, the shading I can be determined at each vertex of a polygon. At the intersection of each scanline with the edges of a polygon, one may then interpolate colors, first along the edges AB and DC of the polygon rasterized and then between the two intersections [42]. Gouraud interpolation would thus be performed alongside depth interpolation (Chapter 22).

Phong Shading

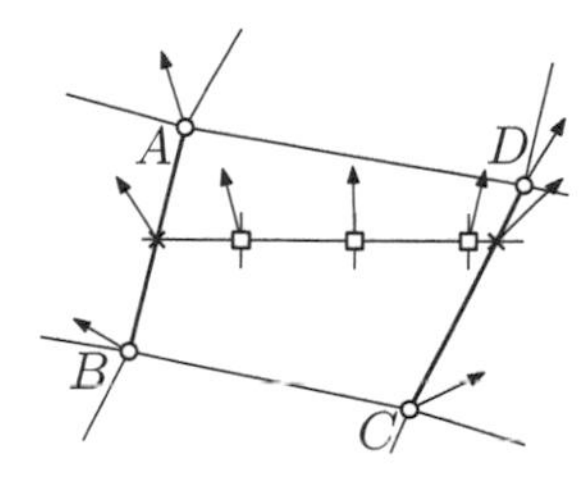

Alternatively, the normal directions can be interpolated along the edges of a polygon and then along each scanline [80]. The interpolated normal is used to shade the pixel. Even though *Phong shading* still has no physical basis, the rendering obtained considerably hides Mach bands, the tendency of the human visual system to accentuate lines of difference in shading.

21.5 Exercises

21.1 Write a program that renders a shaded sphere orthographically projected onto a raster image. Generate three images with different Phong coefficients.

21.2 Repeat Exercise 21.1 using a sphere coarsely triangulated along longitude and latitude lines. Compare the shading obtained when the image is generated using Gouraud shading with that obtained under Phong shading.

21.3 Modify the start-up code labeled "capoman" to implement a simple game. The objective of the game is for the player to run over three prey objects while navigating a maze within a time limit. Read Appendix B and then tackle the following brief milestones.

- Add a set of outer walls in addition to a minimum of two interior walls. Both outer and inner walls must have a "thickness" and so should be modeled by more than just one polygon (see Chapter 26). They should also have a "top" since we are looking down on the maze. Try first to use very few polygons, and then confirm after you add the flashlight that the lighting looks reasonable. If it does not, subdivide the wall polygons into a higher-density mesh.
- Add three objects (e.g., cylindrical prey—see § C.10) moving in the free space (between the walls). The path of the cylinders may

be a parametric function in time $[x = x(t), y = y(t)]$ with or without tangent continuity.

- Simulate each cylinder's illumination of the path in front of it using a flashlight (study the sample code described in § B.8).

21.4 This is a continuation of Exercise 21.3.

The objective is for the player to run over the prey (the three cylinders) within a time limit (say 20 seconds).

1. Introduce a player controlled by the keys (a/s/d/w). Choose your own object for the player (e.g., assemble a few **glutSolidCube**-s—see § C.10).
2. Add controls to ensure the player does not penetrate walls.
3. Add a counter showing the number of prey so far hunted.
4. Animate the capturing of a cylinder by displaying it spinning "in the air." The game should not be interrupted while the animation is in progress.
5. The cylinders should always take the path directed away from the player.

21.5 This is a continuation of Exercise 21.4.

Solve Exercise 12.6 and then modify your work in Exercise 21.4 to render shadows for the player and for the prey from an omnidirectional light at an affine point. Do not render either the shadows of the walls or any shadows from the flashlights.

22 Raster-Based Visibility

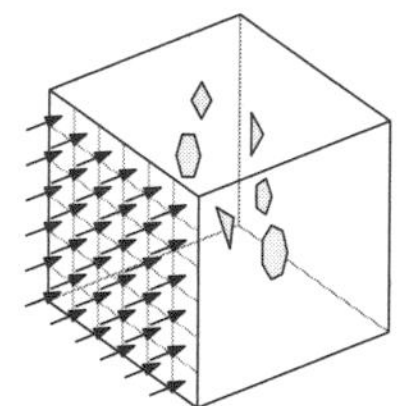

We wish to render a raster image of a set of polygons. The polygons have already been transformed (§ 15.5) and clipped (§ 5.4) to the unit cube. We lay a grid over one side (xy) of the cube and define a set of rays at the pixels induced by the grid. Each pixel conceptually spawns a ray (Chapter 23), and we wish to solve the batch intersection problem: Find the intersection of n rays with polygons of e total edges in time significantly lower than $\Theta(ne)$. The need to find the answer at only a discrete set of samples simplifies the problem considerably compared to a vector solution (Chapter 32). The two algorithms described, Z-buffer and scanline, take advantage of this discretization in addition to the fact that the discrete samples are equally spaced on the grid.

22.1 Pixels on a Scanline

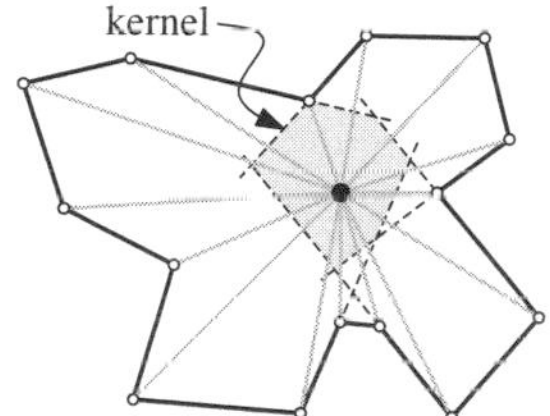

In the polygon-point containment problem discussed in Chapter 20 we determined whether *one* point lies inside *one* polygon. The next natural question to consider is the improvement that can be achieved if the input consists of many points. If the p-sided polygon is star-shaped (all interior and boundary points are visible from a point in the polygon's *kernel*—the intersection of the half-spaces defining the boundary), we would search in a radial list of the vertices [in time $O(\log p)$] to determine whether the query point is inside the polygon. This strategy is also applicable to convex polygons, which are a subset of star-shaped polygons.

Even if the convex or star-shaped polygon will be queried at many points, it is unlikely that searching in a sorted list would produce practical gains. One way for making gains when batch searching using a large set of points is to exploit the structure of the points. Since the points arise as pixels, they are arranged on a grid. If the grid lines are parallel to the x- and y-axes, a set of pixels on a single line parallel to the x-axis are said to form a *scanline*. It is then more efficient to determine the first and the last pixel on the scanline to which the edges of the polygons rasterize and to fill each consecutive set of pixels by the index (or color/shading) of the polygon.

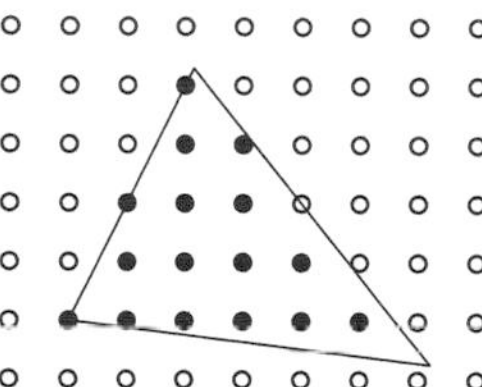

22.2 The Z-Buffer

Just as a raster image consists of a 2D array of color (shading) values, the *Z-buffer* is a 2D array of depth values. The purpose of saving these depth values

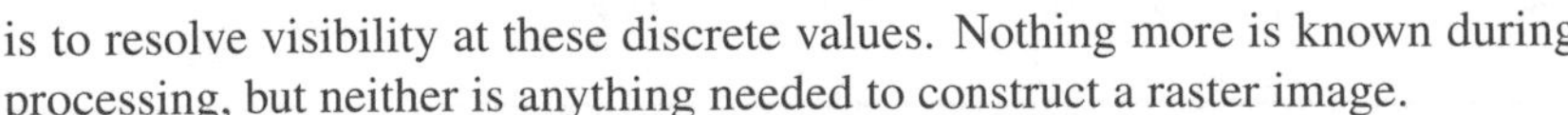

is to resolve visibility at these discrete values. Nothing more is known during processing, but neither is anything needed to construct a raster image.

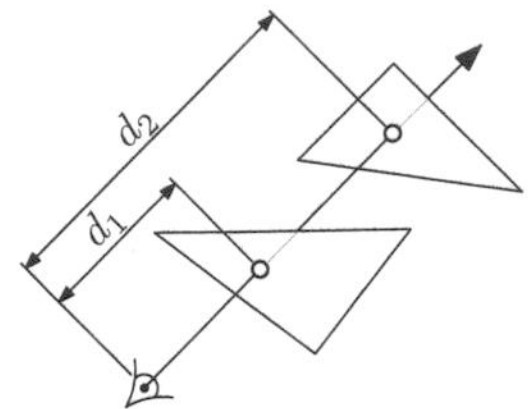

To resolve visibility, one iterates over the polygons and rasterizes each polygon while determining its depth (z) at each pixel. If the depth is smaller (closer to the viewer) than the z value already stored at that pixel, both the depth and the color at that pixel are updated. If the depth is larger (farther than the viewer), neither is updated. To rasterize a polygon, the intersection of its edges with scanline is determined. Each pair of consecutive intersections define a set of adjacent pixels that can be handled by modifying the polygon's z value using the polygon's slope in the plane of its intersection with the scanline.

22.3 The Active Edge List

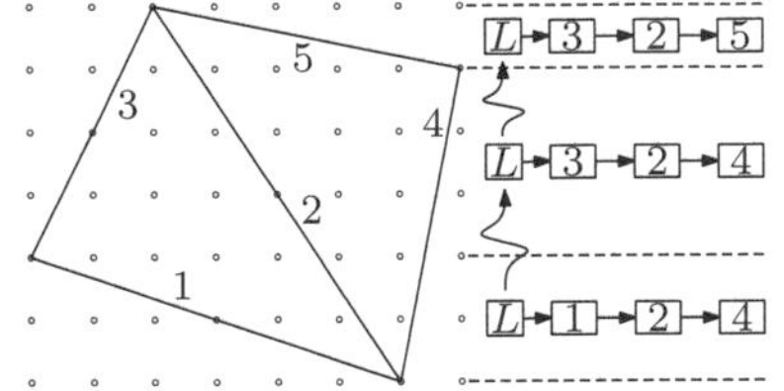

The *active edge list* exploits the following observation. If the consecutive sets of pixels that need to be written are known at one scanline, subsequent sets can be modified by adding a constant factor for each bounding edge. The factor is simply the product of the slope of the line by the distance between the scanlines (perhaps unity).

In the adjacent figure the scanlines can be partitioned into three sets. For each consecutive scanlines in one set, the intersection of the polygons' edges remains identical, even if the locations of the beginning and end of intersections vary. Three active edge lists need to be created for the three distinct sets, but the three lists are so similar that it is only necessary to create the first list. The subsequent ones can be determined by updating from a prior list and by consulting the edge table.

22.4 The Edge Table

The active edge list is updated incrementally at each scanline with the new location of the intersections. Occasionally, it is also necessary to insert or delete list items when the endpoint of an edge is encountered. If the endpoints are sorted by y value into a list, the changes can be made in the order they arise in the list [83].

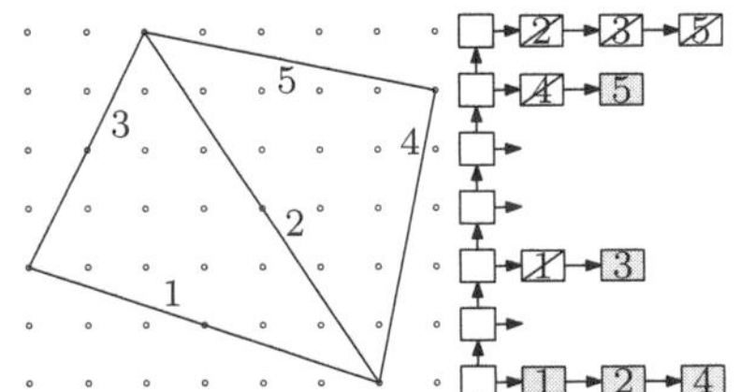

Because the change in the active edge list triggered by the edge table is minuscule (an edge is inserted, an edge is deleted, or two edges swap order), it is sufficient to resort an already nearly sorted list using bin sorting. Also, because a raster image has only finitely many scanlines, a table with as many entries as there are scanlines is created and the endpoints of the edges are inserted into the table at the entry corresponding to their scanline [113]. Each entry is also flagged to indicate whether an edge will be inserted or deleted when that scanline is processed. The resulting *edge table* is consulted at each scanline to determine if any updates need to be made to the active edge list. The use of an edge table and an active edge list makes it possible to reuse the set of depth values for each scanline.

22.5 The Graphics Pipeline

We consider the graphics pipeline one last time and discuss the appropriate stages for visibility and for shading.

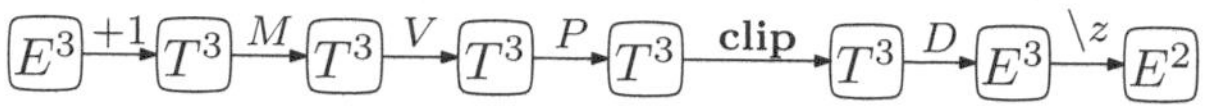

$$E^3 \xrightarrow{+1} T^3 \xrightarrow{M} T^3 \xrightarrow{V} T^3 \xrightarrow{P} T^3 \xrightarrow{\textbf{clip}} T^3 \xrightarrow{D} E^3 \xrightarrow{\backslash z} E^2$$

The natural stage to perform raster visibility is in E^3 following perspective divide. At that point perspective has been performed, all projectors are parallel, the polygons have been clipped, and the polygons have been mapped back from T^3 to E^3. We also assume that the value of the depth z will be discarded at the same stage.

By specifying that the geometric data lies in E^3 or in T^3, we implicitly meant that they are in $E^3(\mathbb{R})$ or in $T^3(\mathbb{R})$: The geometric objects are parameterized by a number type capturing the reals in the corresponding space. To distinguish the final stage, in which a set of pixels with integer coordinates in the Euclidean plane is considered, we explicitly identify that stage as $E^2(\mathbb{N})$.

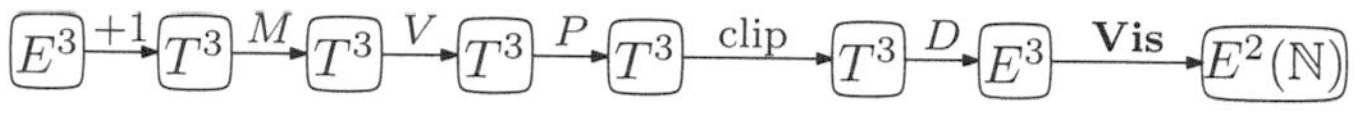

$$E^3 \xrightarrow{+1} T^3 \xrightarrow{M} T^3 \xrightarrow{V} T^3 \xrightarrow{P} T^3 \xrightarrow{\text{clip}} T^3 \xrightarrow{D} E^3 \xrightarrow{\textbf{Vis}} E^2(\mathbb{N})$$

Shading, on the other hand, can only be performed under accurate angles. The first transformation in the pipeline that destroys angles is the perspective transformation P. That is also why shading must be performed before P is applied:

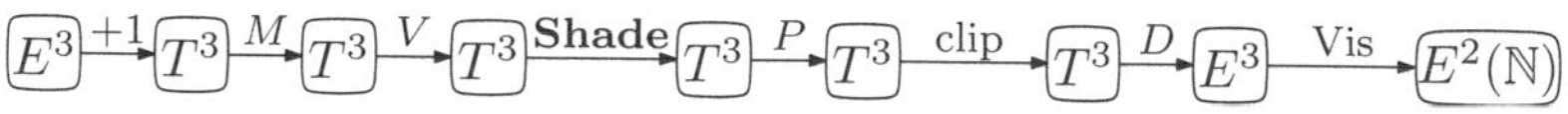

$$E^3 \xrightarrow{+1} T^3 \xrightarrow{M} T^3 \xrightarrow{V} T^3 \xrightarrow{\textbf{Shade}} T^3 \xrightarrow{P} T^3 \xrightarrow{\text{clip}} T^3 \xrightarrow{D} E^3 \xrightarrow{\text{Vis}} E^2(\mathbb{N})$$

This explains that even though shading is applied only to attributes of the geometry rather than to the geometry itself (much as we will see in § 29.2), the stage at which shading is computed must be specified within the pipeline.

The discussions on the Z-buffer and scanline visibility conclude the sequence of operations that need to be performed to produce a raster image (§ 19.2) from a set of polygons. In the most common scenario these polygons describe a solid (Chapters 26 and 27) in Euclidean space (Chapter 3). The polygons are mapped to projective space and then projected on an image plane (Chapter 11). After clipping (Chapter 15), perspective divide (Chapter 12) is applied and the polygons are back in Euclidean space. Their orthogonal projection is the input to raster visibility.

22.6 Exercises

22.1 a. Implement a system that generates a random n-sided convex polygon.

b. Implement a point-inside-a-convex-polygon system that searches in the sorted list of the polygon's vertices and reports whether the query point is inside the polygon.

22.2 a. Implement a system that generates a random n-sided star-shaped polygon.

b. Implement a system that computes the kernel of a star-shaped polygon and, if the interior of the kernel is not empty, reports a point inside the kernel (say the centroid).

c. Implement a point-inside-a-star-shaped-polygon system that searches in the sorted list of the polygon's vertices and reports whether the query point is inside the polygon.

22.3 Implement a system that rasterizes a convex polygon using an edge table and an active edge list.

22.4 Compare your solution for Exercises 22.1 and 22.3. Is there a value of n (the number of sides of the polygon) at which you believe binary searching will take over the edge table in actual time?

23 Ray Tracing

Light illuminates scenes by the collective impinging of a large number of photons that start at a light source. We perceive views of scenes by the arrival on our retinas of a subset of photons reflected from the scenes we observe. Ray tracing is an illumination model that makes it possible to simulate glossy and transparent surfaces by recursively constructing then casting rays at surface intersections [118]. The tracing of rays is performed in the direction opposite to that traversed by the photons.

23.1 Ray Casting

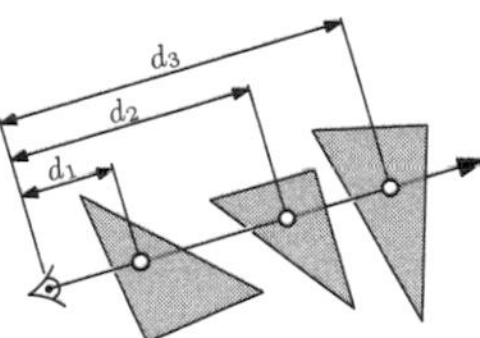

Casting a ray in (Euclidean) space refers to generating a single ray and intersecting it with the scene geometry. The nearest intersection determines the surface visible, which is then shaded (Chapter 21) to generate an image.

Spawning Rays from the Viewer

In the Z-buffer algorithm a raster image is generated by iterating over the polygons in a scene and determining the depth of each polygon at each pixel at which the polygon is rasterized. An alternative pixel-based approach is possible. A ray is generated for each pixel in the raster image and the first visible scene polygon is determined and shaded.

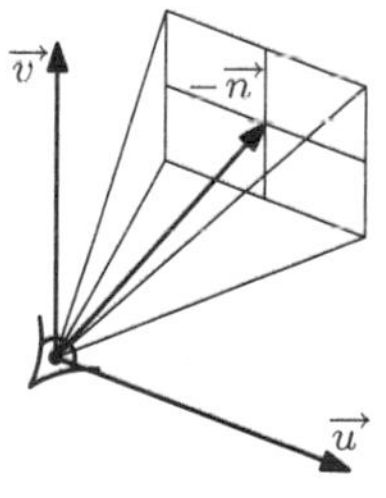

To find the set of rays that need to be cast to generate an image, the three vectors defining an orthogonal view transformation (§ 4.7) are determined from a view direction and an up vector. The view angle and the resolution of the image are then used to determine the increments on the image plane (which is parallel to $\vec{u}$ and $\vec{v}$).

The primitive operation performed once a ray is initialized will be to cast the ray by perhaps launching the ray using a function such as the following:

```
Color_3f ray_cast(const Scene& scene, const Ray_E3d& ray)
```

23.2 Ray-Polygon Intersection

If the scene is defined using a set of polygons, determining whether a ray intersects one polygon can be done by intersecting the ray with the plane carrying the polygon then determining whether the intersection point lies inside the polygon. Ray-polygon intersection may be used for both convex and concave

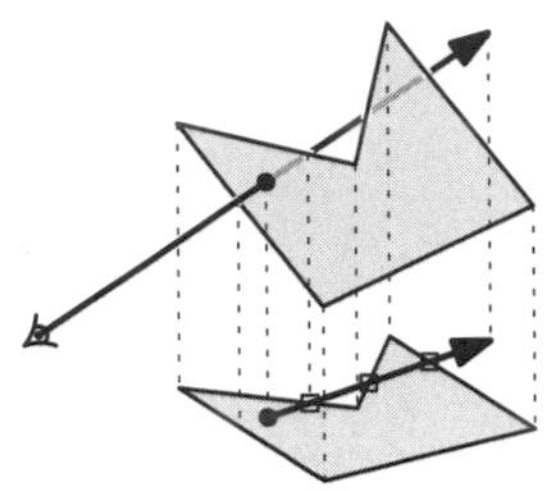

polygons (§ 20.3). Yet determining whether the intersection point lies inside the polygon cannot be performed in 3D space. The difficulty is numerical. If a ray is issued from the intersection point, all its intersections with the edges in 3D will most likely not be detected. It is thus necessary to project the polygon and the ray on an arbitrary plane and to perform the point-inside-a-polygon test in the 2D projection. It is easiest to perform the projection on one of the coordinate planes, since the projection is then simply a matter of discarding one coordinate, but it is crucial to choose that plane such that its normal is the dominant axis (§ 3.7) to the plane carrying the polygon. At the extreme, the polygon is parallel to, say, the xz-plane ($\pm y$-dominant) and the projection must be done on that plane.

23.3 Reflection

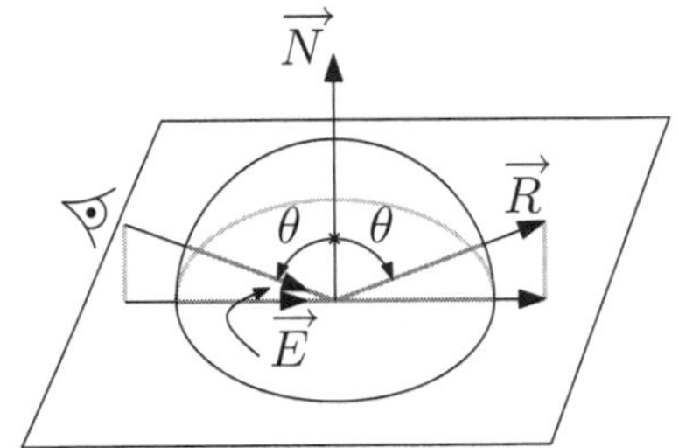

In addition to specular shading (§ 21.2), one can determine the surface that would be seen if the ray incoming from the eye were to be reflected on the surface. The shading of that surface is used to calculate the shading in the direction of the eye. The ray $\overrightarrow{R}$ is the reflection of the eye ray $\overrightarrow{E}$ at the point of intersection of $\overrightarrow{E}$ with the surface. If the data type Direction_E3 (§ 3.2) is used to capture the normalized vector of the ray, then the vector part of the reflected ray can be determined simply by computing

$$\overrightarrow{R} = \overrightarrow{E} + 2\overrightarrow{N}.$$

The direction in which rays are traced (that of $\overrightarrow{E}$ and $\overrightarrow{R}$) is called in computer graphics, by convention, the *forward* direction of the ray, even though that is the opposite of the direction of the simulated photon.

23.4 Refraction

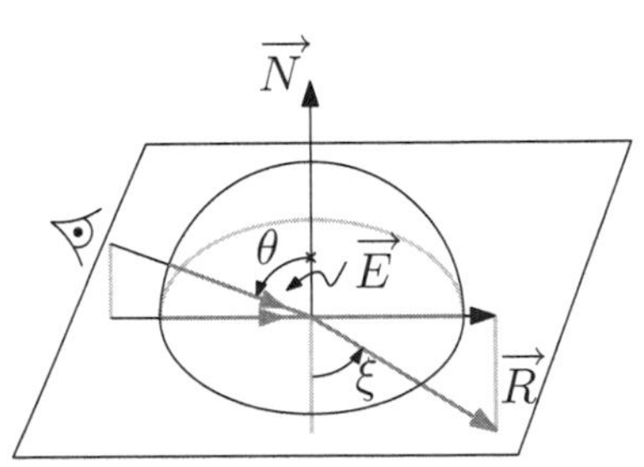

Refraction is another visual effect that can easily be simulated by ray tracing. When an incoming ray $\overrightarrow{E}$ traverses from one medium to another, its angle of incidence θ and its angle of refraction ξ are related by Snell's law. If the speed of light is v_θ in the source medium and v_ξ in the target medium, the angle of refraction can be determined by

$$\frac{\sin(\theta)}{\sin(\xi)} = \frac{v_\theta}{v_\xi}.$$

23.5 Recursion

The distinction between ray casting and ray tracing is the following. *Ray casting* is the process of finding the intersection of a single ray with a scene and returning either the nearest surface intersected or the lighting at that surface.

Ray tracing is the illumination model resulting from casting successive generations of rays starting from one ray spawned at one sample point (pixel) in the image.

Reflection and refraction are simulated by casting additional rays. Since these two rays would be implemented using two function invocations while the same ray-casting function is executing, ray tracing is compellingly implemented using recursion.

23.6 Exercises

23.1 Implement a system for generating ray-traced images of a set of spheres.

23.2 Implement a system for generating ray-traced images of a set of triangles.

23.3 Plücker coordinates were discussed for lines in P^3 in §12.4, but an identical formulation is suitable for representing lines in oriented projective space T^3. One only has to ensure that the order of the two points defining a line is maintained [104]. The significant additional benefit is that one line then partitions the set of *lines* in T^3 into three sets: those that lie on one side or the other and those that intersect it. Write a function that performs this classification.

23.4 Extend Exercise 23.3 by writing a function that determines whether a ray intersects a triangle in space using Plücker line coordinates.

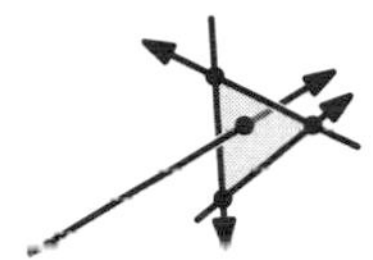

23.5 Repeat Exercise 23.2, but use your solution for Exercise 23.4 to determine quickly whether a line intersects a triangle. If you isolate (and ignore) the time needed for shading, how do the two implementations compare in running time? The coordinates for the lines bounding the polygons must naturally be cached.

23.6 Repeat Exercise 23.1, but add reflection, refraction, and shadows.

23.7 Repeat Exercise 23.2, but add reflection, refraction, and shadows.

Part V

Tree and Graph Drawing

24 Tree Drawing

Humans frequently visualize trees while discussing algorithms. Such visualization aids the comprehension of algorithms on trees and facilitates debugging implementations of tree algorithms. This chapter discusses two algorithms for drawing trees.

It is useful to make a distinction between drawing a tree and finding an embedding. Finding an embedding refers only to associating (2D Euclidean) points to nodes, whereas drawing refers in particular to the process of generating the drawing, though finding the embedding may also be understood to be part of the drawing process. *Rank* embedding is discussed in § 24.2 and *Tidier* embedding in § 24.3. Section 24.4 discusses the drawing proper.

24.1 Binary Search Trees

The embedding algorithms are parameterized by a type Node that captures a node in a binary search tree. For simplicity we implement binary search trees using a single class, although in practice distinguishing the notion of a tree by implementing a second class may be desirable. A tree class would act as a wrapper that delegates client requests to the root of the tree. Using a single Node class means that an empty tree would be represented directly by a NULL pointer, or no object at all, rather than indirectly to a NULL root through a pointer to a tree object.

A *binary* tree is one in which each node has zero, one, or two children. Each node in a tree will usually store some information, including a *key*. Assuming keys are distinct, two keys a and b can be ordered such that exactly one of $a < b$ and $b < a$ holds. A binary *search* tree is one where the key of a left child is smaller than the key of its parent and the key of a right child is greater than that of its parent.

The two terms "left" child and "right" child are entrenched in computing, but they lack generality since they assume that the keys can be mapped to a point on the Euclidean line. As we encounter binary trees for arbitrary geometries in Chapters 28 and 29, and since we would like to be able to use the same routines under any geometry, we assume instead that a node has a *negative* child and a *positive* child. Considering here the specific and simple example of the geometry of E^1 (§ 2.5), inserting a node in a binary search tree consists of making comparisons at a path from the root of the tree to a leaf node. At each node R with key R_k, the key N_k of the inserted node N is compared with R_k. If $N_k - R_k < 0$, the node N is inserted recursively at the negative child, or a

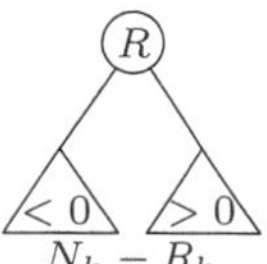

negative child is created if none exists—and likewise for the positive branch if $N_k - R_k > 0$.

If a node is implemented using the following three data items,

```
template<typename T>
class Node
{
  T i;
  Node *negative_child;
  Node *positive_child;
```

then the following code inserts an item of type T:

```
  void insert(const T& item)
  {
  bool difference_is_negative = (item < i);
  if( difference_is_negative )
    if(negative_child)
    negative_child->insert(item);
    else
    negative_child = new Node<T>(item);
  else
    if(positive_child)
    positive_child->insert(item);
    else
    positive_child = new Node<T>(item);
  }
```

Any class with pointer data members usually requires a nondefault copy constructor, an assignment operator, and a destructor. An example of the first for the Node class is shown below.

```
  Node::Node(const Node<T>& mynode)
  {
  i = mynode.i;
  negative_child = mynode.negative_child ?
    new Node<T>( mynode->negative_child ) : NULL;
  positive_child = mynode.positive_child ?
    new Node<T>( mynode->positive_child ) : NULL;
  }
```

An associative container, **std**::map, is used to return the embedding data—a point in E^2 is associated with each node in the tree—before an embedding function is called. The client code follows.

```
  typedef Node<int> Node_type;
  typedef Point_E2<float> Point_E2f;
  typedef std::map<const Node_type*, Point_E2f > My_map;

  int numbers[] = {5,3,4,6,7};
  Node_type * N = new Node_type(numbers[0]);
  int i = 0;
  while( ++i!=5 )
    N->insert(numbers[i]);

  My_map M = embed_tree_by_rank_E2<Node_type, float >(N);
```

24.2 Tree Drawing by Rank

It is reasonable to expect that the nodes of a search tree will be drawn such that all nodes at the same depth (the number of edges from the root) appear at the same y-coordinate, with the root of the tree at the largest y value.

If the *rank* of a node is its index in an in-order traversal of the tree, then one could use the rank of a node for its x-coordinate. In addition to being simple, such a *rank drawing* [60] of trees ensures that the sorted order of the nodes can be easily seen in the drawing.

The function embed_tree_by_rank_E2 takes a pointer to the root of a tree and returns a map from node pointers to Point_E2 objects. The function is parameterized by two types. The Node_type and the type used as parameter to the point class (itself generic).

```
template<typename Node_type, typename T>
std::map<const Node_type*, Point_E2<T> >
embed_tree_by_rank_E2( const Node_type* mytree )
{
  std::map<const Node_type*, Point_E2<T> > M;
  if( mytree )
  {
  int current_rank = 0;
  assign_rank_and_depth<Node_type, T>(mytree, current_rank, 0, M);
  }
  return M;
}
```

Simply modifying the implementation of Node to include a Point_E2 object would be a poor design choice since it strongly binds the embedding routine with one given node class and disallows reuse. Using an associative container temporarily uses additional storage but decouples the drawing routine from the tree.

Another parameterized function will assign the rank and the depth to nodes in the tree. It will also construct a point object and set the association. At each recursive call, the depth is decremented by a constant. As it is customary to draw trees with the root at the top, the depth parameter is decremented. The rank variable, which is passed by reference and so the same variable is shared throughout the tree traversal, is incremented by one at each node. The recursive routine is an in-order traversal; at any subtree rooted at a node the negative (left) subtree is visited first, the subtree root next, and the positive (right) subtree is visited last. In-order traversal ensures that the node with the smallest key is assigned an x-coordinate of 0 with the remaining nodes assigned x-coordinates in their respective rank.

```
template<typename Node, typename T>
void
assign_rank_and_depth(
        const Node* mynode,
        int& rank, int depth,
        std::map<const Node*, Point_E2<T> > & M )
```

```
{
  Node * N = mynode->get_negative_child();
  Node * P = mynode->get_positive_child();

  if( N ) assign_rank_and_depth( N, rank, depth-2, M );
  M[mynode] = Point_E2<T>(rank, depth);
  rank += 1;
  if( P ) assign_rank_and_depth( P, rank, depth-2, M );
}
```

The only two functions that are invoked on the Node class are

```
get_negative_child();
get_positive_child();
```

and these constitute the only coupling required between the embedding routine and the tree. Since the client of the embedding routine should not be required to inspect the source to see that these two functions are needed, the documentation of embed_tree_by_rank_E2 should indicate the necessity of the presence of these two functions. Notice also that the actual node class must be known before the compiler can verify that the two functions are indeed present; there is currently no way in C++ to declare such substitutability requirements explicitly.

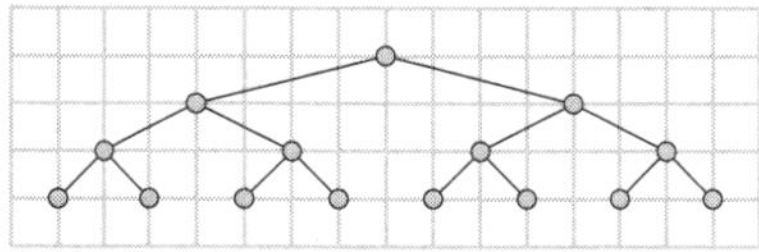

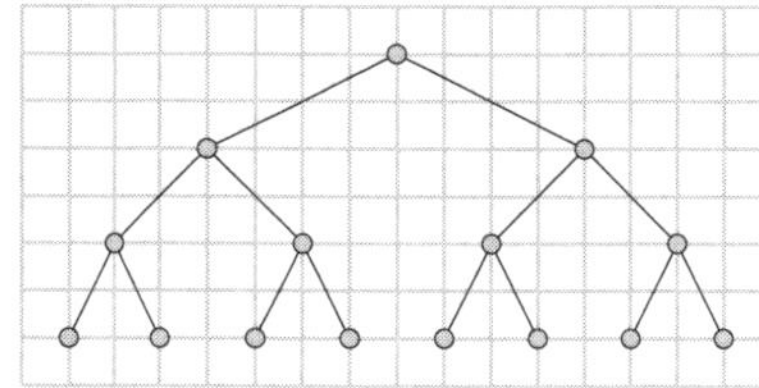

The adjacent figures illustrate two drawings of a complete binary tree. In one the y values are decremented at each level by twice the decrement in the other (the unit is the rank used for the x-coordinate). Because the bottom figure is more appealing, the expression $\mathrm{depth}-2$ appears in the code above. One could, of course, subsequently apply a transformation, but it is easy to avoid bothering.

Several desirable features of tree drawing are satisfied by a rank-based algorithm.

1. The tree drawing is *layered*, that is, nodes at the same depth appear at the same level (y value) in the drawing.

2. The drawing has no edge crossings.

3. The order property of the tree appears in the drawing: Not only are nodes to the left (right) of an ancestor in the negative (positive) side of that ancestor, but nodes also appear strictly in their sorted order from left to right.

But perhaps this last constraint is too strict. It is relaxed in the following *tidier* tree drawing, while additional constraints are introduced.

24.3 Tidier Tree Drawing

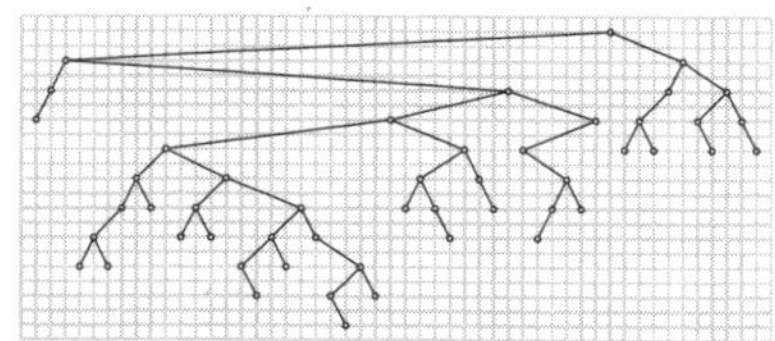

A crucial piece of information is duplicated in rank-based tree drawings: The order of nodes is evident from the drawing and it is redundant to duplicate that information in the x values. Insisting on matching the order with x values often results in unnecessarily long edges.

Tidier drawing replaces the third feature of rank-based drawing with the following two criteria:

1. Two identical subtrees are drawn in an identical manner regardless of where they appear in the tree.

2. The drawing is symmetric; if any subtree is replaced by its logical opposite, the drawing of that subtree will simply be the reflected image of the original drawing. This means that nodes with two children are equidistant from the two children.

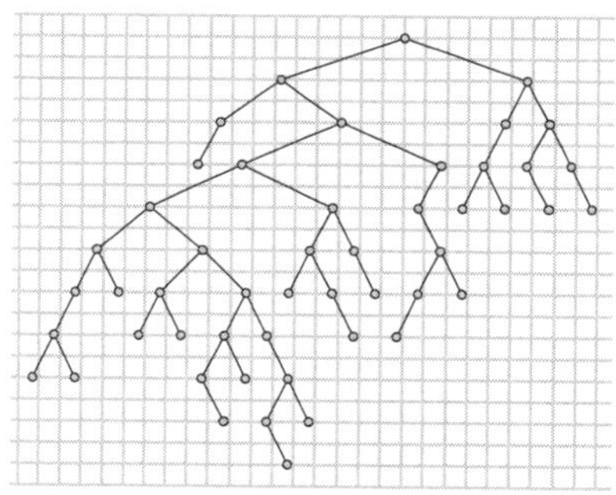

These criteria are satisfied by the following tree-drawing algorithm, named tidier [84] in contrast to the earlier *tidy* tree drawing [117]. Suppose that the right contour of the left subtree and the left contour of the right subtree of a given node have been recursively determined. The two subtrees can be moved as close as possible to each other while maintaining a minimum separation between each two nodes that lie at the same depth in the two subtrees.

Once the minimum separation is established, the root of a subtree is placed halfway in the x direction between the roots of its two subtrees. Because of this symmetry, a tidier drawing uses only one offset variable that determines the x separation between a node and either of its two children. The parent of one or two leaf nodes will have a unit offset.

Should the final coordinates be computed in the same tree traversal that determines separation? Because the processing of a left subtree is always complete before processing the corresponding right subtree has begun, determining the coordinates is possible during the separation traversal (given one thread of computation). But merely touching every node of a right subtree would lead to a quadratic-time algorithm—as can be verified by counting the steps for a perfect binary tree (a complete binary tree with $2^n - 1$ nodes).

For this reason a linear-time implementation requires two traversals, one post-order (or bottom-up) and the second pre-order (or top-down) [117, 84]. In the first traversal the offset at each node is determined and in the second the x-coordinates are calculated as the sum of the (signed) offsets of a node's ancestors. The root of the tree is assigned $x = 0$. Nodes in the left subtree may appear to the right of the root, and vice versa.

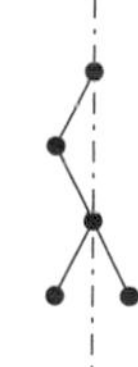

There is no need to store explicitly either the contour nodes themselves or their position relative to a subtree's root. Since the offsets of a node's descendants are known when the offset at that node is being determined, the x-position of a contour node can be determined relative to the root of the subtree. At each level that is common between the left and the right subtrees, the two offsets are calculated and the offset at the root increased to accommodate them if needed. Most of the time, child nodes are also the contour nodes. When joining two subtrees that differ in height, however, an additional contour link needs to be determined.

Define the leftmost extreme node of a subtree as the node at the lowest level in that subtree that appears leftmost in a drawing, and likewise for the rightmost node—but note that a single node may be both the leftmost and the rightmost extreme node.

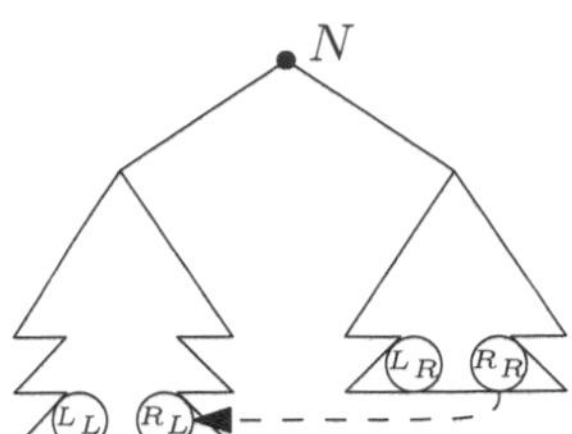

The leftmost and the rightmost extreme nodes are calculated as a side effect of the post-order traversal function. At the conclusion of the post-order traversal, the two extreme nodes of the left subtree (L_L and R_L) and the two extreme nodes of the right subtree (L_R and R_R) will be known. So it is sufficient to return the two required extreme nodes among these four nodes. If one subtree is taller, its two extreme nodes are those returned. If the two subtrees have the same height, the leftmost of the left subtree and the rightmost of the right subtree are returned.

The post-order function has a second side effect; it also determines the contour links (shown dotted). These are stored in an associative container Contour that saves contour links for only those nodes that need one. The offset of such contour links also needs to be saved and the embedding routine may store it either along with the otherwise unused offset of a leaf node (R_R in the figure) or as an additional data attribute along with contour links. The latter approach is slightly wasteful but results in a cleaner system.

The readability of the code can be further improved using the following two helper functions that isolate determining the next nodes on the left and right contours [22].

```
next_left(v)
if v has a child then
    return leftmost child of v
else
    return contour_link(v)
end if
```

```
next_right(v)
if v has a child then
    return rightmost child of v
else
    return contour_link(v)
end if
```

Since contour links are needed only at leaf nodes (which have two NULL pointers), Reingold and Tilford overload child pointers of a leaf as the left and right contour links. They add a Boolean flag to signal that the meaning of these pointers has changed. Doing so saves a little storage but otherwise unnecessarily complicates the implementation. Modifying nodes in the tree is also precluded if we insist that the tree be passed as a **const** parameter. Observe also that updating, however temporarily, the structure of the tree precludes reading the tree from multiple processing threads.

To see that the algorithm is linear in the number of nodes, consider that there is a "zipper" shuttering the two subtrees of *each* node in the tree. Were this shuttering to proceed to the higher of the two subtrees, the algorithm would run in quadratic time—as can be seen by considering a degenerate tree consisting of a linked list. But the shuttering proceeds only to the smaller of the two subtrees. The dominant step of the algorithm is one such zippering unit. Each node in the tree may participate as the (say) left node in the shuttering, but then another node would subsequently become the leftmost node on that level, precluding that a node be used twice.

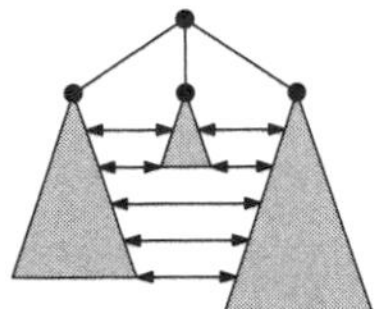

The algorithm can be generalized to ordered trees with more than two children [32], but the modification is not as simple as may seem at first sight because two subtrees whose roots are not adjacent may need to be separated [22].

24.4 Tree Drawing

To draw a tree, we first determine the bounding box of its embedding. Since any order of traversal will do, we use whichever is offered underneath by the associative container.

```
template<typename T>
Bbox_E2<T>
get_Bbox( const My_map & M )
{
  Bbox_E2<T> bbox;
  My_map::const_iterator mi = M.begin();
  while( mi != M.end() )
  bbox += mi++−>second;
  return bbox;
}
```

We could draw the nodes (as circles) and the edges of the tree in one traversal, but that would require determining the points of intersection of lines and circles. It is easier to rely on overlaying objects (§ A.4) and draw edges first and nodes second in two traversals. The recursive function shown below for drawing tree edges fetches the embedding from the associative container M.

```
template<typename T>
void
draw_tree_edges(
    Postscript<T> & PS,
    const Node_type * mynode,
    My_map & M )
{
  if(mynode−>get_negative_child()) {
  Node_type * N = mynode−>get_negative_child();
  PS.draw( Segment_E2<T>( M[mynode], M[N] ) );
  draw_tree_edges( PS, N, M );
  }
  if(mynode−>get_positive_child()) {
  Node_type * P = mynode−>get_positive_child();
  PS.draw( Segment_E2<T>( M[mynode], M[P] ) );
  draw_tree_edges( PS, P, M );
  }
}
```

Figure 24.1 shows the rank-based drawing of a tree. The tree is generated by shuffling the integers 1–100 and inserting the resulting sequence into a binary search tree.

A tidier drawing of the same tree is shown in Figure 24.2.

24.5 Exercises

24.1 Implement the tidier drawing of binary trees described in § 24.3.

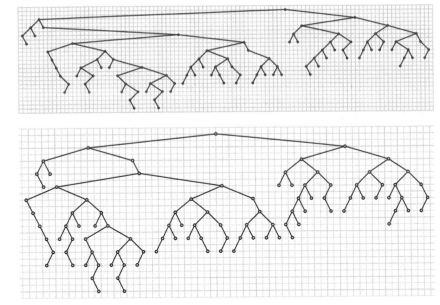

Figure 24.1
A rank drawing of a 100-node tree

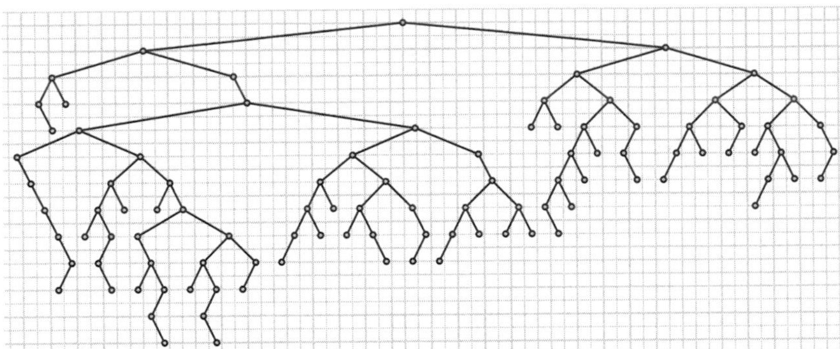

Figure 24.2
A tidier drawing of a 100-node tree

24.2 Even though the rank-based drawing of a tree shown in Figure 24.1 is considerably wider than the tidier drawing shown in Figure 24.2, it has the potentially significant advantage that all nodes are drawn at grid points—and therefore have integer coordinates. Yet it would appear that the tidier drawing could be somewhat salvaged: By merely translating by a constant fraction, a significant subset of the nodes could be made to have an integer x-coordinate. Modify your implementation of the tidier tree drawing so that some leaf node at the largest depth in the tree is translated to have an integer x. Deduce experimentally the percentage of nodes that would then have integer coordinates when the same translation is applied on all nodes.

24.3 Considering that the number of nodes at a given level in a tree increases exponentially, it would appear that a sphere is more suitable than the Euclidean plane for drawing trees. Develop a system that renders the root at the "north pole" and its children recursively on the adjacent lines of latitude. Your system must naturally be interactive since otherwise the tree would have to be again distorted when projected on the Euclidean plane.

24.4 Observe that the tidier drawing of a large tree has a skewed aspect ratio; the figure is too wide compared to its height. The clarity of the drawing decreases because edges near the root of the tree appear almost like horizontal edges. Devise and implement a heuristic to fix this problem. Consider here that edges that are nearly vertical are still acceptable.

25 Graph Drawing

Before discussing two methods for graph drawing in § 25.2 and § 25.3, we spend a moment debating the design of classes for graphs. The two main design constraints are those of genericity and type safety. Genericity implies that the set of graph classes we implement would be reusable in other applications with minimal effort. Type safety guarantees that the compiler has the ability to confirm that absurd type constructs are flagged as such. By necessity, the discussion in § 25.1 is only a sketch. The sheer number of natural questions one can ask about a graph makes the design of a library a topic for a book-length treatment. Four graph libraries have been designed and implemented as platforms suitable for building vast systems: the Stanford Graphbase [61], Sedgewick's [93], LEDA [67], and the Boost Graph Library [100]. If only a handful of graph functions will be needed in a system, then developing one's own graph toolkit is likely the more suitable approach. But systems that depend on a large number of graph functions should be built by choosing from these four libraries.

25.1 A Sparse Graph Implementation

To choose a design for a graph and associated classes, we decide first what the client-side code would look like. In the example, we assume that the nodes represent cities and that the edges represent flights between the cities.

Three classes for a graph, for a node, and for an edge are needed. Operations for inserting and deleting nodes and edges will be performed through a graph object. Thus, nodes and edges do not know the graph to which they belong, but a graph stores lists of its nodes and edges. The client-side code may look as follows:

```
void graph_setup()
{
  My_graph G;
  My_node *n1 = G.insert_node( City( "Cairo", 15e6 ) );
  My_node *n2 = G.insert_node( City( "Alexandria", 2e6) );

  My_edge *e1 = G.insert_edge( n1, n2, Flight( 225, 'f' ) );

  assert( G.are_adjacent(n1, n2) == e1 );
  assert( !G.are_adjacent(n2, n1) );
}
```

To make the implementation reusable, the node, the edge, and the graph classes are parameterized. The actual types are City and Flight.

```
struct City
{
  std::string city_name;
  int population;
  ...
};
```

```
struct Flight
{
  float distance;
  char type;
  ...
};
```

These two types are collected in a traits class [71, 112] (see § 18.2) that acts as a function from the generic types Node_type and Edge_type to the actual types City and Flight. The graph class is parameterized by My_graph_traits to instantiate an actual class. The actual node and edge classes can be instantiated either directly (as is done for the graph class) or, as below, by extracting the two types that are in any case needed in the graph class.

```
struct My_graph_traits
{
  typedef City Node_type;
  typedef Flight Edge_type;
};

typedef Graph<My_graph_traits> My_graph;
typedef My_graph::My_node  My_node;
typedef My_graph::My_edge  My_edge;
```

We assume that the graph is directed; hence, we distinguish between the source and the target of each edge. Each node object stores one list for the outgoing edges and another list for the incoming edges. This makes it possible to iterate over only those edges that have their source at the node or over only those that have their target at the node. A node also stores an instance of the template node parameter type, which is **typedef**ed for convenience.

```
template<typename Graph_traits> struct Node
{
  typedef typename Graph_traits::Node_type  Parameter_node_type;
  typedef typename Graph_traits::Edge_type  Parameter_edge_type;

  typedef Edge<Graph_traits>  My_edge;

  std::list<My_edge*> sourceOf; // node is source of these edges
  std::list<My_edge*> targetOf; // node is target of these edges

  Parameter_node_type nt;

  Node(const Parameter_node_type& _nt)
    : sourceOf(), targetOf(), nt(_nt) {}

  virtual ~Node() { }
  ...
};
```

To determine whether two given nodes are adjacent clients would normally send messages to the graph object. The latter in turn delegates the request to the first of the two nodes. It is sufficient for that node object to check the list of edges of which it is source.

```
template<typename Graph_traits> struct Node
{
  ...
  My_edge*
  is_adjacent(Node *n2) const
  {
    typedef typename std::list<My_edge*>::const_iterator Edge_iterator;
    for( Edge_iterator eit=sourceOf.begin(); eit!=sourceOf.end(); ++eit )
      if( (*eit)->target() == n2 ) // directed graph
        return *eit;
    return NULL;
  }
};
```

Three member variables are needed in an edge object, for the source and target nodes as well as for the template edge parameter type.

```
template<typename Graph_traits> struct Edge
{
  typedef typename Graph_traits::Node_type  Parameter_node_type;
  typedef typename Graph_traits::Edge_type  Parameter_edge_type;

  typedef Node<Graph_traits>  My_node;

  My_node * _source;
  My_node * _target;

  Parameter_edge_type et;

  Edge() {}
  Edge(My_node* s, My_node* t, const Parameter_edge_type& et)
    : _source(s), _target(t), et(et) {}

  virtual ~Edge() { }

  const Parameter_edge_type info() const { return et; }
  Parameter_edge_type& info() { return et; }

  My_node * source() const { return _source; }
  My_node * target() const { return _target; }
};
```

Each graph object stores lists of its associated nodes and edges.

```
template<typename Graph_traits> struct Graph
{
  typedef typename Graph_traits::Node_type  Parameter_node_type;
  typedef typename Graph_traits::Edge_type  Parameter_edge_type;
```

```
typedef Node<Graph_traits> My_node;
typedef Edge<Graph_traits> My_edge;

std::list<My_node*> nodes;
std::list<My_edge*> edges;

Graph() {}
virtual ~Graph() { clear(); }
```

The function for inserting a node expects an (optional) instance of the node parameter.

```
typedef Node<Graph_traits> My_node;
typedef Edge<Graph_traits> My_edge;

std::list<My_node*> nodes;
std::list<My_edge*> edges;

Graph() {}
virtual ~Graph() { clear(); }

virtual My_node* insert_node(
                    const Parameter_node_type& nt =
                    Parameter_node_type())
{
  My_node* mynode = new My_node( nt );
  nodes.push_back(mynode);
  return mynode;
}
```

Likewise, the function for inserting an edge expects the edge parameter as well as the source and target nodes.

```
virtual My_edge* insert_edge(
                    My_node* source, My_node* target,
                    const Parameter_edge_type& et =
                    Parameter_edge_type() )
{
  My_edge* newedge = new My_edge(source, target, et);

  source->sourceOf.push_back(newedge);
  target->targetOf.push_back(newedge);

  edges.push_back(newedge);
  return newedge;
}
```

25.2 Barycentric Graph Drawing

A classical and versatile method for graph embedding proceeds by iteration from an initial position. The user selects three or more nodes in the graph forming a cycle and maps each node to a point in the Euclidean plane E^2. The location of the remaining nodes is determined iteratively. The method is guaranteed to produce a planar drawing for planar graphs [111].

Non-pegged nodes of the graph gradually move to their final position. The terminating criterion is some threshold below which motion would in any case be imperceptible. At each iteration, each node is moved to the barycenter of its neighbors. All nodes are assumed to have unit weight. The barycenter was discussed in Chapter 13 as the centroid of three masses, but determining the center of mass can be readily generalized to more than three points. A new set of coordinates is calculated in a first iteration over the nodes before the new coordinates are copied in a second iteration while verifying the termination condition.

```
bool barycentric_iteration(My_graph& G, double threshold)
{
  const Point_E2d ORIGIN(0,0);
  std::map<My_node*, Point_E2d> M;
  typedef std::list<My_node*>::const_iterator NIT;
  typedef std::list<My_edge*>::const_iterator EIT;
  for(NIT nit = G.nodes.begin(); nit != G.nodes.end(); ++nit)
  {
    if( (*nit)->info().is_free ) {
      Vector_E2d sum(0,0);
      int degree = 0;

      std::list<My_edge*> &Ls = (*nit)->sourceOf;
      for(EIT eit = Ls.begin(); eit != Ls.end(); ++eit, ++degree)
        sum = sum + ((*eit)->target()->info().coords - ORIGIN);

      std::list<My_edge*> &Lt = (*nit)->targetOf;
      for(EIT eit = Lt.begin(); eit != Lt.end(); ++eit, ++degree)
        sum = sum + ((*eit)->source()->info().coords - ORIGIN);

      M[*nit] = Point_E2d(ORIGIN + (sum/double(degree)));
    }
  }

  bool still_moving = false;
  for(NIT nit = G.nodes.begin(); nit != G.nodes.end(); ++nit)
  {
    if((*nit)->info().is_free) {
      if( squared_distance((*nit)->info().coords, M[*nit]) > threshold )
        still_moving = true;

      (*nit)->info().set_coords( M[*nit] );
    }
  }
  return still_moving;
}
```

The iterations shown in Figure 25.1 suggest the two dominant features of barycentric graph drawing. If the graph is planar, the final drawing is guaranteed to be planar (edges meet only at nodes), and the ratio of edge distances in the final drawing can be arbitrarily large.

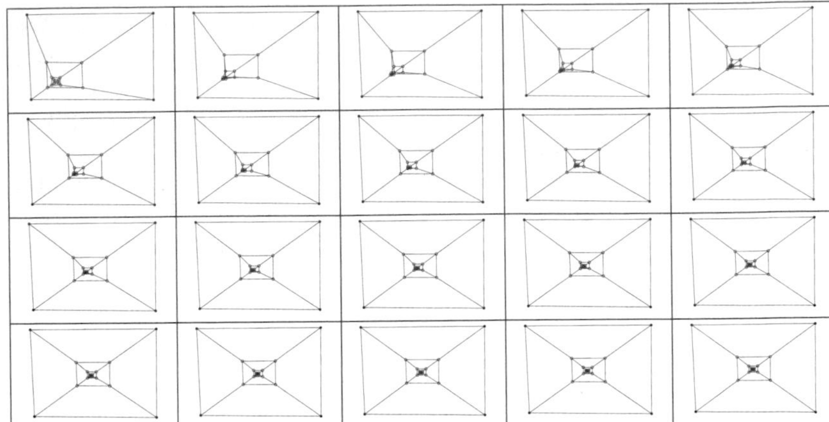

Figure 25.1
Steps in barycentric graph drawing

25.3 Force-Directed Graph Drawing

An alternative for embedding a graph in the plane is to assume that each node carries an electric charge (all nodes are either positive or negative) and that each edge is a spring. The idea is that the node scattering resulting from the electric charges will be counteracted by the spring forces to produce an acceptable drawing. Two nodes attract only if they are joined by an edge, but nodes exert a repulsive force on all other nodes.

The electric force, according to Coulomb's law, is proportional to the two charges and is inversely proportional to the square of the distance. If all nodes are assumed to have a unit charge, it is only necessary to choose empirically a constant of proportionality that determines the strength of the repulsion force. Since the application to graphs is in any case only heuristic, one can experiment with the proportionality and replace $1/r^2$ with either $1/r$ or $1/r^3$, where r is the distance separating two given nodes.

The spring force, according to Hook's law, is proportional to the (signed) difference between the rest and the actual lengths of the spring. Figure 25.2 shows the initial stages of the resulting graph animation. For this figure springs have the same length at rest and the electric force follows an inverse square distance law.

Convergence is slower than in the barycentric method. Figure 25.3 shows a sequence in the late stages of the drawing.

The exercises pursue a few related possibilities for graph drawing.

25.4 Exercises

25.1 Given the relative merits of the two graph-drawing algorithms discussed, it is natural to consider the following heuristic. Blend the two start-up

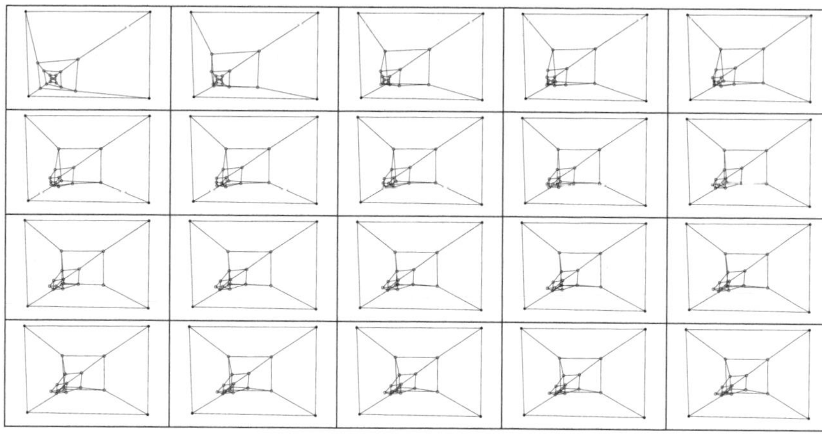

Figure 25.2
Initial steps in force-directed graph drawing

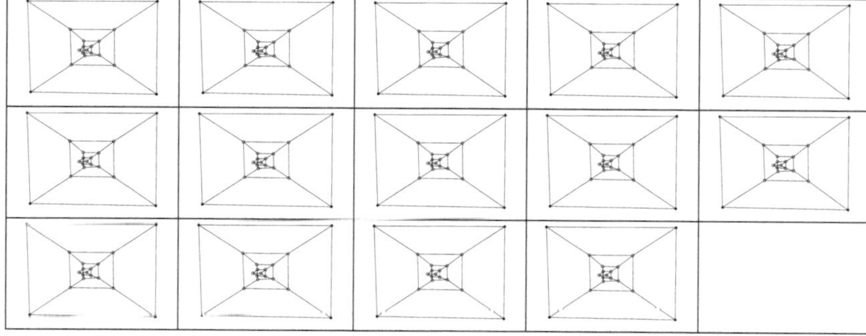

Figure 25.3
Late steps in force-directed graph drawing

programs labeled "barycentric" and "forces" so that the force-directed method is used for the first half of the free nodes and the barycentric method takes over for the second half. Report on your observations.

25.2 Solve Exercise 25.1, but add a slider that makes it possible for the user to choose the fraction of the free nodes after which the barycentric method takes over.

25.3 Force-directed graph drawing depends on the stiffness of springs and on the constant used for the electric repulsion. Modify the start-up code labeled "forces" to provide the user with sliders that make it possible to manipulate the two constants in some judicious range.

25.4 Consider the following heuristic for a variation on the force-directed method for drawing graphs. Instead of deciding to stop iterations for all nodes in the graph, modify the start-up program labeled "forces" and implement a force-directed program that will stop only those nodes that have moved in the prior iteration less than a given threshold. As the number of nodes that become fixed increases, your program will also increase the stiffness of the springs attached to edges and reduce the rest length of springs. Write a brief summary of your opinion on the merit of this approach.

25.5 Modify the start-up code labeled "forces" to make it possible for the user to manipulate the nodes of the graph as the graph is being rendered.

25.6 Modify the start-up code labeled "barycentric" to read a 3D solid (see Chapter 26), to construct a graph corresponding to the nodes and edges of the solid, and to draw the graph.

25.7 Apply force-directed rendering to nodes of a binary tree (Chapter 24) while constraining each node to lie on a line that corresponds to its level in the tree.

Part VI

Geometric and Solid Modeling

26 Boundary Representations

What is the most natural way to model solids? The answer depends on whom one asks. A wood worker or a machinist will suggest that one should start from elementary building blocks that are added or from which space is subtracted (or drilled)—the topic of Chapter 30. A tailor will suggest instead that the boundary of a solid provides an adequate abstraction. Such boundary representations, frequently abbreviated as *b-reps*, are discussed in this chapter. In the context of this text, the most versatile method is binary partitioning, which is discussed in Chapters 28 and 29.

26.1 Face Sets

If a solid in question is enveloped by a planar graph, the elements of the graph—vertices, edges, and faces—are suitable for modeling the solid. Generating shaded drawings of a solid only requires the ability to rasterize a given facet, but the notion of an edge need not be represented explicitly. This leads to the first useful boundary representation for a solid, *face sets*. It is simplest if all faces are constrained to be triangles. In that case it is sufficient to store the coordinates of the vertices for each triangle, but the number of vertices for each facet remains implicit.

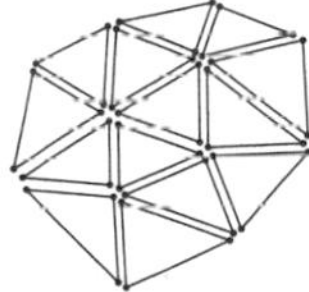

```
class Tri_face_set {
  vector<Point_E3d> P;
public:
  ...
};
```

The coordinates defining the first face are 0, 1, and 2. Those defining the second face are 3, 4, and 5, and so on. We can remove the restriction that the faces be triangles by storing the number of vertices (or edges) for each face. It is also convenient to store the (redundant) running total of the number of edges to avoid computing the total multiple times.

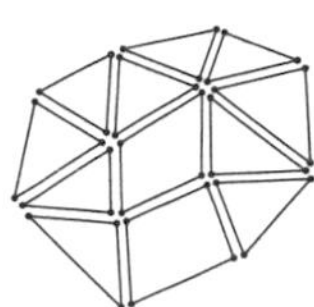

```
class Face_set {
  vector<int> num_sides;
  vector<int> point_index;
  vector<Point_E3d> P;
public:
  ...
};
```

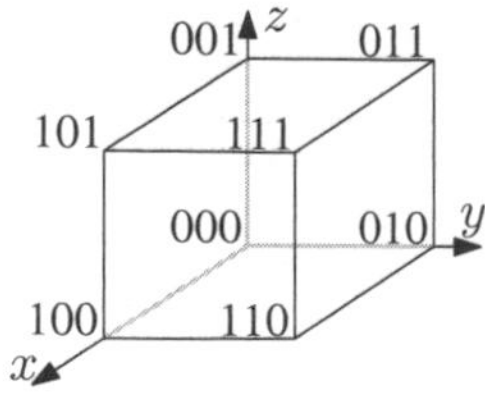

Either of the structures above would need to be completed by ensuring that at least normal vectors for either faces or vertices and material properties are stored. But a serious flaw makes the use of face sets an unattractive proposition in practice. The first few lines of a file that would be loaded into one of the structures above representing a cube will illustrate the difficulty:

```
6
0 4 8 12 16 20
0.0 0.0 1.0   1.0 0.0 1.0   1.0 1.0 1.0   0.0 1.0 1.0
1.0 0.0 1.0   1.0 0.0 0.0   1.0 1.0 0.0   1.0 1.0 1.0
...
```

One significant drawback in using the face set format is that it is frequently necessary to know whether two faces are adjacent. It is possible to conclude from the above data that the two vertices 1.0 0.0 1.0 and 1.0 1.0 1.0 are common between the top and the front faces, and, hence, that these two faces share an edge between the two vertices, but doing so requires relying on a geometric comparison to conclude topological, or connectivity, information, which is exactly what the discussion in Chapter 7 suggested should be avoided. Relying on coordinates for connectivity information is bound to lead to incorrect conclusions because it in turn relies on comparisons between floating point numbers.

26.2 Indexed Face Sets

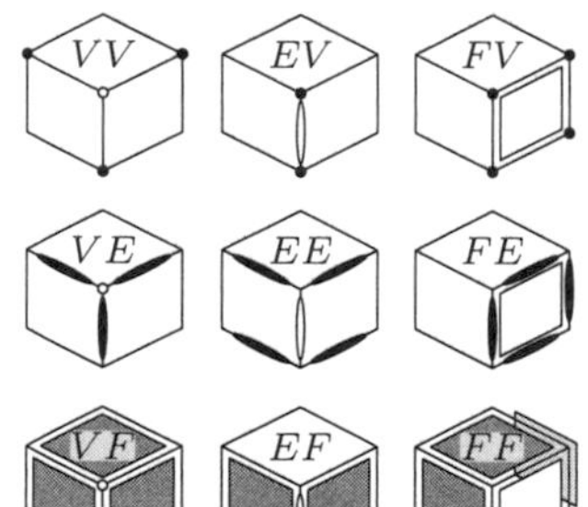

The adjacent figure, which is due to Baer, Eastman, and Henrion [2] and to Weiler [114], illustrates the nine potential (and not necessarily mutually exclusive) links between the sets of vertices, edges, and faces describing a solid. We may choose, for instance, for each vertex object to store pointers to the (ordered list of) adjacent vertices. The notions of an edge and of a face would be wholly lacking. Or we could choose to represent edges and vertices, and store links from each edge to its source and target vertices. The face set representation does not match the FV representation, for in face sets vertices are not captured topologically, but only geometrically. The objective from choosing one representation or another is to provide an easy mechanism for implementing iterators over the elements of the solid. The VE representation would, for instance, make it possible to iterate over the edges adjacent to each vertex.

The FV representation is sufficient for representing arbitrary solids unambiguously. This *indexed face set* representation stores face-to-vertex adjacencies in addition to vertex coordinates. Thus, an indexed face set representation makes it possible to test whether two polygons are adjacent without relying on geometric comparisons.

```
class Indexed_face_set1 {
private:
  vector<Point_E3d> V;
  vector<int> vertices;
  ...
};
```

The above code illustrates one method for capturing polygons with an arbitrary number of sides. Whereas previously the indices of the starting and ending vertices were saved, we now use a special index value, such as -1, to separate the indices of vertices on different polygons.

The normal vectors to the polygons can be computed from the coordinates of the vertices defining each polygon, but recomputing the normals from vertex coordinates is unattractive. On the one hand, it would be wasteful to determine the normal vectors repeatedly, especially since they are frequently needed. Precision is another reason. Normal vectors can be determined more accurately if they are directly generated from the modeling application. In particular, if a curved surface is modeled, using first-generation normal vectors is preferable over recomputing them as a second generation from the first-generation vertex coordinates.

For these reasons most implementations are likely to cache the polygon normal vectors. This requires introducing the notion of a polygon explicitly in an IFS data structure. Doing so also makes it possible to store information such as material properties.

```
class Polygon {
  friend class Indexed_face_set2;
  vector<int> vertex_indices;
  Direction_E3d normal;
public:
  ...
};
```

```
class Indexed_face_set2 {
private:
  vector<Point_E3d> V;
  vector<Polygon> polygons;
  ...
};
```

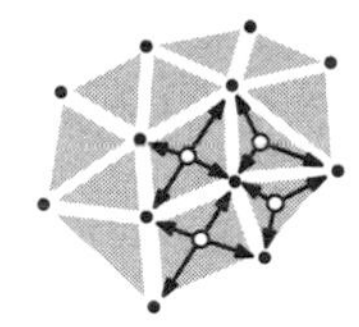

The Object File Format (OFF) is a popular format for describing solids using an indexed face set. An OFF file is identified by the appearance of its acronym on the first line, followed by the number of vertices V, of faces F, and of edges E, in that order. The data consist of V lines containing the vertex coordinates followed by F lines containing the cardinality of each face and the indices of its vertices. Historically, OFF files also included information about the edges, hence the presence of their count in the header, but at this time it appears that no edge data exist in the currently used OFF files. Yet, perhaps to accommodate potential ancient readers, the header is kept intact.

```
OFF
4 4 6

 1.0  1.0  1.0
-1.0  1.0 -1.0
-1.0 -1.0  1.0
 1.0 -1.0 -1.0

3 0 1 2
3 1 0 3
3 0 2 3
3 2 1 3
```

In the file above four sets of vertex coordinates and four sets of face data describe the solid. This solid, a regular tetrahedron, is discussed next.

26.3 Platonic Solids

A polygon whose sides are equal and whose angles are also equal is termed a *regular polygon*. A solid whose boundary is composed of identical regular polygons and whose vertices have identical adjacencies is termed a *regular polyhedron* or a *Platonic solid* [25].

The Regular Tetrahedron

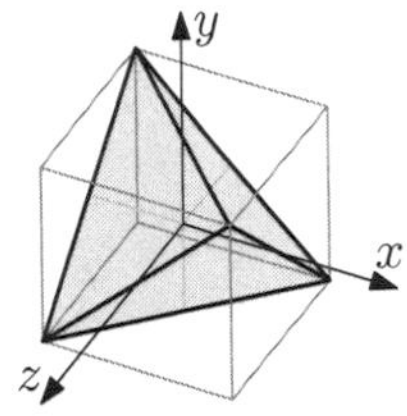

Since the diagonals of the faces of a cube are equal, a regular tetrahedron can evidently be inscribed in a cube. The four points $(1, 1, 1)$, $(1, -1, -1)$, $(-1, 1, -1)$, and $(-1, -1, 1)$ can then be used to model a regular tetrahedron. These coordinates are the simplest to devise, but the most convenient orientation to use a tetrahedron as a solid model is perhaps for one of its faces to be parallel to one of the coordinate planes. This may be effected by, for instance, applying a rotation about the axis $(-1, 0, 1)$ by an angle equal to the one between (§ 1.6) the two vectors $(0, 1, 0)$ and $(-1, 1, -1)$.

It is interesting, if only as a diversion, to discuss the set of a tetrahedron's *axes of symmetry*. What are the axes and angles or, alternatively, quaternions (Chapter 10) about which the present tetrahedron can be rotated such that its vertices coincide again with the original orientation? It is easy to see that a rotation of $\pm 2\pi/3$ about the axis from the origin to any of the tetrahedron's four vertices is an axis of symmetry. Likewise, a rotation of π about any of the three major axes will return this tetrahedron to its initial position. Exercise 26.4 pursues this direction further.

The Cube and the Regular Octahedron

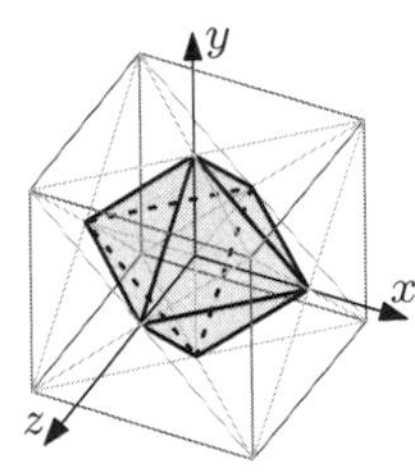

We already encountered one canonical cube with coordinates $(\pm 1, \pm 1, \pm 1)$ (and hence centered at the origin) in § 15.1. The same cube can be used to devise the coordinates of the regular octahedron. Whereas each face of a cube is defined by four vertices and each vertex is adjacent to three faces, each face of a regular octahedron is defined by three vertices and each vertex is adjacent to four faces. The vertices of a regular octahedron are described by $(\pm 1, 0, 0)$, $(0, \pm 1, 0)$, and $(0, 0, \pm 1)$.

The Regular Icosahedron and the Regular Dodecahedron

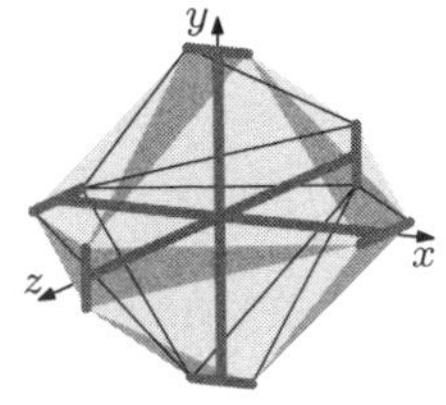

The regular icosahedron has 12 vertices and 20 faces. An elegant way can be devised for finding the coordinates of its vertices [5]. Consider first that an icosahedron is created by forking pairs of vertices at each vertex of a regular octahedron as shown in the figure. The vertices of the icosahedron are $(\pm 1, 0, \pm t)$, $(\pm t, \pm 1, 0)$, and $(0, \pm t, \pm 1)$.

When t is small, the larger eight triangles are already equilateral by construction. So now we seek a value for t that would make the (darker-shaded) triangles also equilateral. The length of one side is that between, say, $(t, 1, 0)$

and $(0, t, 1)$, or $\sqrt{2t^2 - 2t + 2}$. The length of the other is $2t$. Equating, we find

$$t^2 + t - 1 = 0.$$

We recall for a moment the golden ratio and ask for the value of t such that the two rectangles in the figure have the same aspect ratio. This yields

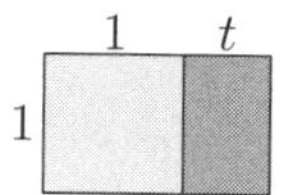

$$\frac{1+t}{1} = \frac{1}{t} \implies t^2 + t - 1 = 0 \implies t = \frac{1 \pm \sqrt{5}}{2}.$$

The positive value for t, approximately 0.61803, is the coordinate we need. The connection between the golden ratio and the icosahedron was first made, according to Coxeter [27], by Luca Pacioli in 1509 as the 12th property of the "divine proportion."

The regular octahedron can be constructed by connecting the centroids of the regular icosahedron, as will be clear in § 26.5.

Only Five Platonic Solids Exist

We can see why only five Platonic solids can exist by considering the Schläfli symbol, named after the Swiss mathematician Ludwig Schläfli (1814–1895). Suppose that the symbol $\{p\}$ identifies a regular p-sided polygon in the plane and that the symbol $\{p, q\}$ identifies a Platonic solid having p-sided faces and q-sided vertices. That is, p vertices (and edges) are adjacent to each face and q faces (and edges) are adjacent to each vertex [25].

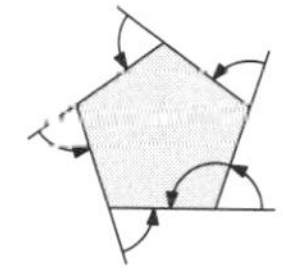

The interior angle at a vertex of $\{p\}$ is $(1 - 2/p)\pi$. To see that that is the case, consider the turn at each vertex needed by a turtle traversing the boundary. The sum of all turns must be one full turn, or 2π. Thus, each exterior angle of $\{p\}$ is $(2/p)\pi$ and each interior angle is $(1 - 2/p)\pi$.

Looking at a vertex of $\{p, q\}$, we notice that the sum of the interior angles of all adjacent faces must be less than 2π, for otherwise the solid has no curvature at that vertex. This sum is q times the interior angles, and so

$$q(1 - \frac{2}{p})\pi < 2\pi,$$

which leads to

$$\frac{2}{p} + \frac{2}{q} > 1 \implies (p-2)(q-2) < 4.$$

The only values for p and q that satisfy this equation are those for the Platonic solids. Their names are derived from the number of faces (*icosa* = 12 and *dodeca* = 20).

Platonic Solid	Schläfli Symbol $\{p, q\}$	V	E	F
Regular Tetrahedron	$\{3, 3\}$	4	6	4
Regular Hexahedron (Cube)	$\{4, 3\}$	8	12	6
Regular Octahedron	$\{3, 4\}$	6	12	8
Regular Dodecahedron	$\{5, 3\}$	20	30	12
Regular Icosahedron	$\{3, 5\}$	12	30	20

26.4 Euler Equation

The number of vertices, edges, and faces of a solid are related by Euler's equation:

$$V - E + F = 2.$$

V	0	1	2	3	3
E	0	0	1	2	3
F	0	1	1	1	2

A simple way to be convinced that this equation holds is to consider incrementally constructing the planar map overlaid on a solid. When starting from nothing, we assume by convention that the three values V, E, and F are initially 0 and that a vertex cannot exist by itself, but that a face must also accompany it. The first step creates a vertex and a face, leaving the equation invariant. Inserting an edge and a vertex adds 1 to each of E and V, and closing a face keeps V invariant but increments E and F by one each.

The Euler equation applies to one (connected) solid. If n solids are captured, the number of vertices, edges, and faces is related by $V - E + F = 2n$. This form is convenient because it then also holds before any objects are created.

26.5 Graphs of Solids

Schlegel Diagrams and Duals of Solids

Consider constructing the surface of a solid from a rubber-like material and then puncturing one face and stretching the material until the edges of that face are adjacent to an outer face in a planar map. A Schlegel diagram of the solid, shown in Figure 26.1 for Platonic solids, is a planar drawing of the solid in which all edges are traced as straight lines.

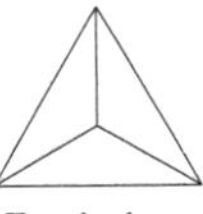

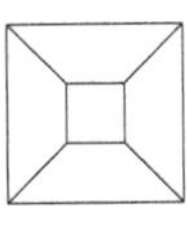

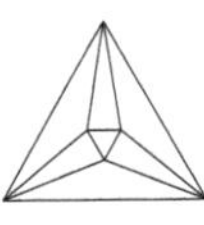

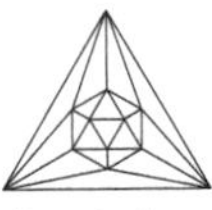

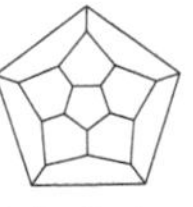

Figure 26.1
Schlegel diagram of Platonic solids—vertex graphs

If face, rather than vertex, adjacencies are modeled, then the resulting graphs in Figure 26.2 can be used for coloring the faces such that no two adjacent faces have the same color. The duality of Platonic solids is also now more evident. The face graph of the octahedron is the vertex graph (Schlegel diagram) of the cube—and likewise for the dodecahedron and the icosahedron. The tetrahedron is the dual of itself.

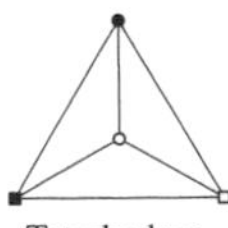

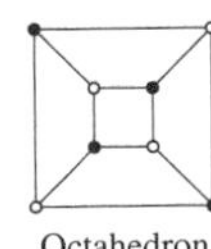

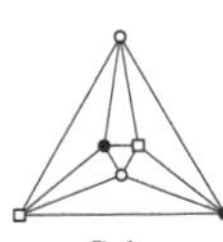

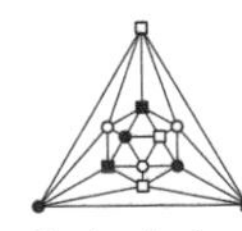

Icosahedron

Figure 26.2
The use of the face graphs of Platonic solids for coloring

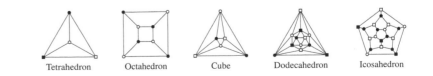

Regular Plane Tilings

How many ways are there to tile the Euclidean plane using a regular polygon? It is convenient to use once again the Schläfli symbol. The symbol $\{p, q\}$ will represent a tiling in which each polygon has p equal sides and each vertex is adjacent to q polygons. Since a regular polygon with p sides has interior angle $\pi - 2\pi/p$ and since the angle at each vertex is $2\pi/q$, we conclude from the equality of these two values that $pq - 2p - 2q = 0$. Since a triangle has the smallest number of sides, $p \geq 3$, and since the interior angle in a regular polygon is strictly less than π, $q \geq 3$. The latter constraint imposes that $p \leq 6$. Since solving for $p = 5$ leads to a fractional q, the only possible tilings, shown in Figure 26.3, are $\{4, 4\}$, $\{3, 6\}$, and $\{6, 3\}$.

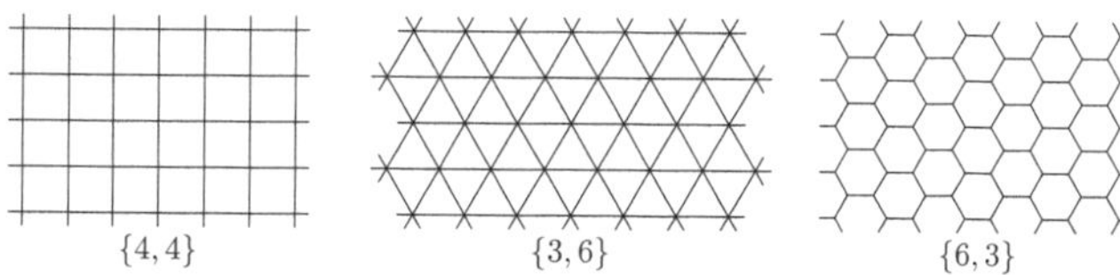

Figure 26.3
Regular tilings of the Euclidean plane

26.6 Exercises

26.1 Implement a (command-line) program that takes a single argument n and generates an OFF file for a uniform n-sided prism. A *prism* is a polyhedron formed by connecting the edges of two regular n-gons ($\{p\}$; see § 26.3) by n rectangles. A *uniform* prism is a prism in which the rectangles become squares [27]. Orient the polygons such that the interior of the polyhedron lies on the negative side of their carrying plane.

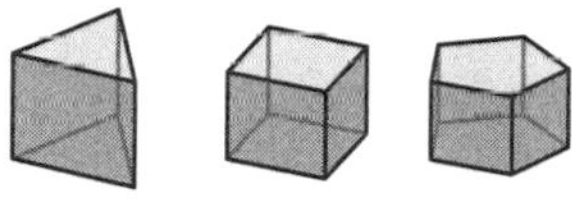

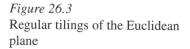

26.2 Show examples of why one cannot define a Platonic solid as one whose boundary consists of regular polygons and whose vertices have the same geometry and connectivity.

26.3 Implement a (command-line) program that takes a single argument n and generates an OFF file for a uniform n-sided antiprism. An *antiprism* is a polyhedron formed by connecting the edges of two regular n-gons by $2n$ isoceles triangles. A *uniform* antiprism is an antiprism in which the triangles are equilateral [27]. Ensure that the polygons describing the boundary are consistently oriented.

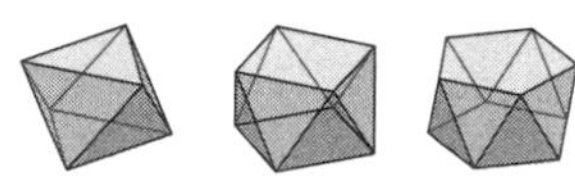

26.4 Implement a system that visualizes a tetrahedron's axes of symmetry by

1. devising the quaternions (how many are there?) corresponding to the positions of the tetrahedron that match its rest position. Ignore transformations that include a reflection.
2. rotating the tetrahedron from one position to a random other position at a uniform speed, resting, and then repeating.

Distinguish among the four vertices by rendering a different feature at each vertex.

26.5 Consider that the planar map in Figure 26.4 is overlaid by a second map that is inserted incrementally. Write the change in V, E, and F at each of the first few steps.

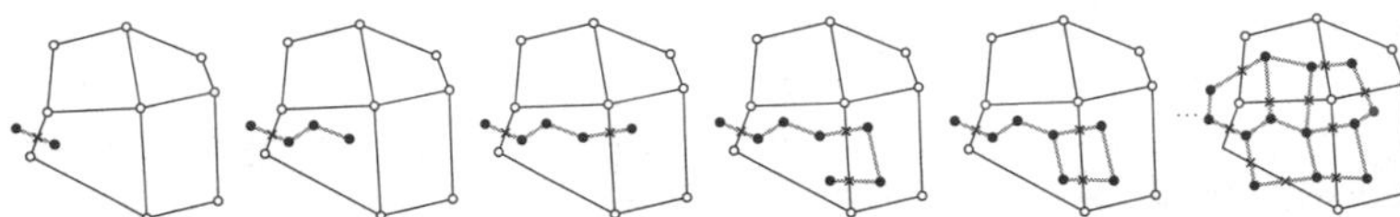

Figure 26.4
The overlay of two planar maps satisfies the Euler equation.

26.6 Exercise 9.4 discussed one way of partitioning the sphere into regions. But regular polyhedra suggest that a different method is possible [110].

1. Modify the start-up code labeled "regular_spherical_model"' to display the five possible regular tessellations of the sphere.
2. Implement a method that ensures each edge is drawn only once.

26.7 Use the class GLimageWriter in the code accompanying this text to generate an animation that illustrates how a regular octahedron can morph to a regular icosahedron (§ 26.3) when the appropriate proportion (the golden ratio) is reached.

26.8 Use the class GLimageWriter in the code accompanying this text as well as your solution to Exercise 26.4 to generate an animation that passes through all permutations and that returns to the initial position (the animation would thus appear seamless as an infinite loop). Is it possible to pass through each permutation exactly once?

27 The Halfedge Data Structure and Euler Operators

If an application only needs to visualize a solid, then indexed face sets are adequate; nested iterations over the set of faces and then over the set of vertices defining a face suffice. But IFS representations are not adequate if one wishes to make incremental modifications to the set of vertices, edges, and faces defining a solid. This chapter discusses the halfedge data structure and a set of accompanying algorithms, Euler operators, that make such modifications possible.

27.1 Introduction

Consider an artist who models an object by starting from a cube and who wishes to modify the cube such that one of its corners is *taper*ed. The modification involves replacing the selected vertex with three vertices and replacing the vertex in the definition of each of the three polygons with two of the three vertices. Indexed face sets make it possible to determine the vertices adjacent to a particular face, but there is no provision to find the faces adjacent to a particular vertex. A taper operation could still be implemented, but one must then search in the set of all faces for those that are adjacent to the selected vertex. Tapering could be performed in time proportional to the number of adjacent faces if those adjacent faces are stored for each vertex.

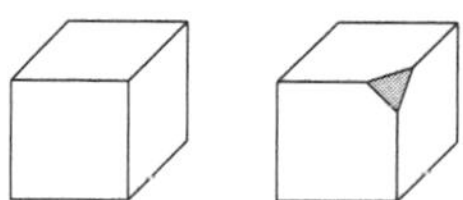

```
struct Vertex {
    Point_E3d coordinates;
    list<Face*> adjacent_faces;
};
```

As suggested in § 26.2, vertex objects are now explicitly represented and store point coordinates and face adjacency information. A second example will suggest that edges also need to be promoted. One could reasonably wish to apply a tapering operation on edges similar to the one applied on vertices. As with vertex tapering, an implementation could implicitly represent the edge being tapered by storing its two adjacent vertices, but to this inconvenience would be added the inefficiency of searching for the two faces among those adjacent to the two vertices for the ones adjacent to the edge. In this case also we could benefit from storing extra connectivity information. Edges are promoted and each edge is explicitly represented by an object. In addition to referencing its two adjacent vertices along one (arbitrary) orientation, an edge would also reference its two adjacent faces. This leads to an edge object such as the one defined as follows:

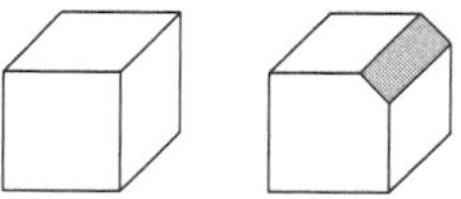

```
struct Edge {
    Vertex *source, *target;
    Face *left_face, *right_face;
};
```

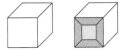

Likewise, we may wish to provide an *inset* operation on faces. But since vertex and edge objects now store information about their adjacencies, implementing an inset operation on a quadrangular face would not be just a matter of replacing one face by five others; performing the updates on adjacent vertex and edge objects will be necessary, which suggests that storing the information in a data structure such as the one below may be adequate.

```
struct Face {
    list<Vertex*> adjacent_vertices;
    list<Edge*> adjacent_edges;
};
```

The set of connectivity information in the structures for a vertex, an edge, and a face above is sufficient for tapering and other "surgical" operations on the surface of the solid, but it suffers from several serious drawbacks:

1. The total storage requirements for the vertex and face objects (along with their accompanying lists of pointers) is not constant. In a system that allocates and deallocates memory while creating and destroying such objects over long sessions, significant memory segmentation could occur.
2. Given a vertex and an adjacent face, it is not possible to query in constant time for the preceding and succeeding vertices and faces around the face and the vertex, respectively. Since no maximum is imposed on either the number of vertices defining a face or the number of faces adjacent to one vertex, the overhead imposed in manipulating objects with high connectivity could be significant.
3. The orientation of an edge agrees with one of its two adjacent faces but not the other. By necessity, such an asymmetry percolates throughout the algorithms and systems built on this design outline. Equally seriously, the effort required to maintain systems is unnecessarily increased and the maintenance relies on programmers reading documentation carefully.

27.2 Halfedge Data Structure

The *halfedge data structure* is a variation on the design sketch above [65]. HEDS solves the three problems:

1. Each object has constant space requirements.
2. Lists of adjacencies can be reconstructed in time proportional to their size.

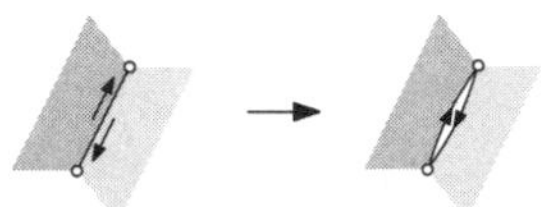

3. An edge is split into two twin halfedges, making it possible to refer simply to either orientation by paying a small price in storage duplication.

The first description of a data structure with equivalent power, the *winged-edge data structure* [7], satisfied the first constraint (significantly, since the implementation was in FORTRAN) as well as the second.

```
namespace HEDS {
  struct Vertex {
    Point_E3d coordinates;
    Halfedge *out_neighbor;
  };
  struct Halfedge {
    Vertex *source, *target;
    Face *adjacent_face;
    Halfedge *succ, *pred;
    Halfedge *twin;
  };
  struct Face {
    Halfedge *adjacent_edge;
    Direction_E3d normal;
  };

  struct Solid {
    vector<Vertex*> vertices;
    vector<Halfedge*> edges;
    vector<Face*> faces;
    ...
  };
}
```

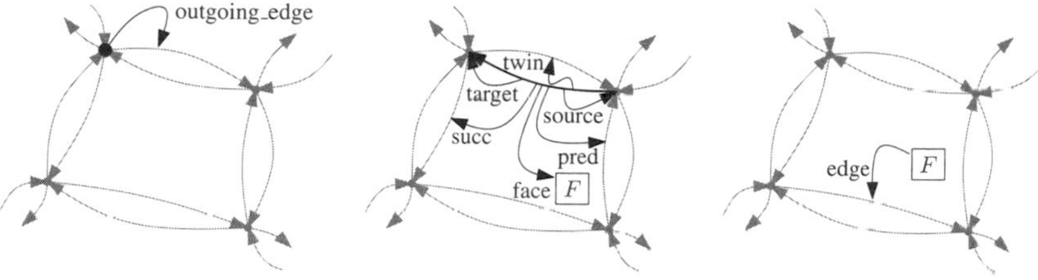

Figure 27.1
The halfedge data structure

The halfedge data structure, illustrated in Figure 27.1, makes it possible to

1. traverse the list of vertices or the edges of a face in time proportional to their cardinality.
2. traverse the edges outgoing from or incoming to a vertex as well as the adjacent faces in time proportional to their cardinality.
3. traverse the faces adjacent to a given face in time proportional to their cardinality.

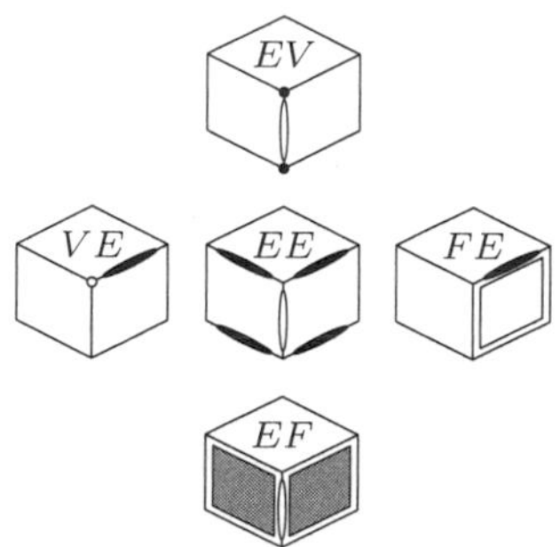

Vertex and face objects are light: Each stores a single pointer to an adjacent halfedge. The data structure is, on the other hand, halfedge-centric (hence the name). Each halfedge stores pointers to the two adjacent vertices and the single adjacent face as well as pointers to the succeeding and preceding halfedges in the same face. The crucial twin halfedge pointers link each two halfedges that are adjacent (in opposite order) to the same pair of vertices.

We can repeat the subset of a figure shown on page 238 that is captured in a HEDS—the edges reference adjacent vertices, edges, and faces, and each face and vertex each reference one halfedge.

Note that this extensive set of connections is sufficient, but not necessary. A solid consisting of E edges needs $12E$ pointers, or $96E$ bytes on a 64-bit machine, for the edges alone. This can be partially remedied by discarding either the source or the target of a halfedge as well as either each successor or predecessor vertex, but this comes at the price of two indirect memory accesses each time a discarded pointer is needed.

Linking Twins and Outgoing Halfedges

The extensive set of pointers will normally not be saved to disk, but a format such as the indexed face set (§ 26.2) will be used. The successor and predecessor pointers for halfedges can be initialized during reading. The face adjacent to each halfedge and one halfedge adjacent to each face can also be initialized then. Two sets of pointers remain: the pointer from a halfedge to its twin and the pointer at each vertex identifying one outgoing halfedge.

Neither operation can be performed while the data are read from disk since the requisite objects would not have been created (and therefore their addresses would still be unknown). Once the disk file is read, linking twin pointers can be performed by iterating over the halfedges and saving in an associative container (e.g., **std**::map) pointers to the halfedge objects indexed by a tuple (e.g. **std**::**pair**) combining the source and target vertex. A second iteration over the halfedges determines whether the twin key (target, source) has been inserted and points to the halfedge associated with that key if it exists.

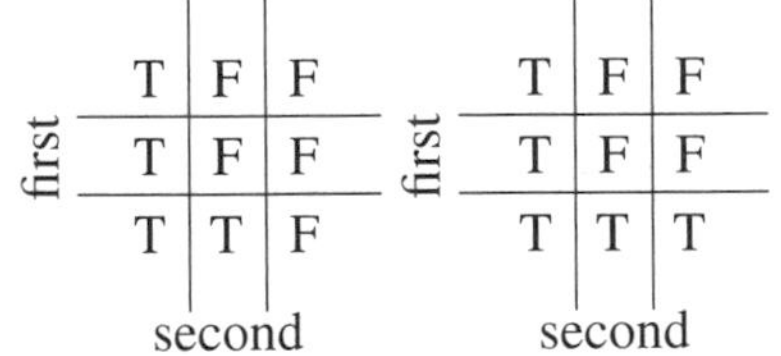

lexicographic order   **std**::**pair** order

Associating a halfedge pointer with a pair of vertex pointers requires establishing a total order on the latter. Since no such pair could be equal (no two halfedges could have the same source and the same target vertices), there is—or should be—no concern in this case regarding collisions between keys. It is worth observing that the comparison operator implemented by the Standard Template Library is subtly different from lexicographic ordering; although in this case either comparison operator is equally suitable.

Likewise, setting outgoing halfedge pointers can be performed by two iterations—the first indexing the halfedges by their source vertex as key (with duplicates allowed) and the second extracting any of the stored halfedge pointers among those indexed by a given vertex. Alternatively, an associative container not allowing duplicates may be used. In that case only the first halfedge with a given vertex is stored. Storage is then proportional to the number of vertices rather than to the number of halfedges. These initializations are the topic of Exercises 27.4 and 27.5.

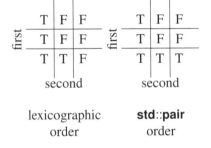

Manifold Topology

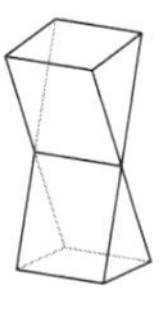

The halfedge data structure does not make it possible to handle objects in which more than two halfedges are adjacent to the same pair of vertices or in which the neighborhood of a vertex is not equivalent to a disk. To handle non-manifold topologies another data structure such as the radial edge [115] or selective geometric complexes [88] needs to be considered instead.

27.3 Euler Operators

The halfedge data structure is attractive partly because it provides these query abilities, but particularly because it is possible to implement operators that locally modify the boundary, or surface, of the solid [7, 34, 66, 86]. Each such *Euler operator* can be executed in time proportional to the local complexity.

In some sense HEDS are the generalization to two-dimensional surfaces of (the one-dimensional) linked lists. From a software engineering perspective, just as it would be unwise to allow a client programmer who needs an implementation of a linked list to operate directly on the list pointers, it would also be unwise to expose the HEDS internal pointers to client programmers. It is in this spirit that a HEDS implementation should provide a set of functions that act as an intermediate interface.

An *Euler operator* is a function that operates on a halfedge data structure by adding and/or deleting vertices, halfedges, and/or faces so that Euler's equation ($V - E + F = 2$) remains valid on the manipulated solid (or planar map) after the operation completes, although the equation will, in general, not be satisfied during the execution of the function.

The following acronyms are used for the operators:

m : make	**k** : kill	**v** : vertex	**e** : edge	**f** : face

The mev operator, for example, stands for make-edge-vertex. It makes, or creates, a vertex and an edge. Its opposite, kev, the acronym for kill-edge-vertex, destroys a vertex and an edge. In both cases Euler's equation is left invariant.

The following is a partial list of Euler operators:

- **mev** make-edge-vertex; **kev** kill-edge-vertex
- **mef** make-edge-face; **kef** kill-edge-face
- **mvfs** make-vertex-face-solid; **kvfs** kill-vertex-face-solid
- **split_edge**; **join_edge**

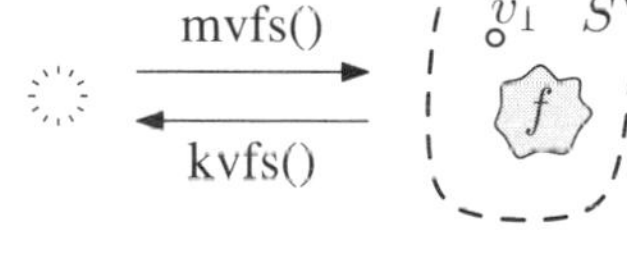

The first step in constructing a solid, and the last step in destroying one, are special. Initially, no objects exist. A vertex is created along with a face, but with no edges. An example for locally modifying a solid (or a planar map) using Euler operators is illustrated in Figure 27.2.

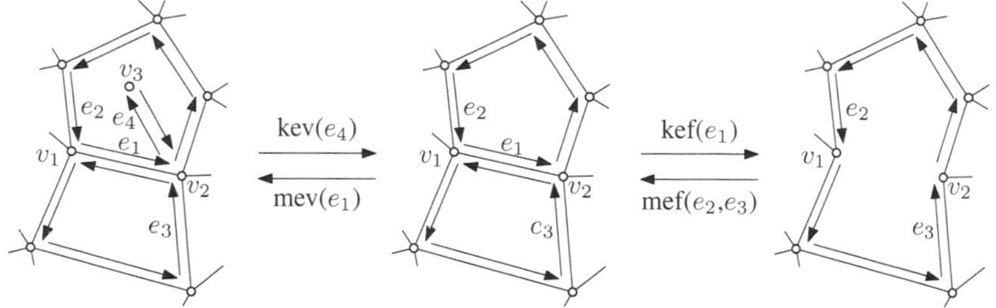

Figure 27.2
A subset of Euler operators

The two operations split_edge and join_edge have the same effect as mev and kev (adding or removing one vertex and one edge), but the interface is distinct and hence distinct functions are needed. We will need the first in § 32.2.

These operators are called *Euler operators* because they keep Euler's equation invariant. Note that E is the number of edges and not that of halfedges.

A tetrahedron, for example, has four vertices, six edges, and four faces; indeed $4 - 6 + 4 = 2$.

Parameters of Euler Operators

Another advantage of using halfedges is the conciseness of the parameter list for Euler operators. Passing one halfedge as parameter to a mev operator, for instance, unambiguously defines where the vertex and the edge should be inserted. The objects created are then returned from the mev function. Two halfedges need to be passed to mef, identifying that their two target (or source) vertices should be joined by an edge.

Using Euler Operators to Construct a Cube

Euler operators act as an intermediate software layer that hides the pointers from HEDS clients. They would be used to manipulate the surface of a solid—one already satisfying Euler's equation, but they can also be used to construct elementary solids.

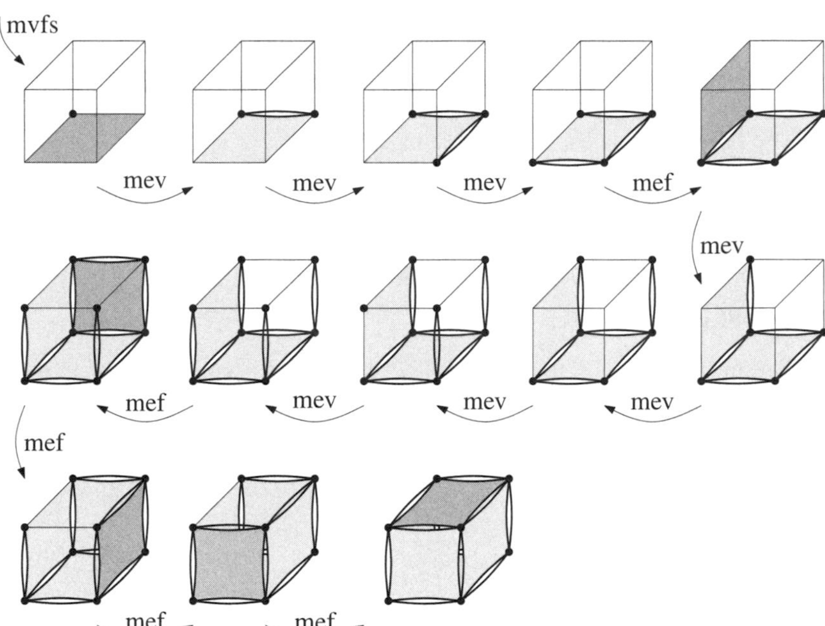

Figure 27.3 Using Euler operators to construct a cube

In intermediate stages the solid constructed will satisfy Euler's equation, but it will not be a sensible solid. Before the last operator in Figure 27.3 is applied, for example, the solid is valid topologically, although it has one face between six vertices. Since these vertices are not coplanar, the face thus defined is not planar. Such unrealistic geometries are alright during the construction of a solid as long as the result is valid also geometrically at the end of the sequence.

And so the designer may find it useful to use several strata for the HEDS and its accompanying operators. The first, or bottom-most, layer of functions directly manipulates the pointers between the vertices, edges, and faces. A layer on top implements the Euler operators and maintains the topology, but not the geometry. This second layer will prevent clients from accessing functions belonging to the first layer. A third layer will consist of a set of functions that each maps to a sequence of Euler operators. The construction of a cube above is an example of one function belonging to that third layer. Other primitive operations (from the viewpoint of the user of a solid modeling system) will be those of extrusion, sweep, tapering, etc. and will also belong to the third layer. This separation is useful conceptually, but it is possible for two layers or all three layers to be coalesced into member functions of a single class.

27.4 Solids of Arbitrary Genus

A set of vertices, edges, and faces suffices to represent a genus-1 object, such as a donut, but the standard Euler equation no longer holds. A slightly modified version

$$V - E + F = 2(S - H) + R$$

is needed. This *Euler–Poincaré* equation also ties the number of holes H and the number of rings R. S is the number of solids.

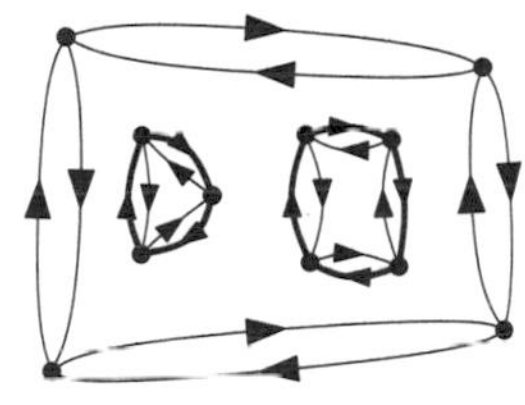

A *ring* is a cycle of edges partially defining a face. Each face is described by a single outer ring in addition to an arbitrary number of inner rings. The orientation of halfedges is always such that travelers along a halfedge have the face to their left. The outer ring will be oriented counterclockwise and the inner rings clockwise as seen by an observer located outside the solid.

A HEDS implementation incorporating the notion of a ring might be

```
namespace HEDS2 {
  class Vertex {
    Point_E3d coordinates;
    Halfedge *out_neighbor;
  };
  class Halfedge {
    Vertex *source, *target;
    Face *adjacent_face;
    Halfedge *succ, *pred;
    Halfedge *twin;
  };
  class Ring {
    Halfedge *adjacent_edge;
    Ring *succ, *pred;
  };
  class Face {
    Halfedge *outer_halfedge;
    Ring * first_inner_ring;
    Direction_E3d normal;
  };

  class Solid {
  private:
    vector<Vertex *> vertices;
    vector<Halfedge *> edges;
    vector<Ring *> rings;
    vector<Face *> faces;
  public:
    mev(...);
    ...
  };
}
```

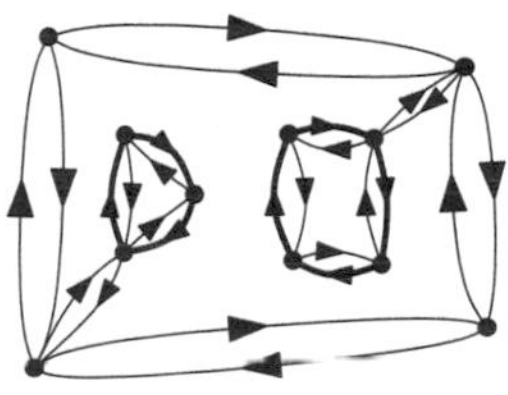

Faces with holes could be captured while keeping the notion of a ring implicit. To that effect a pair of halfedges would connect a vertex on the outer ring to a vertex on each inner ring.

Two more pairs of operators are needed to manipulate a HEDS with rings and holes. The complete set of operators becomes

- **mev** make-edge-vertex; **kev** kill-edge-vertex
- **mef** make-edge-face; **kef** kill-edge-face
- **mvfs** make-vertex-face-solid; **kvfs** kill-vertex-face-solid
- **split_edge**; *join_edge*
- **kemr** kill-edge-make-ring; **mekr** make-edge-kill-ring
- **kfmrh** kill-face-make-ring-hole; **kfmrh** make-face-kill-ring-hole

Using Euler Operators to Construct a Genus-1 Object

Figure 27.4 shows an example illustrating the new set of operators. Assuming that a solid cube is represented using the HEDS2 set of classes and is built using the operators previously outlined, the steps in the figure show the incremental construction of a hole in the cube.

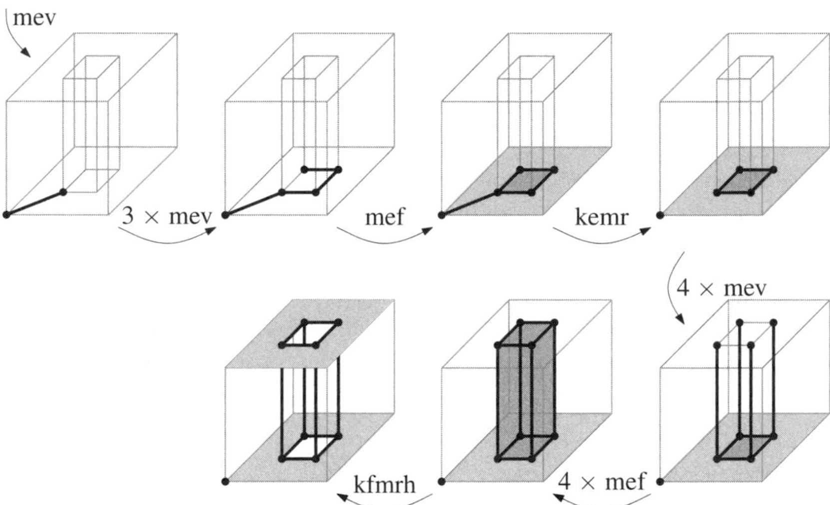

Figure 27.4
Complete set of Euler operators

27.5 Exercises

27.1 Devise the sequence of Euler operators needed to morph a regular octahedron into an icosahedron as discussed in § 26.3. Illustrate the steps for one of the eight vertices.

27.2 The smallest polygon, a single triangle formed by three vertices, three halfedges, and one polygon, is not a legal solid since it does not satisfy Euler's equation and does not enclose a finite volume. One can define a zero-volume solid consisting of a triangular area by using *two* triangles defined by the same vertices in opposite order. Provide a sequence of Euler operators that constructs such a solid, dubbed a triangular *lamina*.

27.3 Show how the operations of tapering a vertex, tapering an edge, insetting a face, and extruding a face can be reduced to a sequence of Euler operators.

27.4 Modify the start-up code labeled "select-polygon" to link twin halfedges.

27.5 Modify the start-up code labeled "select-polygon" to set the outgoing halfedge pointers at each vertex.

27.6 After completing Exercises 27.4 and 27.5, implement a minimal set of Euler operators (mvfs, mev, kev, mef, kef).

27.7 The user interface of a geometric modeler is one of its most important components, and the usability of a modeler is often one of its main evaluation criteria.

It is common practice to precede development with a *storyboard* outlining the steps the user takes when operating a program [82]. Devise three different storyboards for a geometric modeler that makes it possible to taper a vertex, to inset a face, to taper an edge, and to extrude a face. The three storyboards should represent rather different user interface policies, but each set should be consistent. For instance, will the user select the data first and then select the function (taper, etc.) that will be applied, or will the function be selected first?

27.8 Choose one of the storyboards you designed in Exercise 27.7 and then modify the start-up code labeled "select-polygon" to implement these modeling operations.

27.9 Ask a friend to participate in testing the usability of your implementation of Exercise 27.8. After providing minimal assistance, observe how your friend is interacting with the program and write a few lines suggesting improvements you could make to your user interface.

27.10 The duality between the cube and the octahedron discussed in Chapter 26 can be taken one step further using Euler operators. Generate a sequence of images (using the class GLimageWriter in the start-up code) that show the result of applying vertex tapering simultaneously on the cube's eight vertices. When the three vertices generated at each vertex meet at the cube's edge, the triangles become hexagons, four vertices of which eventually meet to become the octahedron's vertices. For many more ideas of possible animations, see Holden's *Shapes, Space, and Symmetry* [53].

28 BSP Trees in Euclidean and Spherical Geometries

Chapters 26 and 27 discussed how point sets can be represented using their boundary. This and the next chapter discuss an alternative method for representing point sets that relies on the recursive partitioning of space. We define in this case a standard set of operations on the point set (point containment and Boolean operations). The algorithms can be dissected into combinatorial parts, discussed in Chapter 29, and geometric parts, discussed in this chapter. This illustrates how a set of geometric classes can be plugged into a generic combinatorial algorithm to produce a concrete algorithm operating on the geometry in question.

Representing solids as CSG trees (Chapter 30) makes it trivially easy to perform Boolean operations, but point containment queries are difficult. Representing solids using their boundary (Chapters 26 and 27) makes point containment queries relatively easy, but Boolean operations are involved. Does a representation exist that makes both operations simple? Binary space partitioning (BSP) trees are such a representation.

28.1 Labeled-Leaf Binary Search Trees

Consider constructing a binary search tree in which the keys in interior nodes are points in the Euclidean line E^1.

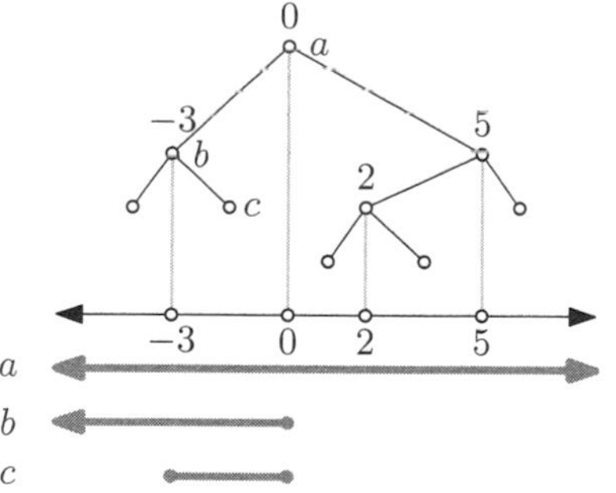

Because the tree satisfies the search property, points in a left subtree are smaller than (have an x-coordinate that is smaller than) the root of the subtree and points in a right subtree are greater than the root.

The binary partitioning defined by the tree naturally, but implicitly, defines a region for each node. The root node a is defined on the region $(-\infty, +\infty)$, node b is defined on the region $(-\infty, 0)$, and node c is defined on the region $(-3, 0)$. The resulting structure is termed a *binary space partitioning tree*, or BSP tree [38].

We will initially call the two subtrees "left" and "right" subtrees. Doing so is fine for one-dimensional geometries, but since our objective is ultimately to write a single implementation for an arbitrary dimension, we will soon move to naming the two subtrees the "negative" and the "positive" subtrees, paralleling the convention taken for the comparison predicate in § 2.5.

We now observe that the leaf nodes partition E^1 into regions and that we can represent a function on E^1 by storing the function in the leaf nodes. To represent a set of segments in E^1, for instance, we simply store a Boolean flag at each leaf node. A leaf is labeled **true** whenever the segment it captures is

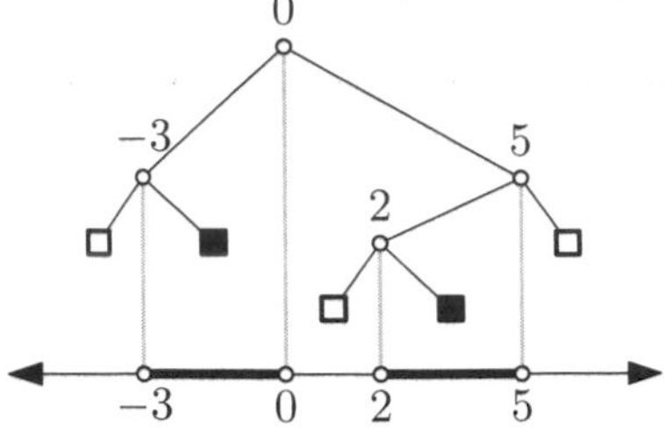

in the set represented and is labeled **false** otherwise. The resulting structure is called a *labeled-leaf binary search tree* [109].

Observe that in a labeled-leaf BSP tree the interior nodes are distinguished from the leaf nodes. An interior node stores a point in the Euclidean line, whereas a leaf node stores a Boolean flag.

Even though the segments are represented only implicitly in the tree, we can find the bounds of the segment defined by a node by inspecting the path from the root to that node. The deepest interior node with a right child on the path defines the left endpoint and the deepest interior node with a left child on the path defines the right endpoint.

28.2 Operations on Segments in E^1

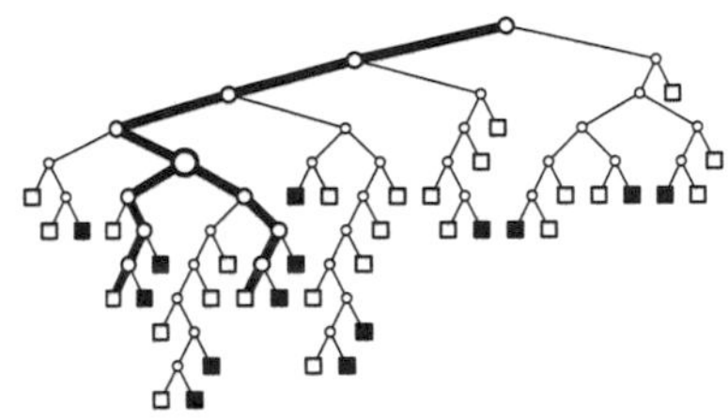

Point Containment

To determine whether a given point in E^1 is in the set captured by a labeled-leaf BSP tree, we search for the point in the tree. The answer is provided by the Boolean flag at the leaf. If the point searched is not itself a label for an interior node in the tree, the search path is a single path from the root to a leaf. But if the point searched is encountered on the path, recursion forks and both subtrees are searched.

If both recursive forks return an identical answer (either **true** or **false**), that is the answer sought. But the two recursive forks need not return the same answer, which suggests that a simple Boolean return flag is insufficient. The return type is an enumeration of one of *three* flags:

```
enum Set_membership
{
    INSIDE_SET = −1,
    ON_SET_BOUNDARY,
    OUTSIDE_SET
};
```

The function BSP_node::classify, the predicate that determines point containment, is discussed in § 29.5.

Boolean Operations on Segments in E^1

In contrast to an ordinary binary search tree on E^1, which makes it possible to insert a point, a labeled-leaf BSP tree makes it possible to insert a segment. To do so, the segment is *split* at the root of the tree. The two new fragment segments are recursively inserted in the two subtrees.

Figure 28.1 shows an example for determining the Boolean union operation between the segments AB and CD. We start from an empty tree consisting of a single node labeled **false**. First AB is inserted, followed by CD. When CD is inserted at the root node B, CD is split at B, but because both C and D lie on the positive (right) side of the splitter B, the segment is inserted intact in the positive (right) subtree of B. When CD reaches the leaf node at the right child of B, a new subtree is constructed for the segment CD. Indeed, the tree for AB itself was constructed in the same vein; a subtree is constructed for AB and that subtree is appended in the initially empty BSP tree.

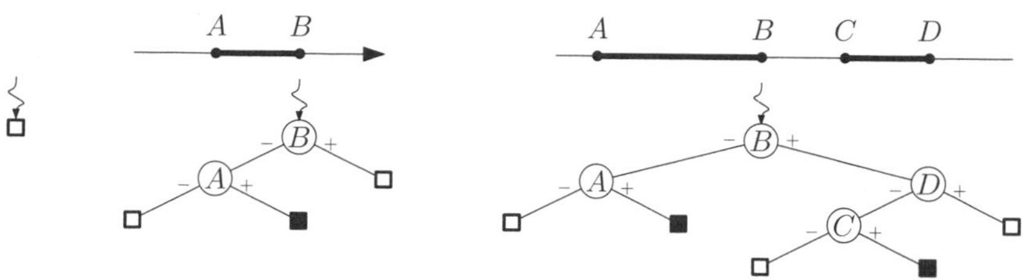

Figure 28.1
Incremental construction of the union of two segments

It is easy to see how intersection and difference operations can be performed; the particular Boolean operation desired is recursively invoked at the root. A difference operation would flip the Boolean flag from **true** to **false** for each leaf node reached and an intersection operation would flip the Boolean flag from **true** to **false** for each leaf node *not* reached.

Simple as they are, operations on sets in E^1 exhibit many (but not all) of the characteristics of the operations in other geometries and higher dimensions. The two operations we consider next are how a subtree is constructed and how a segment is split.

Subtree Construction

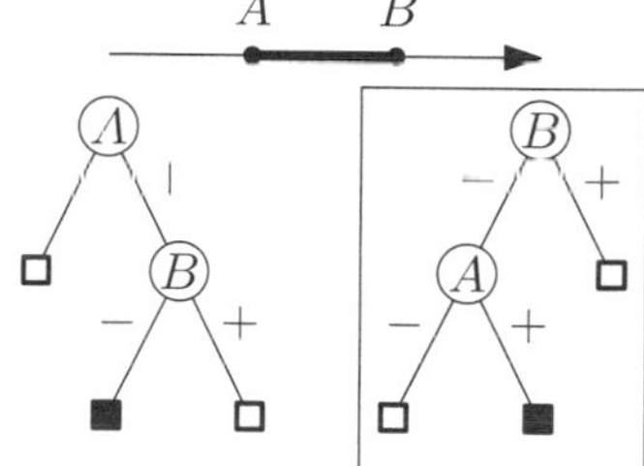

Evidently, the subtree for a segment AB consists of five nodes—two interior splitting nodes and three leaf nodes—but which of the two possible trees should be adopted? We take the convention of using the larger of the splitting points at the root. Neither is elegant; if a set is reflected before insertion into a BSP tree, we cannot hope for the two trees to be the mirror image of one another. But symmetry is missing in any case. Compare, for instance, the tree resulting from $CD \cup AB$ with the one resulting from $AB \cup CD$.

Necessity of Maintaining Neighborhoods

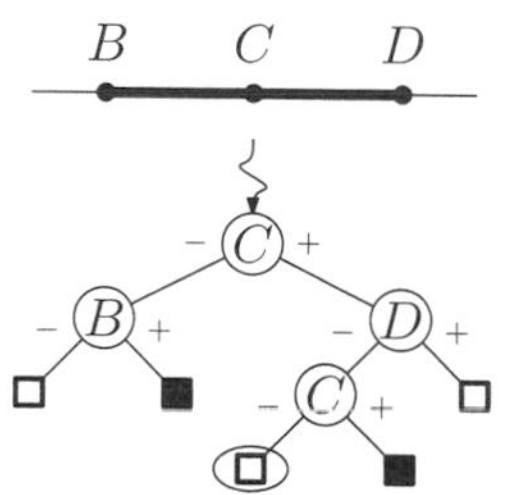

The boundaries of the segments need, of course, not be distinct. Now consider what would happen if we attempt to construct the tree corresponding to the expression $BC \cup CD$.

Splitting CD at C produces just one fragment and a subtree corresponding to that fragment is inserted at the positive (right) child of C. But now the tree has *two* interior nodes labeled C. The tree also has a leaf node defined on an empty segment, for the points in it must be simultaneously larger and smaller than C.

To avoid this waste of time and space, we define a wrapper object for segments in E^1. An object of type BSP_E1_segment consists of an instance of Segment_E1 (§ 2.5) in addition to two Boolean flags. Recall that Segment_E1 in turn stores two points in E^1, a source and a target. The two Boolean flags capture whether the segment represented is open or closed on either side. Initially, the inserted segment is closed at both sides. But if a segment such as CD is split at C, the source flag becomes "open."

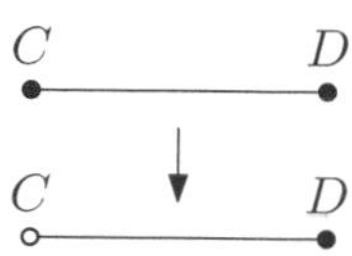

Splitting a Segment in E^1

The splitting operation takes a segment and a splitter and returns two segments, one of which may be empty. A sketch of BSP_E1_segment follows:

```
template<typename NT>
class BSP_E1_segment
{
  bool is_empty_var;

  Segment_E1<NT> segment;
  bool  source_is_closed, target_is_closed;
  ...
};
```

We stop for a moment and notice how the types are simplified in E^1. Each segment is determined by two points and the splitter is likewise a point. In general, the segment is a *convex polytope* and the splitter is a *hyperplane*. The generic (combinatorial) BSP data structures and algorithms discussed in Chapter 29 use these more general terms. An instantiation to a given geometry will map the generic types to concrete types. In E^2, for instance, the convex polytope is a convex polygon and the hyperplane is a line.

Five cases arise when splitting a segment in E^1. The splitting point may lie on either side of the segment, may lie inside the segment, or it may coincide with one of the two boundaries.

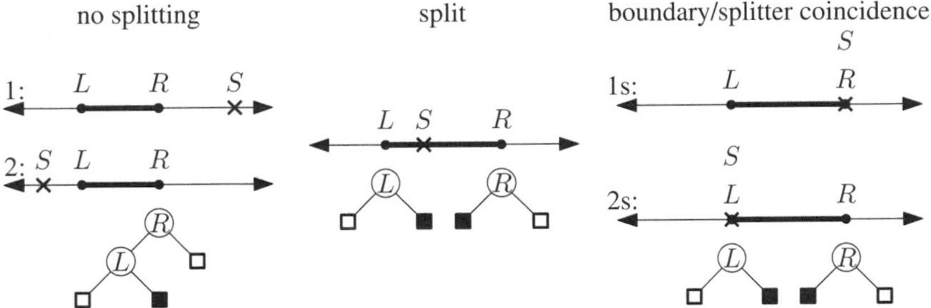

Figure 28.2
Splitting a segment in E^1

We noticed that should the splitter coincide with one of the two boundaries, that boundary should be flagged as open. If the splitter falls inside a segment, both segments need to be flagged as open at the splitting point. As illustrated in Figure 28.2, constructing a subtree to be attached at a leaf will convert a segment into a subtree while using only the one or two boundaries that remain flagged as closed. Yet a segment can encounter splitters at both endpoints. In that case the resulting subtree is a single node.

28.3 Operations on Segments in S^1

Representing a Segment in S^1

A spherical segment AB is represented using a BSP tree rooted at B. The root node partitions S^1 into B^+ and B^- (see § 8.2). The latter set is likewise partitioned by the node A into B^-A^+ and B^-A^-. The three leaf nodes correspond to the three segments in S^1.

An instance of BSP_S1_segment represents a segment in S^1 in the context of BSP trees. In addition to storing the segment represented, two Boolean flags signal whether the segment is the entire set of points in S^1 or the empty set.

```
template<typename NT>
class BSP_S1_segment
{
    bool is_empty_var;
    bool is_full_var;

    Segment_S1<NT> segment;
    bool  source_is_closed, target_is_closed;
    ...
};
```

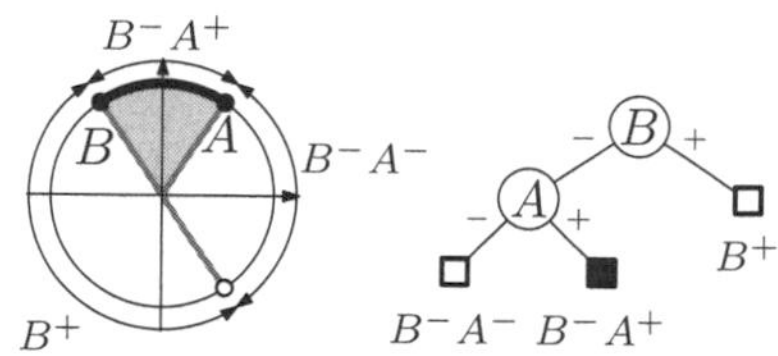

Splitting a Segment in S^1

We look at the four configurations that can arise when splitting a segment by a point in S^1 as the contrast with E^1 may be illuminating. In cases a and c in Figure 28.3 the segment is not split, whereas cases b and d require a split. If the segment is split by either the point (b) or its antipode (d), both fragments are marked as open at the splitting point. Cases a1 and a3 (which are not mutually exclusive) do not cause a split, yet the resulting segment is marked as open at one or both endpoints. Symmetric cases arise when the segment lies on the closed positive halfspace defined by the splitting point (c).

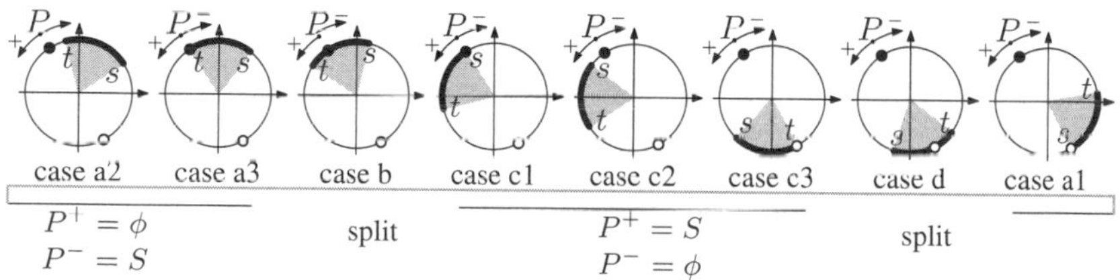

Figure 28.3
Splitting a segment in S^1

Handling Degeneracies

Handling degeneracies correctly in S^1 yields unexpected results. Consider using the three points $A(1,1)$, $B(-1,1)$, and $C(-1,-1)$ to construct two segments AB and BC.

Figure 28.4 shows the evaluation of $AB \cup BC$. While BC is percolating down the tree, it is flagged as open at B and lands in the positive branch of B. Since only one of the two endpoints is closed (C), only that endpoint induces a new interior splitting node.

By contrast, Figure 28.5 shows the evaluation of the expression $BC \cup AB$ (for the same set of points), AB is classified as lying in the negative side of C, but because A is the antipode of C, the segment AB is flagged as open at A. When AB subsequently reaches B, it is classified as open at B. Thus, AB

Figure 28.4
Asymmetry of constructing a tree in S^1—$AB \cup BC$

reaches the negative branch of B with both endpoints open and only has the effect (on union) of converting the leaf node for the region l from **false** to **true**. Since both children of B now have the label **true**, one may reduce the tree by replacing the subtree at B with a single **true** leaf flag.

Figure 28.5
Asymmetry of constructing a tree in S^1—$BC \cup AB$

Likewise, if after evaluating $AB \cup CD$, we evaluate $[AB \cup CD] \cup BC$, neither C nor D induces a new partitioning node (they are the antipodes of A and B, respectively). And so inserting BC only has the effect of modifying the leaf labels.

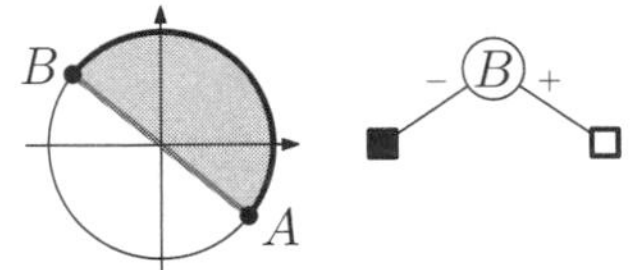

A library for one-dimensional spherical geometry could be designed and used while allowing for segments to exceed half of S^1. This is the reason we do not assert when constructing a segment that the target lies in the positive halfspace defined by the source. This constraint becomes required when dealing with BSP trees. An instance of BSP_S1_segment can capture all of S^1, or half or less of S^1, but it cannot capture a segment that is strictly larger than half and smaller than full. Aside from this constraint, constructing a subtree in S^1 is nearly identical to the details for E^1, although we do need to confirm that the two segment endpoints are not antipodal, for then a subtree with only three nodes is needed.

28.4 Operations on Convex Polygons in E^2

Point sets with linear boundaries in E^2 can be represented using BSP trees. Interior nodes store splitting lines and each leaf node stores a Boolean flag.

A Polygon in E^2

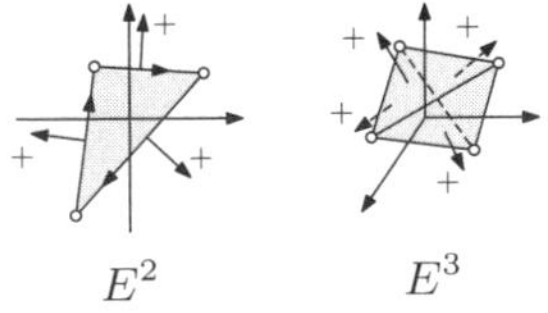

It is convenient here to take polygons in the plane as oriented clockwise, rather than the traditional counterclockwise orientation. There are two reasons for this choice. A convex polygon in 2D is bounded by a set of oriented halfspaces (lines) just as a solid in 3D is bounded by a set of oriented planes. Since the 3D solid is on the negative side of its defining oriented planes, it is more consistent (and will help a generic implementation) if the inside of a polygon is also on the negative, or right, side of an oriented line in the plane.

The second justification for this unorthodox orientation is that we would like to compute visibility by projecting from either E^2 to S^1 or from E^3 to S^2

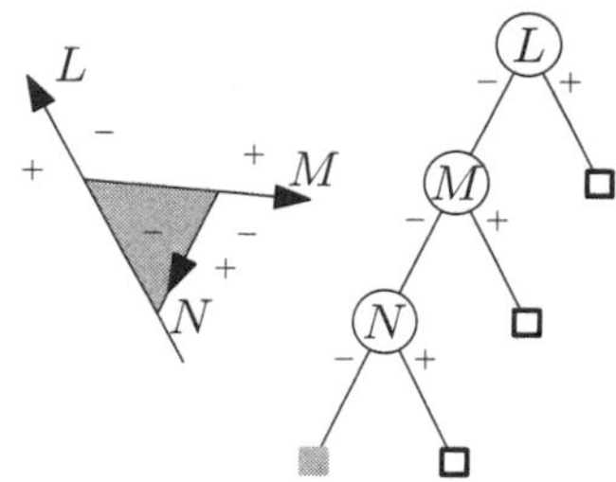

(Part VII). In either scenario the center of projection will lie outside the solids and in both cases we would like the center of projection to lie on the positive side of at least one bounding halfspace.

As would be expected to parallel the two 1D geometries, representing a convex polygon in E^2 requires saving whether each of its bounding lines is open or closed. As a polygon is split at interior nodes of a tree, the edges resulting from a split are flagged as open neighborhoods. The subtree construction routine subsequently only uses those boundaries that remain closed; each of the boundaries flagged as open would have already been represented at an ancestor node in the tree.

```
template<typename NT>
class BSP_E2_convex_polygon
{
  bool is_empty_var;

  std::vector<Segment_E2<NT> > lines;
  std::vector<bool> line_is_closed;
  ...
};
```

The first few lines of an implementation of a convex polygon show that segments, rather than lines, in E^2 are used to define the polygon. The reason this may be necessary is that we may define a line from two points and subsequently find that a test for point-line incidence fails. Such tests are exactly those that make it possible to discard empty nodes in a BSP tree. Exercise 28.4 asks you to confirm whether floating point on your machine would be reliable for this purpose.

Splitting a Polygon by a Line in E^2

We need a new splitting algorithm for two reasons. On one side the representation of the polygon is dual to the traditional one; it is based on bounding lines rather than the polygon vertices. More importantly, the algorithm needs to return two polygon fragments that correctly include the neighborhood information. This information is linked with line boundaries, which must be interleaved while the point-based sidedness tests are performed.

Describing a splitting algorithm that operates on edges rather than on vertices will seem to be a step backward—at least if one adopts the now-established thinking that operating on vertices is simpler (because that makes it more amenable to a hardware implementation [107]). But simplicity is not necessarily desirable, rather the simplest simplicity that adequately explains problems and that provides a general solution is what we ought to be seeking. Vector visibility (Part VII) will also seem to be a step backward compared to raster methods. Ultimately, it all depends on the specification of the problem studied.

The traditional approach for representing a polygon is to use neither pairs of points (or segments, as just discussed) nor lines. But the obvious approach of representing a polygon using vertices is flawed. As mentioned in § 6.4, failure to keep the algebraic degree of all expressions in a geometric system at a minimum is a recipe for introducing difficulties. If floating point numbers

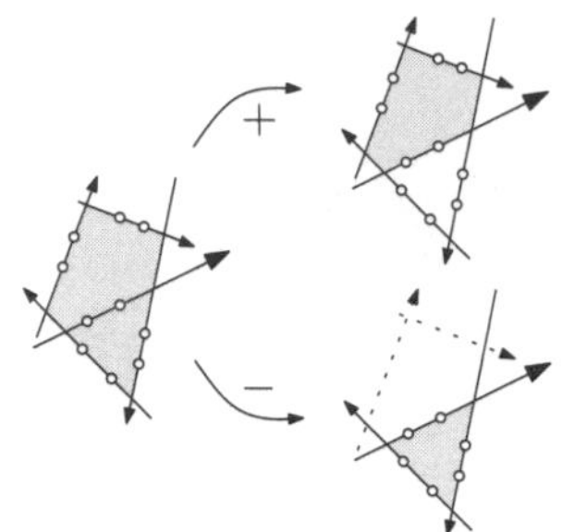

are used, the predicates will be increasingly unreliable. If rational numbers are used, the size of the numbers quickly dominates the storage. Floating point filters help a little, but filtering will only delay, not eliminate, the problem, which is simply encountered with larger inputs.

As shown in the adjacent figure, a polygon is defined using its bounding lines. Each line is in turn defined using two input points. Splitting a polygon by a line consists of inserting the splitting line twice in the circular list defining the polygon to determine the two fragments.

The algorithm shown in Figure 28.6 splits a polygon whose boundary is defined by pairs of points. Since we have in any case to compute the vertices of the polygon to determine their orientation with respect to the splitting line, this is the first step in the code. If all vertices lie on one side of the splitting line, we determine whether any of the bounding lines coincides with the splitting line to construct the Boolean flags correctly then return.

The algorithm then has to handle the 10 cases shown. Consider, for instance, case (f) in the figure. The target of an edge is found to lie on the splitting line. This indicates that the next edge to be inserted in the positive fragment will be colinear with the splitting line and will need to be flagged accordingly. Cases (f) and (h) are not mutually exclusive. Cases (i) and (j) are the classical polygon-clipping steps discussed in § 5.4. C++ code for this function is included in the code accompanying the text. If the interior of the polygon lies to the left (positive) side rather than the right side of the boundary, the same pseudo-code applies. One only has to swap the two lines labeled "insert L" and "insert -L."

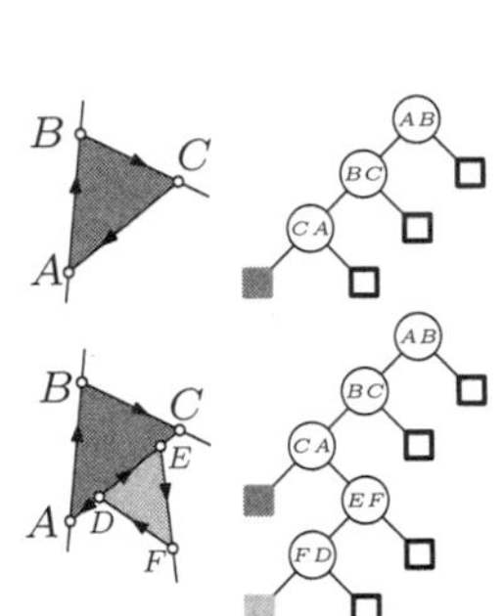

After a polygon has been split by partitioning lines on paths from the root to leaves, a tree is constructed for each resulting fragment at a leaf. The new subtree for those fragments whose boundaries resulted exclusively from splits is simple; it consists of a single node. An example illustrates the need for the split function above. Consider constructing the union of two triangles ABC and DEF such that A, C, D, and E are colinear. When DEF is inserted in the tree, it is split by AB, but only one fragment, whose three sides are labeled as closed, results. The same occurs when DEF is split at BC. Splitting at CA, on the other hand, signals that the edge DE already occurs in the tree and that DE is thus open. For this reason, the tree constructed for the triangle DEF omits the partition line at DE; it would duplicate CA and result in an empty set at a leaf. The code for subtree construction in E^2 is described in § 29.4, and the case when $D = A$ and $E = C$ is discussed further in Exercise 28.4.

28.5 Exercises

28.1 Show the labeled-leaf BSP tree that would result from the following operations in E^1:

- $[1, 5] \cup [2, 7]$
- $[1, 5] \cap [2, 7]$
- $[1, 5] - [2, 7]$
- $[1, 5] - [2, 4]$
- $\{[-3, 0] \cup [2, 5]\} \cup [0, 2]$
- $\{[-3, 0] \cup [2, 5]\} \cap [-3, 5]$

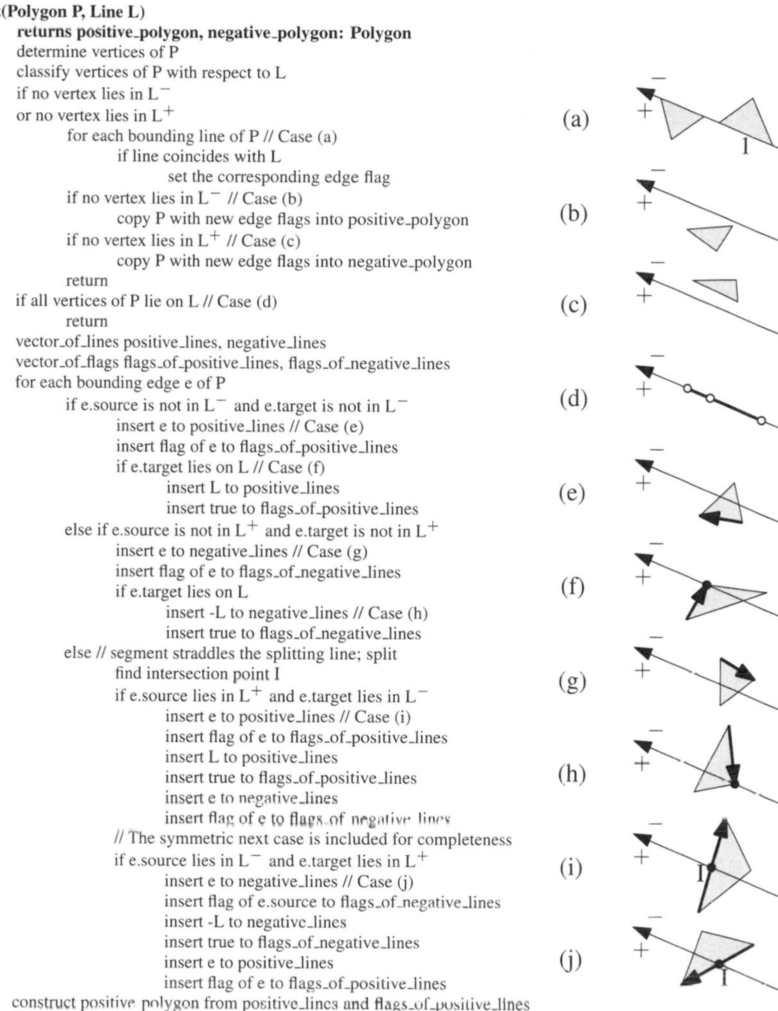

```
Split(Polygon P, Line L)
    returns positive_polygon, negative_polygon: Polygon
    determine vertices of P
    classify vertices of P with respect to L
    if no vertex lies in L⁻
    or no vertex lies in L⁺
        for each bounding line of P // Case (a)
            if line coincides with L
                set the corresponding edge flag
        if no vertex lies in L⁻ // Case (b)
            copy P with new edge flags into positive_polygon
        if no vertex lies in L⁺ // Case (c)
            copy P with new edge flags into negative_polygon
        return
    if all vertices of P lie on L // Case (d)
        return
    vector_of_lines positive_lines, negative_lines
    vector_of_flags flags_of_positive_lines, flags_of_negative_lines
    for each bounding edge e of P
        if e.source is not in L⁻ and e.target is not in L⁻
            insert e to positive_lines // Case (e)
            insert flag of e to flags_of_positive_lines
            if e.target lies on L // Case (f)
                insert L to positive_lines
                insert true to flags_of_positive_lines
        else if e.source is not in L⁺ and e.target is not in L⁺
            insert e to negative_lines // Case (g)
            insert flag of e to flags_of_negative_lines
            if e.target lies on L
                insert -L to negative_lines // Case (h)
                insert true to flags_of_negative_lines
        else // segment straddles the splitting line; split
            find intersection point I
            if e.source lies in L⁺ and e.target lies in L⁻
                insert e to positive_lines // Case (i)
                insert flag of e to flags_of_positive_lines
                insert L to positive_lines
                insert true to flags_of_positive_lines
                insert e to negative_lines
                insert flag of e to flags_of_negative_lines
            // The symmetric next case is included for completeness
            if e.source lies in L⁻ and e.target lies in L⁺
                insert e to negative_lines // Case (j)
                insert flag of e.source to flags_of_negative_lines
                insert -L to negative_lines
                insert true to flags_of_negative_lines
                insert e to positive_lines
                insert flag of e to flags_of_positive_lines
    construct positive_polygon from positive_lines and flags_of_positive_lines
    construct negative_polygon from negative_lines and flags_of_negative_lines
```

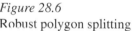

Figure 28.6
Robust polygon splitting

28.2 Implement a system that takes a set of nonoverlapping segments in E^1 and constructs a balanced labeled-leaf BSP tree directly from the segments (without incrementally inserting the segments). Write test routines to confirm that your code is correct.

28.3 The ability to insert a set of segments incrementally into an initially empty tree obviates the need to write the initialization function discussed in Exercise 28.2, yet such a function may still be useful. Predict which approach is faster and confirm using timing tests whether your prediction was accurate. Does the difference make the specialized initialization code worthwhile?

28.4 In § 28.4 pairs of points, rather than lines, were used to define the boundary of a polygon. Determine whether doing so has any benefit on your machine. To do so, read § 7.3 and then write a program that generates pairs of noncoincident points (in a square) and then constructs a line object from the pair of points. If you then check whether both points are indeed incident to the line they have defined, does your machine sometime conclude that that is not the case? If it does, what is the ratio of success to error?

29 Geometry-Free Geometric Computing

Coordinate-free geometric computing was introduced in Part III. Chapter 17 discussed why we may wish to develop a geometric algorithm for one geometry without manipulating coordinates directly, and Chapter 18 discussed how CGAL [36, 23] achieves coordinate freedom.

This chapter illustrates that we go further than coordinate-freedom and achieve geometry-freedom; a geometric algorithm may be designed and developed such that it remains geometry-neutral and thus purely combinatorial. The geometric parts are isolated from the combinatorial parts, and a geometry is used as a plug-in into the combinatorial algorithm to produce a geometric algorithm.

As templates in C++ can be used for data genericity as well as for algorithm genericity [1, 100], a set of classes and predicate functions can be bound to one of several sets of classes and functions. Such multiple instantiations of a single algorithm are shown to be useful in Part VII, where a single problem requires representing and manipulating point sets in different geometries. Another advantage is that the reliability of the resulting code is increased. If the code is written such that it amalgamates its infrastructure in more than one way, the potential locations for a programming error are diminished.

We have already encountered geometry-free, or geometry-neutral, algorithms in Parts I and II. Clipping and splitting operations were described such that the coupling between the operation proper and the underlying geometry is reduced. As a prominent example, polygon clipping should be described in isolation from the Euclidean or the oriented projective geometry in which its final implementation would reside. Ideally, the implementation itself would be generic.

29.1 A Generic BSP Node

As in Chapter 24, we use a single class to capture a binary search tree. The class BSP_node is parameterized by BSP_geometry as well as by boundary and interior type attributes (discussed in § 29.2). Each geometry provides its own actual types.

```
template<
  typename BSP_geometry,
  typename Boundary_attributes,
  typename Interior_attributes>
class BSP_node
{
  typedef typename BSP_geometry::Hyperplane        Hyperplane;
  typedef typename BSP_geometry::Point             Point;
```

```
    typedef typename BSP_geometry::BSP_convex_polytope  BSP_convex_polytope;

    typedef typename BSP_geometry::Sub_hyperplane       Sub_hyperplane;

    typedef BSP_node<BSP_geometry, Boundary_attributes, Interior_attributes> My_BSP_node;
    // hyperplane and boundary_attributes are valid for interior nodes only.
    Hyperplane hyperplane;
    Boundary_attributes boundary_attributes;

    // cell_is_occupied and interior_attributes are valid for leaf nodes only.
    bool cell_is_occupied;
    Interior_attributes interior_attributes;

    BSP_node * positive_child;
    BSP_node * negative_child;
    ...
};
```

For a generic (combinatorial) class to be used, it must be instantiated for a given geometry. Its algorithms (member functions) operate without knowledge of the particular geometry under which they apply. Whenever a combinatorial operation needs geometric data, the geometric subset is delegated to a given geometric component, itself relying on classes for a corresponding geometry.

To use BSP_node for E^2, for instance, it must be instantiated for a class BSP_E2 that acts as a traits class [71, 112] (see § 18.2) as well as an interface to the basic classes of E^2.

```
template<
    typename NT,
    typename Boundary_attributes,
    typename Interior_attributes>
class BSP_E2
{
public:

    typedef Segment_E2<NT>  Hyperplane;
    typedef Point_E2<NT>  Point;
    typedef BSP_E2_convex_polygon<NT>  BSP_convex_polytope;
    typedef BSP_E2_segment<NT>  Sub_hyperplane;

    typedef BSP_E2<NT, Boundary_attributes, Interior_attributes> BSP_geometry;
    typedef BSP_node<BSP_geometry, Boundary_attributes, Interior_attributes>  BSP_node_E2;
    ...
};
```

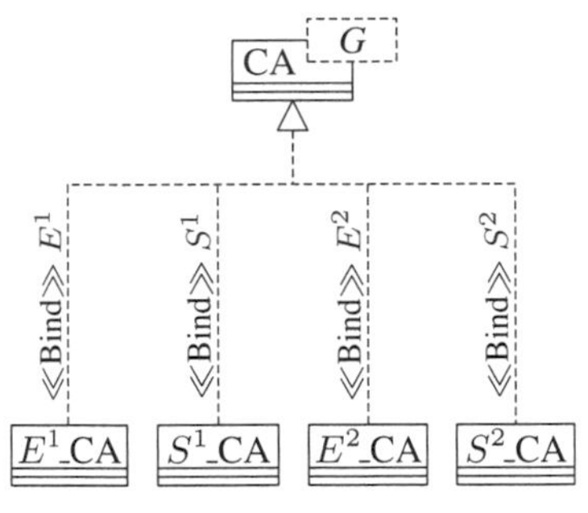

G: Geometry
CA: Combinatorial Algorithm

The adjacent figure illustrates using a conceptual UML diagram how a geometry is used as a plug-in to a combinatorial algorithm to yield one geometric algorithm or another depending on the plug-in.

Each geometry will have its own set of traits. The basic elements are a point and a hyperplane, as well as a function for determining the side of the hyperplane on which the point lies. Leaf nodes represent full-dimensional subsets in the corresponding geometry. Subhyperplanes are discussed in § 29.7.

29.2 Attributes

We often wish in practice to associate some attributes with a point set. In rendering applications we may want to save color and other surface material properties of the point set. In engineering we may want to save the density, strength coefficients, or other properties of the material. The former are examples of *boundary attributes* while the latter are examples of *interior attributes*.

Geometry	Hyper-plane	Point	BSP_convex_polytope	Sub-hyperplane
E^1	Point	Point (in E^1)	Segment	Point
E^2	Line	Point (in E^2)	Convex polygon	Segment †
E^3	Plane	Point (in E^3)	Convex polyhedron	Polygon †
S^1	Point	Point (in S^1)	Segment	Point
S^2	Circle	Point (in S^2)	Convex polygon	Segment

†: The segment or the polygon may be unbounded.

Interior Attributes

Whether an attribute is a boundary or an interior attribute is a relative notion. As we will see in Chapters 31 and 32, for instance, the boundary attribute of a set will become the interior attribute when the same set is projected (from E^2 to S^1 or from E^3 to S^2). Whenever a concrete class is instantiated from the generic class, the two attribute types then also need to be specified.

Interior attributes are only stored for leaf nodes labeled **true**. Since leaf nodes labeled **false** are not in the set, no attribute is maintained for them. In the accompanying code an object, rather than a pointer to an object, is stored, and so a default attribute is used for leaves labeled **false**. Storing the object itself simplifies programming since the client programmer need not worry about memory deallocation. The drawback is that common attributes that occupy significant storage are duplicated, but then the client programmer can define wrappers to share attributes via pointers when needed.

Boundary Attributes

The boundary attributes of a set are stored at interior nodes. As we will see in § 29.7, the boundary is collected from the tree by constructing the subhyperplanes. Because we use in this treatment a single-level binary tree to represent a point set, storing the boundary attributes is inherently limited: Only one attribute is stored at a hyperplane. If the hyperplane is used to construct more than one subhyperplane, all subhyperplanes must share the same attribute.

It is possible to remove this restriction while also adding the ability to represent nonregular point sets (see § 30.5) by using multi-level binary space partitioning. Each interior node for a tree representing a point set in E^2, for instance, would store a tree capturing a point set in E^1 [4, 76]. Thus, an n-dimensional point set would be represented using an n-level binary tree. We only point out to this possibility here, but do not pursue it further. The current framework would also make it possible to be generic in the geometry: an interior node in a binary tree for a geometry G^n would store a binary tree for a point set in the geometry G^{n-1}, where G can be either of the separable spaces E^n or S^n.

Attribute Stability

A Boolean operation is termed *stable* if the attributes of a set before the operation are not altered after the operation is performed. Consider, for example, calculating the difference $AC - BD = AB$ in E^1 for four distinct values A, B, C, and D that are in increasing order. For the operation to be stable, the interior attribute of AB must be that of AC before the operation is performed and the boundary attribute at A must remain unchanged while that for the result AB at B must be that for BD.

29.3 Boolean Operations and Attribute Maintenance

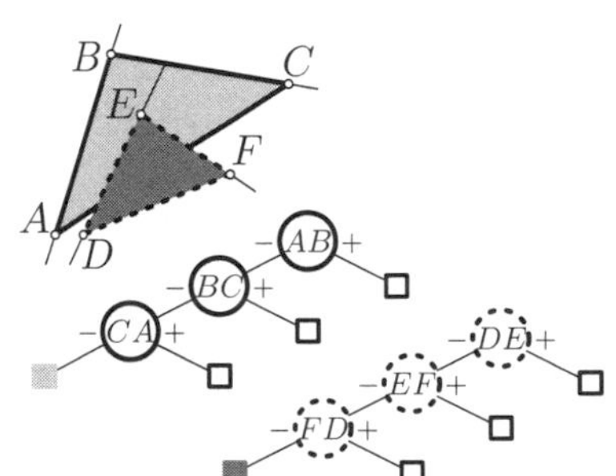

We focus here on computing Boolean operations between a BSP tree and a convex polytope [109], but it is also possible to compute Boolean operations between two BSP trees [72]. In both cases the result of the operation is a BSP tree, which indicates that the interior and the boundary attributes must be saved in the tree; once the operation is completed, no reference to the polytopes that originated the BSP tree is saved. Since the implementation of each Boolean function is only a few lines long, we avoid the obfuscation resulting from combining all functions into one as well as paying at run time the price of testing which operation is performed. As a running example, we consider applying the three functions on the two triangles ABC and DEF shown in the adjacent figure.

Union

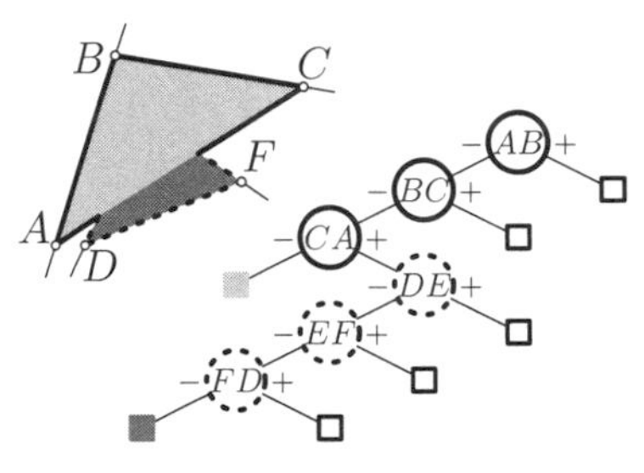

If the convex polytope is at an interior node, it is split into a positive and a negative part. This splitting function is one of the geometric parts of the algorithm and must be delegated to the corresponding geometric component. One or both polytopes after splitting are recursively inserted in the BSP tree. If a polytope reaches a leaf node that is already in the set, the polytope is discarded. If the polytope, on the other hand, reaches an empty leaf node, a subtree is built with the attributes of the inserted polytope and the subtree is appended at the leaf node. Building a subtree is also delegated to a geometric component.

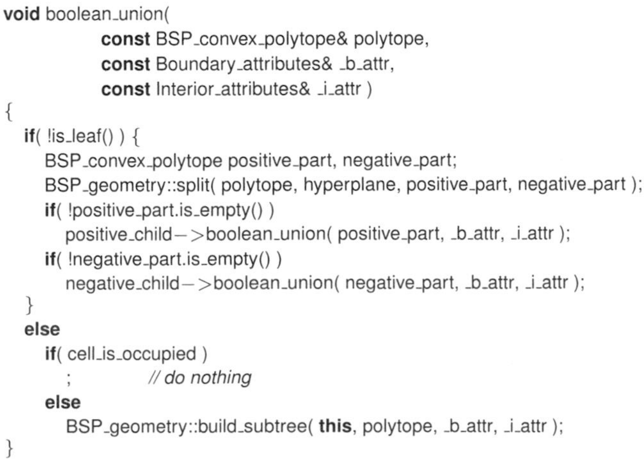

```
void boolean_union(
            const BSP_convex_polytope& polytope,
            const Boundary_attributes& _b_attr,
            const Interior_attributes& _i_attr )
{
  if( !is_leaf() ) {
    BSP_convex_polytope positive_part, negative_part;
    BSP_geometry::split( polytope, hyperplane, positive_part, negative_part );
    if( !positive_part.is_empty() )
      positive_child->boolean_union( positive_part, _b_attr, _i_attr );
    if( !negative_part.is_empty() )
      negative_child->boolean_union( negative_part, _b_attr, _i_attr );
  }
  else
    if( cell_is_occupied )
      ;           // do nothing
    else
      BSP_geometry::build_subtree( this, polytope, _b_attr, _i_attr );
}
```

Intersection

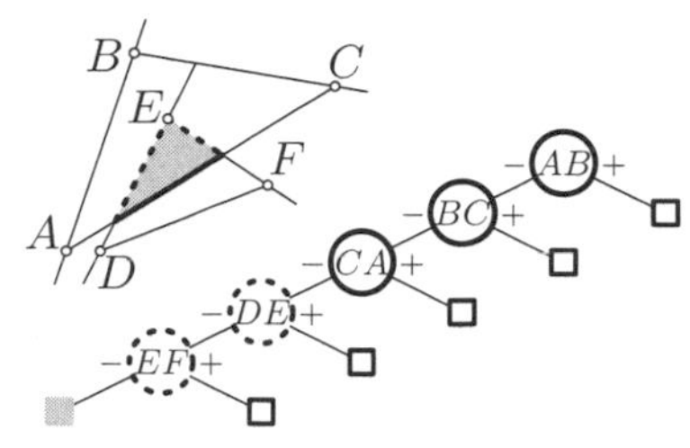

If during an intersection operation a positive part is empty after splitting, the subtree rooted at the positive child is recursively deleted. If it is not empty, it is treated recursively—and likewise for the negative part.

In the base case of the intersection function a subtree is built only at leaf nodes that are already occupied. In that case the boundary attributes of the convex polytope are used, but to maintain interior attribute stability the interior attribute of the leaf node of the BSP tree is used instead.

```
void boolean_intersection(
                const BSP_convex_polytope& polytope,
                const Boundary_attributes& _b_attr )
{
  if( !is_leaf() ) {
    BSP_convex_polytope positive_part, negative_part;
    BSP_geometry::split( polytope, hyperplane, positive_part, negative_part );
    if( positive_part.is_empty() ) {
      delete positive_child;
      positive_child = new BSP_node(false);
    }
    else
      positive_child->boolean_intersection( positive_part, _b_attr );
    if( negative_part.is_empty() ) {
      delete negative_child;
      negative_child = new BSP_node(false);
    }
    else
      negative_child->boolean_intersection( negative_part, _b_attr );
  }
  else
    if( cell_is_occupied )
      BSP_geometry::build_subtree( this, polytope, _b_attr, interior_attributes );
    else
      ;          // do nothing
}
```

Difference

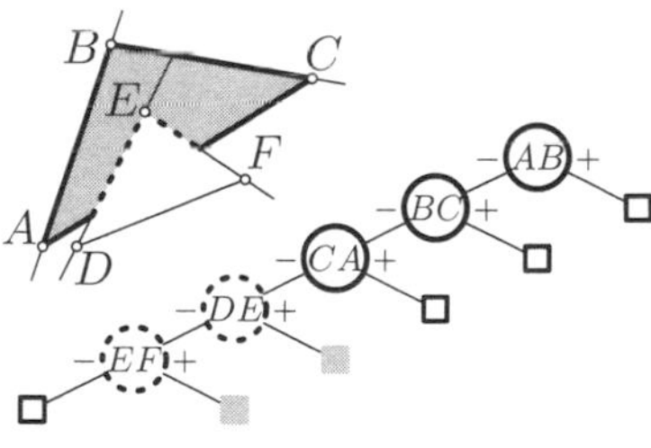

The recursive part of computing the Boolean difference operation is, as with union, a standard example of divide-and-conquer. When a polytope fragment reaches a leaf node that is occupied (already in the set captured), a subtree is constructed at that leaf node. But in this case the subtree construction routine is asked to reverse space: Those leaves that would have been labeled **false** are labeled **true** and the one leaf that would have been labeled **true** is labeled **false**.

```
void boolean_difference(
                const BSP_convex_polytope& polytope,
                const Boundary_attributes& _b_attr )
{
  if( !is_leaf() ) {
    BSP_convex_polytope positive_part, negative_part;
    BSP_geometry::split( polytope, hyperplane, positive_part, negative_part );
    if( !positive_part.is_empty() )
      positive_child->boolean_difference( positive_part, _b_attr );
    if( !negative_part.is_empty() )
      negative_child->boolean_difference( negative_part, _b_attr );
  }
  else
    if( cell_is_occupied )
      BSP_geometry::build_subtree( this, polytope, _b_attr,
                        interior_attributes, true /*reverse space*/ );
    else
      ;          // do nothing (implicitly discard polytope)
}
```

29.4 Subtree Construction in E^2

To construct the subtree, we extract from the convex polygon the *free bounding lines*—those where the polygon remains closed. In this code only the boundary and interior attributes of the convex polygon are involved, unlike the Boolean operations discussed above, where the attributes of the current node as well as the polytope are involved.

```
static void
build_subtree(
        BSP_node_E2 * current_node,
        const BSP_E2_convex_polygon<NT>& P,
        const Boundary_attributes& _boundary_attributes,
        const Interior_attributes& _interior_attributes,
        bool reverseSpace = false )
{
  std::vector<Segment_E2<NT> > bounding_lines = P.get_free_bounding_lines();
  if( bounding_lines.size() > 0 ) {
    typedef typename std::vector<Segment_E2<NT> >::iterator ExtSegIt;
    ExtSegIt lineIt = bounding_lines.begin();
    current_node->set_interior_node( *lineIt, _boundary_attributes );
    BSP_node_E2* current = current_node;
    if( bounding_lines.size() == 1 ) {
      current->positive_child = new BSP_node_E2( reverseSpace );
      if( reverseSpace )
        current->positive_child->interior_attributes = _interior_attributes;
      current->negative_child = new BSP_node_E2( !reverseSpace );
      if( !reverseSpace )
        current->negative_child->interior_attributes = _interior_attributes;
    }
    else {
      do {
        ++lineIt;
        current->positive_child = new BSP_node_E2( reverseSpace );
        if( reverseSpace )
          current->positive_child->interior_attributes = _interior_attributes;
        current->negative_child = new BSP_node_E2(false);
        if(lineIt != bounding_lines.end())
          current->negative_child->set_interior_node( *lineIt, _boundary_attributes );
        else
          current->negative_child->set_leaf_node( !reverseSpace, _interior_attributes );
        current = current->negative_child;
      } while( lineIt != bounding_lines.end() );
    }
  }
}
```

29.5 Point Containment

Determining whether a point is in the point set consists of recursively searching for the point in the binary tree. As long as the point is not incident to a hyperplane stored at an interior node on the search path, the answer will be readily found at the leaf.

If the point is incident to a hyperplane during the search, both subtrees are searched. The positive subtree search determines the positive neighborhood of the point and the negative subtree search determines the negative neighborhood. If both report that the respective neighborhood is in the set and both have the same interior attribute, the point is in the set interior and the attribute is known. If the interior attributes differ, we signal that the neighborhood is heterogeneous by returning the default interior attribute. If the set memberships differ, the point is known to lie on the boundary. We repeat for convenience a figure that we saw on page 256.

```
typedef std::pair<
    Set_membership,
    Interior_attributes > Classification_pair;

Classification_pair
classify( const Point& P ) const
{
    if( is_leaf() )
        return std::make_pair(
                    cell_is_occupied ? INSIDE_SET : OUTSIDE_SET,
                    interior_attributes);
    else {
        const Oriented_side os = oriented_side( hyperplane, P );
        if( os == ON_POSITIVE_SIDE )
            return positive_child->classify( P );
        else if( os == ON_NEGATIVE_SIDE )
            return negative_child->classify( P );
        else {
            const Classification_pair pos = positive_child->classify( P );
            const Classification_pair neg = negative_child->classify( P );
            if( pos == neg )
                return pos;
            else if( pos.first == neg.first )
                return std::make_pair( pos.first, Interior_attributes() );
            else
                return std::make_pair( ON_SET_BOUNDARY, Interior_attributes() );
        }
    }
}
```

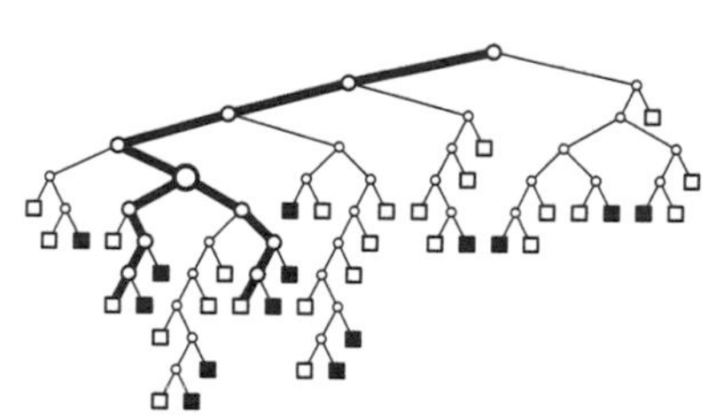

29.6 Collecting the Convex Polytopes

We can collect the point set from the tree by recursively partitioning a certain domain at the root of the tree. The elegance of spherical geometries is revealed when we consider that the universe of S^1 or S^2 can be sent to the function recovering the convex polytopes. Even though a segment in S^1 or a polygon in S^2 is not allowed to be the universal set, it is possible to add a flag to signal the universal set, which is then treated as a special case in the splitting routine.

The difficulty with Euclidean spaces can be solved by defining an *infimaximal* frame [68] that acts as an initially unbounded bounding box. Recursive divisions will be defined as a function of *extended points*, which have the advantage over Euclidean points of simulating ideal points by having at least one coordinate on a large but undetermined value. An implementation based on CGAL, which includes an *extended kernel*, established that infimaximal frames can indeed be used to avoid passing an initial polytope to the root of the tree, but we will be content here with discussing a standalone implementation that expects such an initial polytope.

```
typedef std::pair<BSP_convex_polytope,Interior_attributes> PolyAttr;
typedef std::vector<PolyAttr> Interior_list;

void
get_polytopes(
```

```
            const BSP_convex_polytope & polytope,
            Interior_list & collector ) const
{
    if( is_leaf() ) {
        if( cell_is_occupied )
            collector.push_back(
                std::make_pair( polytope, interior_attributes) );
    }
    else {
        BSP_convex_polytope positive_side;
        BSP_convex_polytope negative_side;
        BSP_geometry::split(
                    polytope, hyperplane,
                    positive_side, negative_side );
        if( negative_child && !negative_side.is_empty() )
            negative_child->get_polytopes(
                                negative_side,
                                collector);
        if( positive_child && !positive_side.is_empty() )
            positive_child->get_polytopes(
                                positive_side,
                                collector);
    }
}
```

The tree is traversed to collect either the interior or the boundary of the set. Collecting the interior occurs at leaf nodes, whereas collecting the boundary, which is discussed next, occurs at interior nodes.

29.7 Collecting the Boundary

To collect the boundary of the point set in the tree we pass a bounding box (for Euclidean geometries) or the universal set (for spherical geometries) to the root of the tree. Recursive splitting yields convex regions at each interior node; the convex polytope at a given node is the intersection of the halfspaces of the ancestors of the node. At any interior node, the intersection of the node's convex region with the node's hyperplane is termed a *subhyperplane*. Subhyperplanes are generic types that map to different concrete types for the different geometries (§ 29.1).

But the linear separation provided by splitting hyperplanes makes it possible to recover the boundary in a special order. If we pass a point to the boundary collection routine, it is possible to determine at each interior node the side on which the point lies. We can then return the boundary of the "far" set (the set opposite the query point) or the boundary of the "near" set (the set lying in the same halfspace as the query point) before the other.

To generate the order from near to far, the halfspace in which the observer lies is traversed first. The near-to-far, or front-to-back, order is particularly appealing as it can be used to construct the view using Boolean operations in a spherical geometry, the topic of Part VII.

```
void boundary(
        Boundary_list & ftb,
        const BSP_convex_polytope & current_cell,
        const Point & observer,
        bool also_get_backfaces )
{
  if( is_leaf() )
    return;

  const Oriented_side s = oriented_side(hyperplane, observer);

  BSP_convex_polytope p_polytope, n_polytope;
  BSP_geometry::split( current_cell, hyperplane, p_polytope, n_polytope );

  if( s == ON_POSITIVE_SIDE ) {
    if( positive_child )
      positive_child−>boundary(ftb, p_polytope, observer, also_get_backfaces);
    Sub_hyperplane sub_hyperplane;
    BSP_geometry::construct_sub_hyperplane(hyperplane, p_polytope, sub_hyperplane);
    ftb.push_back( std::make_pair(sub_hyperplane, boundary_attributes) );
    if( negative_child )
      negative_child−>boundary(ftb, n_polytope, observer, also_get_backfaces);
  }
  else if( s == ON_NEGATIVE_SIDE ) {
    if( negative_child )
      negative_child−>boundary(ftb, n_polytope, observer, also_get_backfaces);
    if( also_get_backfaces ) {
      Sub_hyperplane sub_hyperplane;
      BSP_geometry::construct_sub_hyperplane(hyperplane, n_polytope, sub_hyperplane);
      ftb.push_back( std::make_pair(sub_hyperplane, boundary_attributes) );
    }
    if( positive_child )
      positive_child−>boundary(ftb, p_polytope, observer, also_get_backfaces);
  }
  else { // ON_ORIENTED_BOUNDARY
    // either order will do
    if( negative_child )
      negative_child−>boundary(ftb, n_polytope, observer, also_get_backfaces);
    if( positive_child )
      positive_child−>boundary(ftb, p_polytope, observer, also_get_backfaces);
  }
}
```

The section concludes by discussing how subhyperplanes can be explicitly determined in E^2.

Splitting a Subhyperplane to a Convex Region in E^2

A subhyperplane is a continuous subset of a line in the plane. It may be either a line, a ray, or a segment: The term captures any of the three occurrences. Subhyperplanes are normally represented only implicitly in a BSP tree—in that neither the source nor the target of the segment (with one or both potentially at infinity) is known. Yet determining the two endpoints is occasionally needed. One example is the classical way of drawing a BSP tree in the plane. For the diagram to be intelligible, only those portions of splitting lines that intersect their respective cells are drawn.

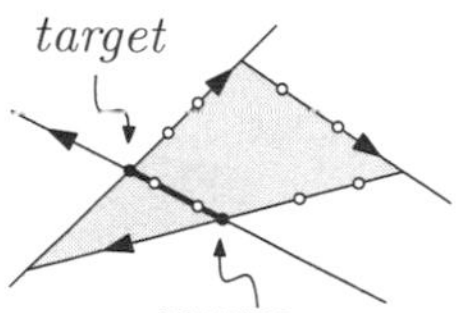

The subhyperplane associated with the root node of a tree is always a line, which can also be seen as a segment with two points at infinity. As before, a bounding box that is passed to the root of the tree is recursively split at interior nodes.

But determining the two endpoints of a subhyperplane (drawn segment) cannot be performed by simply clipping an initial (conceptually infinite)

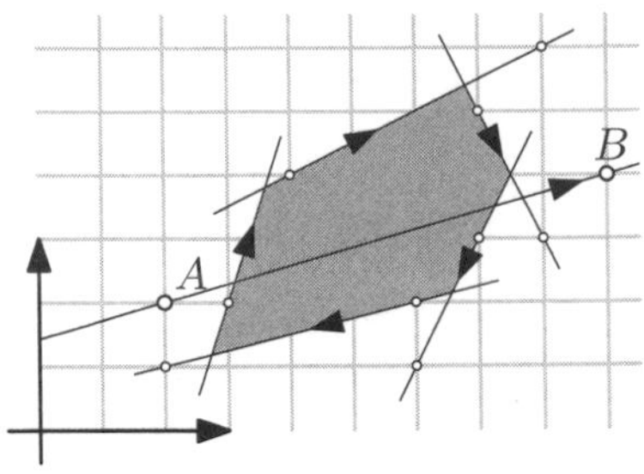

segment by the halfspaces (lines) defining the convex region. Doing so would not put an upper bound on the algebraic degree (§ 6.4) of the two endpoints. In the figure, for example, two unnecessary clip operations could potentially be performed raising the algebraic degree of the resulting points. The subhyperplane remains instead implicitly defined by a carrying line and a bounding cell. The two actual endpoints are determined only if the client code explicitly requests the source or the target of the subhyperplane.

The loop in the following code only determines the two bounding lines that will define the two endpoints, but no clipping is actually performed. Once the two bounding lines are known, two intersection operations will determine the endpoints. The signed distance along the carrying line of the intersection points is maintained. But this step also must be performed with care. Computing the actual distance would not do since it requires a square-root computation. It is also unnecessary to calculate the square of the distance; the projected distance suffices. Choosing between the two axes of projection is made based on the dominant direction (§ 2.4) of the carrying line.

```
template<typename NT>
std::pair<bool,Segment_E2<NT> >
clip_line_to_convex_region(
                const Segment_E2<NT>& carrying_line,
                const std::vector<Segment_E2<NT> >& region)
{
  const NT zero(0);
  NT min_largest  = + std::numeric_limits<NT>::max();
  NT max_smallest = − std::numeric_limits<NT>::max();

  Segment_E2<NT> s_min_largest, s_max_smallest;
  typedef typename std::vector<Segment_E2<NT> >::const_iterator Sci;
  for( Sci s = region.begin(); s != region.end(); ++s )
    if( !are_parallel(*s, carrying_line) )
    {
      bool plus = ( cross_product(carrying_line.get_Vector_E2(),
                        s−>get_Vector_E2()) > zero );

      NT lpd = find_larger_projected_difference(carrying_line, *s);
      if( plus ) {
        if( lpd < min_largest )
        { min_largest = lpd; s_min_largest = *s; }
      } else
        if( lpd > max_smallest )
        { max_smallest = lpd; s_max_smallest = *s; }
    }
    else            // simulate T2
      if( oriented_side(*s, carrying_line.source()) == ON_NEGATIVE_SIDE )
        return std::make_pair(false, Segment_E2<NT>()); // clipped out

  if( min_largest <= max_smallest )
    return std::make_pair(false, Segment_E2<NT>()); // clipped out

  Point_E2<NT> s = intersection_of_lines(carrying_line, s_max_smallest);
  Point_E2<NT> t = intersection_of_lines(carrying_line, s_min_largest);
  return std::make_pair(true, Segment_E2<NT>(s,t));
}
```

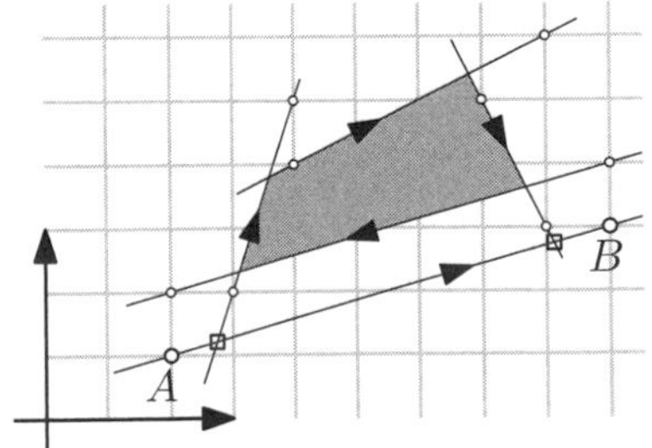

Notice that we cannot simply ignore bounding lines that are parallel to the carrying line. The two endpoints marked by squares would otherwise be reported as the endpoints of the segment when the segment is in fact clipped out by the convex region.

We conclude this section by showing an example for computing Boolean operations on a set of triangles in E^2. The triangles are formed from three vertices chosen randomly inside a square, and each Boolean operation is also chosen randomly between union and difference. Figure 29.1 shows the tree resulting from 25 such triangles, as well as the point set resulting from computing Boolean operations on 25 and from 250 triangles. For clarity, only the subhyperplanes corresponding to the top two levels in the tree are shown in the last case.

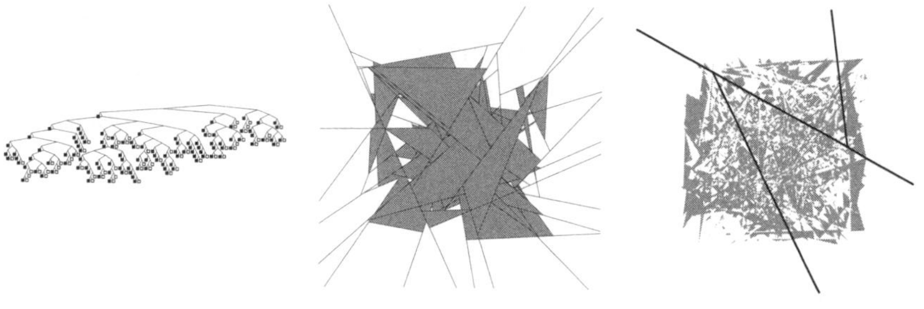

Figure 29.1
A binary tree for 25 triangles is shown on the left; the resulting point set in the middle; and the point set resulting from 250 triangles on the right.

29.8 Developing for Multiple Geometries

The main advantage of developing algorithms by separating combinatorial and geometric steps is the gain in software development and maintenance time. Developers of a geometric library handling g geometries and a algorithms would not need to write ga implementations, but only g geometric infrastructures and a algorithms.

Another advantage of developing algorithms that can be simultaneously used under multiple geometries is the ensuing simplification in unit testing. If one module or unit test is required for each of a algorithms, it would be sufficient to write a tests and ensure that the different geometries are covered in the tests. In this way one can conceivably port software written for Euclidean geometry to spherical geometry or vice versa.

Geometry \ Algorithm	1	2	3	4	5
A				⋆	
B		⋆			⋆
C	⋆				
D			⋆		

We would like in general to layer our geometric systems so that we consider problems one at a time, isolating and developing modules as the project proceeds. Yet as with many geometric systems [52], BSP trees offer an example where the issues become intertwined. Floating point filters (which were briefly alluded to in Chapter 7) do not help with BSP trees [56]. Floating point filters work on the premise that most predicates will make an "easy" decision, where, say, the test between the sidedness of a point and a hyperplane occurs most often when the point is not near coincidence to the hyperplane. Handling inputs of nontrivial size with BSP trees means that the number of predicates where a point will be tested against a hyperplane to which it is indeed incident will grow rapidly. This occurs already for nodes at a very low depth in the binary tree. The result is that floating point filters will fail so often as to render them useless. If the filters trigger computations based on the underlying exact implementation and that underlying implementation continues to require more bits, then we have the recipe for an algorithm that is theoretically correct but useless in practice.

Such a vast gap between theory and practice is an indication of why writing the time and space complexities of an algorithm assuming constant time per real number operation is an incomplete metric. Analysis must also account for the order of constructibility of objects used by the predicates of an algorithm and the ensuing growth in the size of these objects.

Practitioners are also not immune to missing serious issues in their implementations. The current practice is that one or a few carefully selected examples suffice to convince reviewers that an algorithm is sound *and* practical. Yet that is no guarantee that the algorithm can indeed be used for an arbitrary input.

BSP trees make it possible to compute regularized Boolean operations. A more general alternative is to use Nef polyhedra [11, 45], which make it possible to construct and manipulate arbitrary point sets.

29.9 Exercises

29.1 The convex hull of a set of points P in the plane is a minimal sequence of points $C \subset P$ such that points in P are colinear or lie on the left side of each pair of consecutive points in C [83, 74, 30].

This definition can be interpreted in either E^2 or S^2. Write a generic C++ function that computes the convex hull in either geometry. Assume that the points in S^2 all have $z > 0$.

29.2 Generate a PostScript rendering of your solution under both E^2 and S^2 in Exercise 29.1.

29.3 Repeat Exercise 29.1 for Delaunay triangulations [63].

30 Constructive Solid Geometry

Constructive solid geometry (CSG) is computing's equivalent to building solids out of physical primitives. Two primitives can be assembled into a larger one, space can be carved out of a primitive, and, farther from a physical manipulation, the intersection of two objects can be determined.

As with binary space partitioning, also in constructive solid geometry solids are represented using trees. In both a BSP tree and a CSG tree primitives are stored at leaves, but whereas they are defined in a BSP tree implicitly by the intersection of a node's ancestors halfspaces, primitives are defined from elementary building blocks in a CSG tree and Boolean operations are stored in interior nodes to assemble the primitives into the desired compound solid.

30.1 A CSG Example

Consider that we wish to construct a rectangular slab in which four holes will be drilled to accommodate four rivets. Two methods are possible. We could start from a slab and remove four cylinders one after the other, or we could start by building a solid consisting of the four cylinders and remove the resulting solid from the slab. The "remove" operation here is the Boolean difference operation.

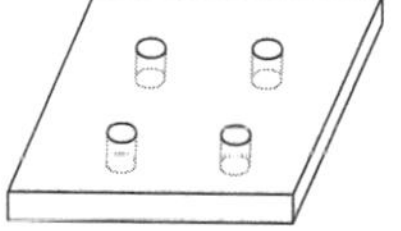

When representing each such negative space using a cylinder, note that the planar facets of the cylinder need not share the planes of the slab. Each cylinder's height should instead be larger than the thickness of the slab to protrude on both sides of the slab.

30.2 CSG Class Diagram

We wish to represent an arbitrary solid that can be constructed by applying Boolean operations from primitive solids. Boolean operations will consist of the set {union, intersection, difference} and primitive solids will consist, say, of the set {cube, cylinder, cone}. A binary tree will represent the solid we are modeling in which the primitive solids are leaf nodes and the interior nodes are Boolean operations. A transformation node also needs to be introduced to make it possible to position a primitive at an arbitrary position in space. Transformation nodes will also be interior nodes.

This suggests the following constraints on the class relationships.

- A node is either a Boolean_operation, a Primitive_solid, or a Transformation.
- An Interior_node is a Boolean_operation or a Transformation.
- A Leaf_node is a Primitive_solid.
- A Transformation_node has a Primitive_solid node as its single child.
- A Boolean_operation has two Interior_node children.

This suggests the inheritance hierarchy in Figure 30.1.

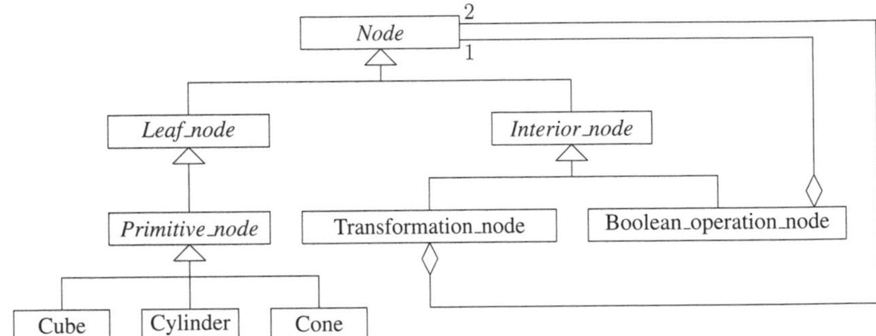

Figure 30.1
CSG inheritance hierarchy—first try

But there is no need to distinguish explicitly between an interior node and a leaf node, and so the hierarchy in Figure 30.2 will be sufficient. Observe that even if the designer imposes that primitives be already positioned at the desired location in space before attaching the corresponding nodes to the tree, a Transformation_node class remains necessary, as we will see shortly.

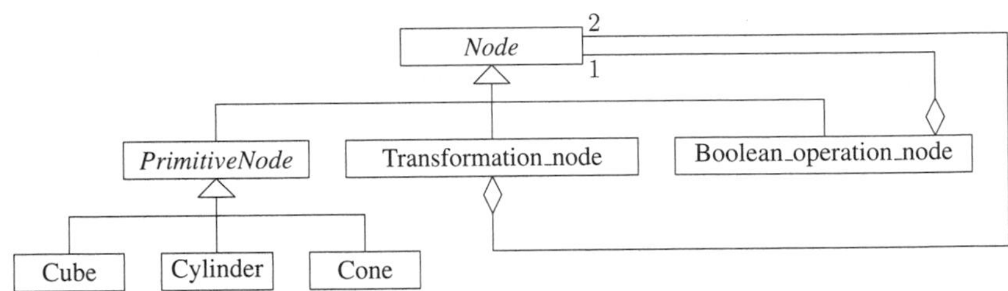

Figure 30.2
CSG inheritance hierarchy

The resulting UML class diagram arises frequently: This *composite design pattern* [39] is encountered in domains from text editors to scene graphs. It is worth iterating the bare version of this pattern in Figure 30.3 to relate the names of its classes to those of a CSG class diagram.

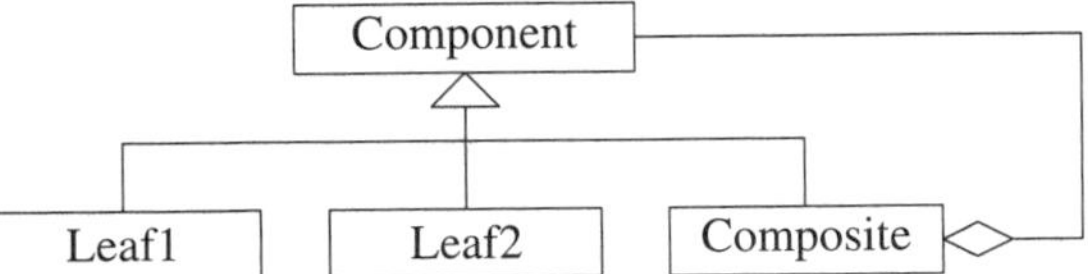

Figure 30.3
The composite design pattern

30.3 CSG Object Diagram

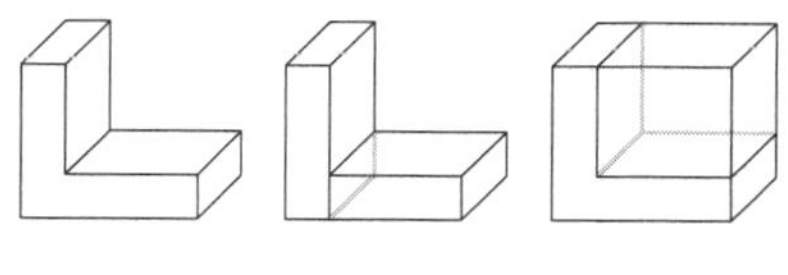

Suppose that we wish to build an L-shaped figure or an angle bracket. The bracket could be constructed in one of several ways, perhaps using union or difference. Choosing to represent the bracket using a union operation leads to the adjacent object diagram. Recall the distinction between class and object diagrams: A class diagram is fixed once the program has been written, whereas an object diagram represents a snapshot for a running program of the machine's state.

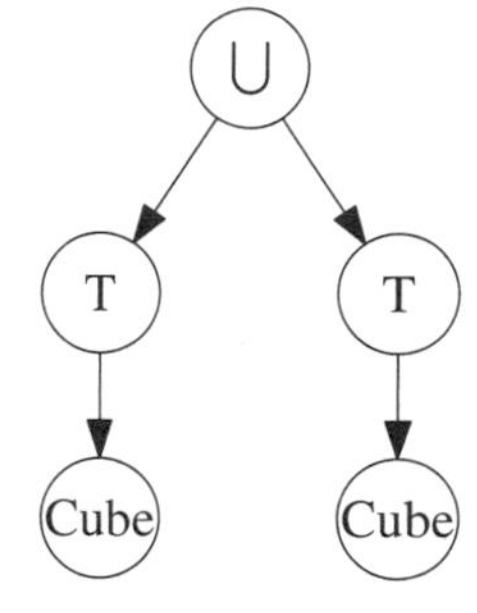

The object diagram that results in this case is a tree, but this need not be the case. If the object being constructed contains some redundancy—four identical wheels in an automobile, an airplane's two wings that are identical by reflection, and so on—one node would be referenced by multiple transformations yielding multiple instances of a subtree. Yet since a part cannot consist of instances of the whole, the graph may not contain cycles. The resulting graph is a directed acyclic graph (DAG). DAGs are also at the core of another modeling schema, scene graphs such as the Virtual Reality Modeling Language (VRML).

30.4 Point Classification

A first application for a CSG system is *point classification*. The CSG solid is queried with a point and is asked whether the point lies inside, outside, or on the surface of the solid. An implementation would define an abstract function Node::classify that will be overridden by the classes in the CSG hierarchy.

```
template<
    typename Point,
    typename Neighborhood,
    typename Transformation>
class Node {
public:
    virtual Neighborhood classify(const Point& p) = 0;
    ...
};
```

As with BSP trees, we can plug some Euclidean geometry at a given dimension into the generic CSG classes. In this way a transformation is to be understood as a type parameter for a concrete transformation class in the chosen geometry. The neighborhood (§ 30.6) class and the point class are also type parameters.

Since the client only has a pointer to the root of the CSG DAG, it is the root node that will first receive the query message. The message is sent down the tree to the leaf (primitive) nodes, each of which will know how to compute a local answer. The answers are then assembled on the way up. Transformation nodes will transform the point on its way down but will not affect the reply. Boolean operation nodes will only pass the request to their children and will apply the particular operation they define on the reply on its way up the DAG.

30.5 Regularization

The difficulty we encounter at this point results from our sloppiness in taking the term *solid* as obvious enough not to need to be defined. To see that the term is not obvious, it is sufficient to consider the problem in one dimension, where a solid becomes a segment in E^1, though it is more convenient to use the term *interval* instead.

An interval in E^1 is defined by two points and consists of the set of points between the two bounds. An interval may be either *closed* or *open* at either endpoint. Boolean expressions can be readily constructed on intervals. An expression such as $[1,3] \cup [3,5]$ evaluates to $[1,5]$. An expression such as $[1,3] - [3,5]$ evaluates to $[1,3[$. The ability to encode open and closed intervals means that it remains easy to evaluate expressions such as $[1,3] \cap [3,5]$, which results in $[3,3]$, and $[0,6] - ([1,3] \cap [3,5])$, which results in the disjoint set $[0,3[\cup]3,6]$.

Although this interpretation of Boolean expressions is intuitive as intervals on the reals, it ceases to be intuitive if we interpret an expression as a (physical) solid. If each interval is meant to capture a cylinder of unit diameter along the x-axis in 3D, for example, it would be natural to interpret an expression such as $[1,3] \cap [3,5]$ not as a volumeless unit disk at the value $x = 3$, but as the empty set. Similarly, we wish to think of the expression $[0,6] - ([1,3] \cap [3,5])$ as resulting in a single cylinder $[0,6]$ rather than in two cylinders equivalent to two semi-open intervals.

This trouble results because the operators that we apply on solids result formally in solids that have no volume or that do not include (a subset of) their boundary, whereas intuitively solids that have no volume have effectively "vanished" and a solid always includes its boundary.

Such an inconsistency between the formal and the intuitive interpretation of operators can be resolved by enforcing these two constraints. This enforcement, or *regularization*, will be applied to the outcome of each Boolean operator. Thus, it is best to replace the formal Boolean operators with more intuitive ones, *regularized Boolean operators*, which are denoted by an additional asterisk.

In one dimension, regularization effectively disallows open intervals and discards intervals whose two endpoints are equal. The cylinder equivalent to $[1,3] \cap [3,5]$ now becomes the empty set and the one equivalent to $[0,6] - ([1,3] \cap [3,5])$ becomes a single cylinder.

30.6 Neighborhood

Neighborhood captures adjacency. The neighborhood of a point $P(x)$ in E^1 is the set of points lying in $(x-\epsilon, x+\epsilon)$ for a small ϵ. Similar notions apply in E^2 and in E^3. To represent whether the neighborhood of a point in one of these two geometries is inside the set in question, one uses a point set defined on S^1 or on S^2, respectively.

Even if the client programmer has no need for the neighborhood at a point and would be content to know whether a given point is inside, outside, or on the boundary of a set, neighborhoods are necessary intermediate results during the propagation of the answer from leaf nodes through the Boolean operators up the tree. To see why, consider in E^1 the value returned from the two operations

$$5 \overset{?}{\in} [1,5] \cup [5,9]$$

and

$$5 \overset{?}{\in} [1,5] \cap [5,9].$$

The number 5 lies in both cases on the boundary of the two intervals, but how could such two values be combined using a union or an intersection operation to yield the required "inside" and "outside" answers, respectively?

The information missing and that needs to be propagated is the neighborhood of the point [86]. In E^1 two additional flags are needed. A left, L, and a right, R, Boolean flag will both be true if the point is strictly in an interval. One or the other will be set to false if the point is in the interval, but its neighborhood on one side is not.

This additional information is in fact all that's needed (and could conceivably be implemented by defining S^0, though that would perhaps be excessive). The interpretation of the pair of Boolean flags is as follows:

1. L=false and R=false $\Longrightarrow$ the point is outside the cylinder.
2. L=false and R=true $\Longrightarrow$ the point is on the boundary (on the left end of the cylinder).
3. L=true and R=false $\Longrightarrow$ the point is on the boundary (on the right end of the cylinder).
4. L=true and R=true $\Longrightarrow$ the point is strictly inside the cylinder.

The implementation of neighborhood in E^1 will reveal the distinction between the difference operator we expect and the subtraction operator in C++. C++ defines **false** – **true** as **true**, whereas we wish for the difference to evaluate to **false**. The problem is avoided by writing the difference as `a && ! b`.

To generalize the notion of neighborhood to two and three dimensions, we define primitives in two dimensions to be a set of polygons and would implement regularized Boolean operations requiring the maintenance of an S^1-neighborhood for the point queried. One way to think of this use of S^1 is by analogy to a pie. Arbitrarily sized slices of the pie can be taken out—they need not be adjacent. If the pie is empty, the point is outside; if the pie is complete, the point is inside; and if the pie is partially complete, the point is on the boundary. The usual robustness problems will apply if one wishes to detect accurately that a point lies on the boundary.

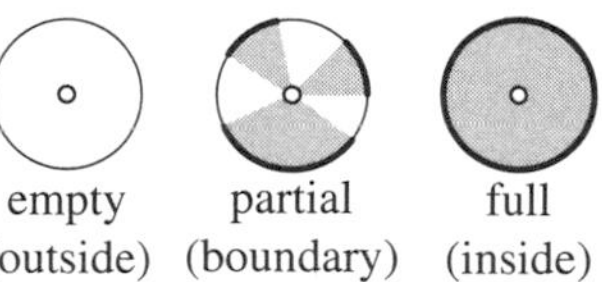

In a rendering application knowing the normal vector is crucial since it makes shading possible. The normal vector to an E^2-solid can be extracted from the S^1-neighborhood.

1. If the point is on an edge, its neighborhood is described by the normal to the edge. The normal is directed outside of the solid being represented.

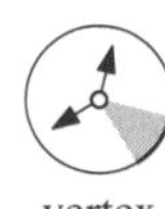

2. If the point lies on a vertex, its neighborhood is described by a set of *pairs* of normals.

In practice, just a single pair of normal vectors will be used to represent a vertex. If more than a pair of normals are needed to represent a vertex, the polygon represented is said not to have manifold topology (§ 27.2) at the vertex.

If we are representing solids in E^3, the data structure needs be augmented yet again. The neighborhood can be one of the following (omitting the cases where it is either a full ball or the empty set):

1. The point is on the face of polygon representing the boundary of the solid. The neighborhood will be described by a single normal vector.
2. The point is on an edge of the solid. The neighborhood is described by a pair of normal vectors. A set of pairs of normals may be allowed. This case is similar to the case of the neighborhood of a point in 2D. In 3D it signifies that multiple faces meet at one edge. A CSG system must be able to handle such a neighborhood as an intermediate result even if the final solid being constructed is not allowed to have such a neighborhood.
3. The point is on a vertex of the solid and the neighborhood is described by a set of normal vectors. These normal vectors describe the polygons adjacent to the vertex. Once again, a set of sets of such normal vectors may be needed, in which case the vertex is degenerate and multiple "spikes" of the solid meet at one point in space. But there is an additional complication: It is possible for one such "spike" to contain a hollow, or a negative space, of a set of normals.

Ray Casting

To determine the intersection of a ray with a CSG solid, the ray is percolated down the tree to the leaves (and transformed along the way at transformation nodes). The intersection of the ray with each solid at a leaf node is determined as an interval on E^1 along the ray with the same origin as the ray. As the intervals are percolated back up the tree, regularized Boolean operations are performed on the intervals, possibly using BSP trees. At the root, the lower boundary of the first interval is reported as the nearest intersection.

30.7 Exercises

30.1 Implement ray casting on a CSG expression in E^2. Choose an appropriate method for computing Boolean operations on intervals in E^1.

30.2 Implement ray casting on a CSG expression in E^3. Also, choose an appropriate method for computing Boolean operations on intervals in E^1.

30.3 A simple way for computing areas and volumes is to use random sampling. Implement a system for a CSG tree in E^2 that estimates the area of the expression by selecting a large set of random points in a box bounding the solid in the plane and classifying each point. The required area is the ratio of points inside the solid over the sum of points inside and outside the solid multiplied by the area sampled. Choose some acceptable tolerance (say 1%) and then determine the number of random points needed for the area to fall within that tolerance from the area calculated exactly (by hand). What is the likelihood for a random point to lie on the boundary of one of the primitives?

30.4 Repeat Exercise 30.3 in E^3.

Part VII

Vector Visibility

31 Visibility from Euclidean to Spherical Spaces

Given an observer and a set of triangles (or other piecewise linear geometric objects) in the Euclidean plane, we wish to determine the subset of the triangles' edges that are visible from the observer. If we assume that the observer is omni-viewing, the view is naturally captured on the surface of a circle centered at the observer. This chapter discusses a solution for this problem, *vector visibility in 2D*. This classical problem in "flatland" has many applications such as rendering scenes consisting of 3D mazes, which can be adequately modeled using a 2D visibility algorithm.

31.1 Incorrectness of Sorting by Distance

Before discussing how one can compute visibility, we spend a moment discussing what cannot be done. It is tempting to assume that one can reduce visibility to sorting, but we will see that sorting is inadequate in 2D, and hence it is also inadequate in 3D.

Suppose the input consists of a set of segments in the plane and one wishes to determine the segments visible from the viewer. It is tempting to assume that the correct near-to-far order from the viewer can be determined by sorting the distances of one sample point on each segment. As the counterexample shown in the adjacent figure suggests, choosing the midpoint of each segment and sorting using the resulting set of points may lead to an incorrect *depth order*.

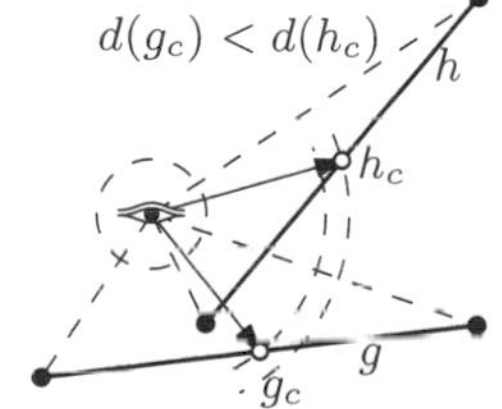

31.2 Representing the Output

2D visibility results in a partitioning of S^1 into spherical segments. Each spherical segment corresponds to a segment in a scene polygon. As discussed in § 28.4, by orienting the polygons clockwise, the circular projection at a viewer located outside the scene polygons will be oriented counterclockwise. The partitioning of S^1 also consists of a set of spherical points. Each point corresponds to the projection of a scene vertex on S^1.

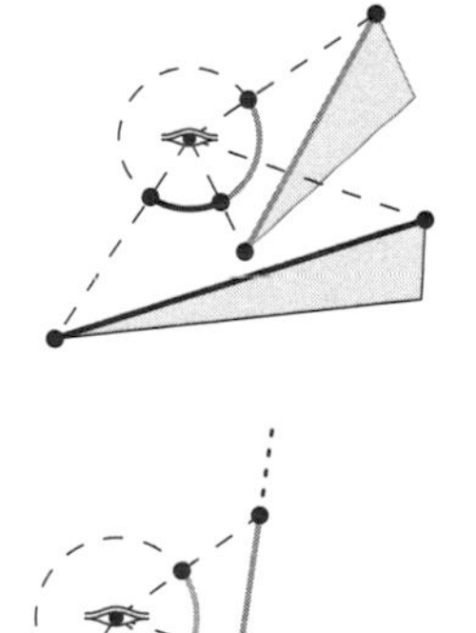

Notice that we are unable to constrain the resulting set of spherical segments to be minimal. As illustrated in the figure, a point E may be induced on an input segment following the construction of the E^2 BSP tree for depth order. Visibility is then determined for a set of three segments and E', the projection of E, will arise in the output. If one wishes, it is possible to purge such extra vertices after visibility is complete.

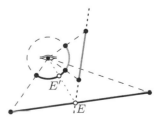

31.3 Data Structures for 2D Visibility

The scene consists of a set of pairs, where each pair joins a triangle with some material type. Since the interior of triangles cannot be seen from an observer situated outside the triangles, it is not necessary to capture the interior material (generically termed attributes in § 29.2). The relevant material is instead that of the boundary of the triangles.

The triangles are inserted in an arbitrary order (though the order can affect time and space requirements [78]) into a BSP tree in E^2. A BSP_node_E2 instance represents the scene in E^2. The view is also represented by an instance of a BSP tree, but one in S^1. The type in that case is BSP_node_S1.

```
template<typename NT, typename Material>
class View_E2
{
public:

    // Types for scene
    typedef std::pair<Triangle_E2<NT>, Material > Triangle_and_material;
    typedef std::vector<Triangle_and_material > Scene;

    // Types for scene as BSP tree
    typedef BSP_E2< NT, Material, char > BSP_geometry_E2;
    typedef typename BSP_geometry_E2::BSP_convex_polytope BSP_E2_convex_polygon_d;
    typedef typename BSP_geometry_E2::Sub_hyperplane BSP_E2_segment_d;
    typedef BSP_node< BSP_geometry_E2, Material, char > BSP_node_E2;
    typedef typename BSP_node_E2::Boundary_list Boundary_list;

    // Types for view tree
    typedef BSP_S1< NT, char, Material > BSP_geometry_S1;
    typedef typename BSP_geometry_S1::BSP_convex_polytope BSP_S1_segment_d;
    typedef typename BSP_geometry_S1::Sub_hyperplane Sub_hyperplane_d;
    typedef BSP_node< BSP_geometry_S1, char, Material > BSP_node_S1;
    typedef typename BSP_node_S1::Interior_list Interior_list;

    ...
};
```

The View_E2 class is parameterized by the number type as well as by the material, or surface attribute, of the scene triangles. Whereas the material is a boundary attribute in E^2, it becomes an interior attribute in S^1. BSP trees provide a function for collecting the interior (§ 29.6) as well as a function for collecting the boundary (§ 29.7). The latter is used for extracting data from the first tree and inserting to the second.

31.4 Computing the View

Computing the view is done in two steps. First the depth order from the observer is established, and then the resulting set of segments is inserted in front-to-back order to construct the view.

To extract the segments on S^1 that together form the view, a special segment is initialized to the complete set of directions (S1_FULL). That set is then recursively divided. The occupied cells among the leaf nodes are collected into a list capturing the view.

```
View_E2(
        const Scene & triangles,
        const Bbox_E2<NT> & bbox,
        const Point_E2<NT> & observer)
{
    get_depth_order( triangles, bbox, observer );
    determine_view( front_to_back_list, observer );

    const BSP_S1_segment_d S1(S1_FULL);
    view_segments = view_tree.get_convex_polytopes(S1);
}
```

31.5 Computing the Depth Order

To determine the depth order, the scene triangles are inserted into the BSP tree in E^2 using Boolean union. At this stage the boundary material of each triangle is also inserted. If the scene is static and the observer moves, the same BSP tree can be used for determining the different depth orders. The front-to-back order is computed by querying the tree from the position of the observer.

```
void
get_depth_order(
            const Scene & triangles,
            const Bbox_E2<NT> & bbox,
            const Point_E2<NT> & observer)
{
    typedef typename Scene::const_Iterator Scene_CI;
    for( Scene_CI ti = triangles.begin(); ti != triangles.end(); ++ti )
        // auto promotion of ti->first
        tree.boolean_union( ti->first, ti->second, 0 );

    const BSP_E2_convex_polygon<NT> initial_convex_polytope(bbox);

    front_to_back_list = tree.get_boundary(
                                initial_convex_polytope,
                                observer );
}
```

31.6 Reducing Visibility to Boolean Operations

After we compute a list of the input segments ordered by depth, visibility is determined by processing the segments from near to far. A *mask* in S^1 could be used to store the set of segments so far encountered. Only those portions of a segment lying outside the mask are visible [55]. If the mask and the view after segment s_{i-1} is processed are M_i and V_i, then the two Boolean operations

$$M_{i+1} = M_i \cup s_i,$$
$$V_{i+1} = M_i - s_i$$

would incrementally update the mask and determine the visible scene fragments.

Yet it is unnecessary to maintain a mask; a BSP tree in S^1 can hold the view and can also act as a mask. Attribute stability (§ 29.2) ensures that the material of invisible fragments does not affect the view and that invisible fragments are quietly discarded. Thus, in the following code a single invocation of Boolean union suffices to determine the view. But first each E^2 segment is projected onto S^1 centered at the observer by two point-subtraction operations.

```
void
determine_view(
            const Boundary_list & front_to_back_list,
            const Point_E2<NT> & observer )
{
  typedef typename Boundary_list::const_iterator Segment_CI;
  for( Segment_CI si = front_to_back_list.begin();
      si != front_to_back_list.end(); ++si )
  {
    Point_S1< NT > src( observer, si−>first.source() );
    Point_S1< NT > tgt( observer, si−>first.target() );

    BSP_S1_segment_d s( src, tgt );

    view_tree.boolean_union( s, 0, si−>second );
  }
}
```

31.7 Example of Vector Visibility

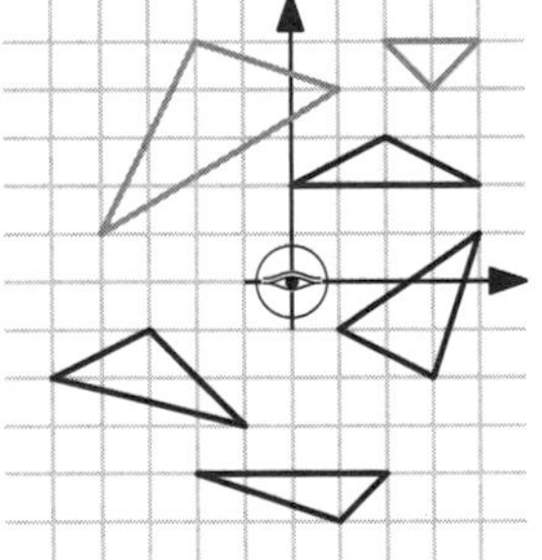

The discussion in 2D may be interesting in its own right, but it is mostly interesting as a prelude for visibility in 3D. Had we only been interested in computing visibility from E^2 to S^1, then either a randomized incremental or a rotational sweep algorithm would have also been adequate [35]. The advantage of the present approach is that the exact same set of functions can be implemented for computing visibility from E^3 to S^2. This is such the case, in fact, that one would be able to implement a single view class that is parameterized with a generic Euclidean geometry and with another generic spherical geometry. But for now we look at determining the 2D view of the set of triangles shown in the adjacent figures.

To render a scene, we construct a View_E2 instance from a scene consisting of a set of triangles, a bounding box, and an observer.

```
const Point_E2d observer(0, 0);
const Bbox_E2d bbox( Point_E2d(−6,−6), Point_E2d(5,6) );
const View_E2d view( triangles, bbox, observer );

render(triangles, "psout/e2_to_s1_scene1.eps", observer, view, bbox);
```

The adjacent figure is then rendered from the triangles in E^2 and the view in S^1.

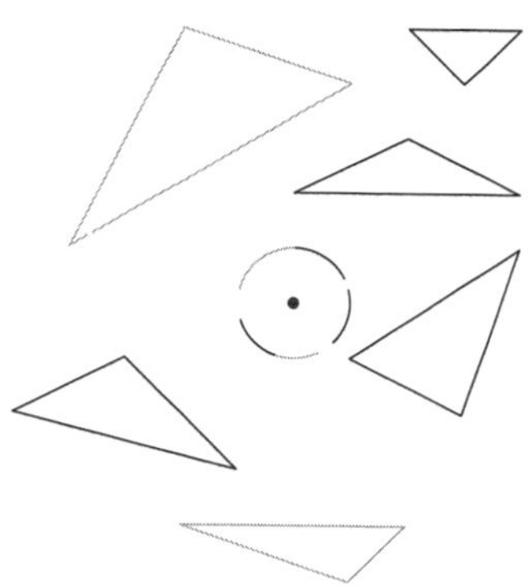

```
void render(const Scene& triangles,
        const string& filename,
        const Point_E2d& observer,
        const View_E2d& view,
        const Bbox_E2d& bbox)
{
  Postscript_d PS(filename, bbox);

  Interior_list view_segments = view.get_view_segments();

  Scene::const_iterator triangle_ci = triangles.begin();
  PS.set_line_width(0.8);
  while( triangle_ci != triangles.end() ) {
    PS.set_gray_stroke( triangle_ci->second );
    PS.draw( triangle_ci->first );
    triangle_ci++;
  }

  PS.set_gray_stroke( 0.0 );
  PS.draw(observer);

  Interior_list::const_iterator ci = view_segments.begin();
  while( ci != view_segments.end() ) {
    PS.set_gray_stroke( ci->second );
    PS.draw( ci->first.get_Segment_S1(), observer, 1.0 );
    ci++;
  }

  PS.close();
}
```

31.8 Exercises

31.1 Give an example to show that the view of two segments in the plane from a viewer may not be determined by sorting the two segments using the closer of their two endpoints.

31.2 Repeat Exercise 31.1 to show that the heuristic of sorting by the distance from the source of each segment (with input triangles consistently oriented) is also flawed.

31.3 Consider the case of an observer moving in a static scene consisting of a set of triangles. Develop an interactive system that uses the class View_E2 to determine the view and that renders a rasterized version of the view interactively as the user drags the mouse to simulate the motion of the observer. Calculate the view direction from the last two distinct points the user has sampled.

32 Visibility in Space

The text concludes with an algorithm for visibility in 3D. This chapter merely pulls together the ingredients, as they have all already been encountered. Discarding invisible sets relies on the use of Boolean algebra on regular sets (Chapters 28 and 29) adapted to spherical geometry (Chapters 8 and 9). By using rational numbers as the number type (Chapter 7) and ensuring that the algebraic degree of intermediate objects and predicates remains low (Chapter 6), it is possible to design and implement a 3D vector visibility system that never fails. The resulting half-edge data structure is constructed using only two Euler operators (Chapter 27).

Using a general and layered solution to this most intriguing and most classical [108] of problems in computer graphics and computer-aided design may be what is missing to spur vector rendering. Raster images are well suited for raster displays, but they are ill-matched to the printed page, which are capable of significantly higher resolutions than can be afforded by raster images. The computation of the *spherical visibility map* [56] may be a suitable starting point for large areas of computer graphics to be restudied with the objective of producing pure vector shaded synthetic imagery aimed at the printed page. The reader interested in a survey of the vast visibility literature is referred to the work just cited for starting pointers.

32.1 Spherical Projection

When bound by a physical medium—a planar canvas, the printed page, or a computer screen—planar projections (§ 11.4) are the appropriate model. But more abstractly, a sphere captures omnidirectional viewing and is the natural projection medium. Spherical visibility is suitable as preprocessing; any planar projection can subsequently be generated by central reprojection onto a planar surface.

A spherical projection is concisely specified by no more than the location of the observer, whereas the observer location, a viewing direction, and the field of view are needed to specify a planar projection.

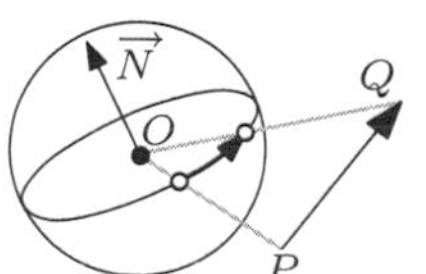

In a spherical projection, a point in E^3 maps to a point in S^2 defined by the direction from the observer to the Euclidean point. A segment in E^3 also maps to a segment in S^2 (unless the observer lies on the segment's carrying line).

Even though we will not perform computation directly on lines in space, their projection on a sphere provides insight. A line in E^3 also projects to a segment in S^2. The endpoints of the image segment are antipodal points and

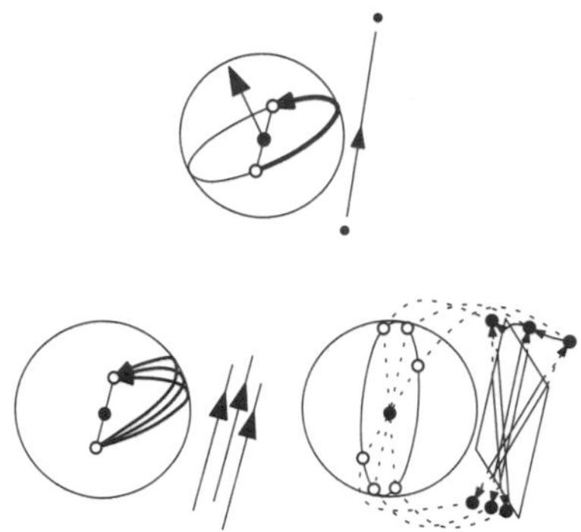

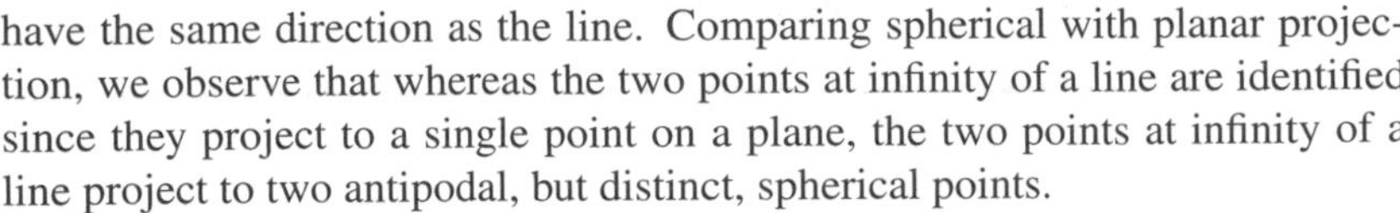

have the same direction as the line. Comparing spherical with planar projection, we observe that whereas the two points at infinity of a line are identified since they project to a single point on a plane, the two points at infinity of a line project to two antipodal, but distinct, spherical points.

Two parallel lines that are coplanar with the observer project to the same spherical segment. A set of parallel lines in general project to distinct spherical segments, but to ones that share the same pair of antipodal spherical points. Multiple sets of parallel lines that are coplanar project to spherical points that are colinear. The ideal line of the plane of the sets of lines projects to a spherical circle defined by the normal direction of the plane. A set of parallel planes have ideal lines that project to the same spherical circle.

32.2 Constructing a DCEL Embedded on S^2

The steps needed for computing visibility in space are identical to those used in the plane—indeed, a generic implementation projecting from E^n to S^{n-1}, for $n = 2, 3$, would be possible, although at the time of this writing the 3D implementation is independent from the present infrastructure. A depth-sorted list of primitives is projected on a sphere centered at the observer and the Boolean union is incrementally computed. As with 2D projection, attribute stability (§ 29.2) ensures that simple union discards invisible features.

After iterating over the scene input, a set of spherical polygons are implicitly represented in the tree. Each edge bounding a polygon is defined by an edge in the scene. Each vertex is defined by two adjacent edges. If the two edges are also adjacent to the same scene vertex, the vertex on the sphere is the projection of that scene vertex. Otherwise it results from the apparent intersection of two scene edges.

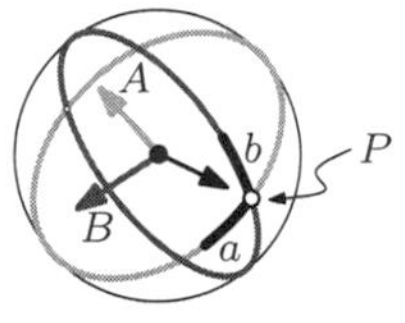

The sequence of vertices of a polygon in S^2 can be obtained from the sequence of spherical circles describing the polygon. If the order in which two spherical circles is determined, then one can define one of the two antipodal intersections to be the point of intersection (§ 9.5). For this convention to hold, we also have to ensure that all spherical polygons are convex. In the figure the point P is uniquely defined by the ordered pair of circles (A, B).

For many applications it would be sufficient simply to extract the resulting set of spherical polygons; each polygon will correspond to a leaf in the BSP tree on S^2. Yet it is more interesting to contruct a doubly connected edge list (Chapter 27). A DCEL stores the adjacency of the resulting spherical image, which can in turn be used if one wishes to refine the output further. One application is the determination of shadows. Each pair of spherical polygons that are adjacent in the DCEL but whose corresponding scene polygons are not adjacent signal a shadow cast from one polygon onto the other. A DCEL is also an ideal representation if visibility is one step in a work flow, for modifying attributes or even changing the position of vertices through a warping effect would be possible without destroying the underlying structure.

To construct a DCEL, a recursive function is passed an initially unpartitioned sphere and a pointer to the root of the tree. The first (root) partition is special and is treated separately. A flag signals that the DCEL is the full sphere. The two endpoints of the subhyperplane corresponding to the root of the tree define two spherical points and the root partition is defined using these two spherical points and their antipodes. A special DCEL function then creates the necessary four halfedges.

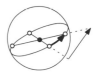

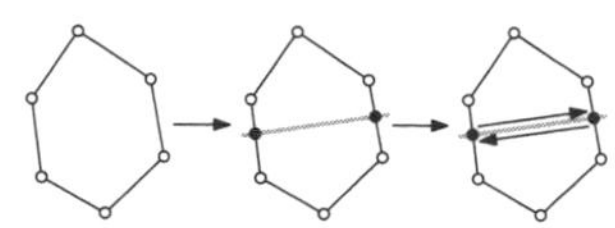

Only two Euler operators are needed after the initial partition is complete. The convex spherical polygon corresponding to each node in the tree is maintained recursively and the face in the DCEL to which it maps is also maintained. As an interior node spawns two children, so is the face in the DCEL also partitioned into two faces. If exact arithmetic is not used, this becomes a numerically delicate operation, for the case of a vertex lying on a partitioning plane needs to be detected. If it is not, a second vertex will be mistakenly inserted near the first one. The splitting plane may intersect two spherical segments in their interior or it may pass through one or two vertices. In the first case the Euler operators split_edge and make_edge_face are used. In the second, only make_edge_face is needed.

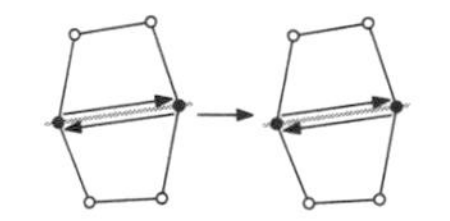

32.3 Exercises

32.1 The accompanying code labeled "inside-sphere-view" may be used for visualizing the vector view of a scene as seen from a viewer. The user interface allows you to manipulate the near clipping plane and the field of view (see Chapter 15). Occasionally the quality of the image deteriorates. Inspect the image to determine the criteria under which it does.

32.2 In § 32.2 a special DCEL function was used for the root partition. Describe instead how this initial partition can be created using only a sequence of Euler operators.

32.3 Given one triangle in E^3, develop a function that constructs a DCEL on S^2 by partitioning it sequentially at the three edges.

32.4 Repeat Exercise 32.3, but implement recursive partitioning and make it possible to handle an arbitrary convex polygon.

32.5 Solve Exercise 32.4 and then construct the spherical view of a set of polygons that do not overlap as seen by an observer.

32.6 Use the start-up program labeled "inside-sphere-view" to visualize the spherical view you constructed in Exercise 32.5.

Appendices

A The PostScript Language

Even if a geometric system is developed with no visualization component as its primary objective, such a component often becomes necessary, if for nothing else, as a stimulant to further development.

The best training in geometric and graphical computing can be achieved by writing raster or vector image files and using a utility to visualize one image or a sequence of images making an animation without relying on prefabricated libraries. Yet *interactive computer graphics* requires the use of libraries for input and output, the subject of the next two appendices, and printing devices require, most often, the use of the PostScript language, the subject of this appendix.

A.1 Introduction

Even if one is only interested in interactive visualization through a library such as OpenGL (see Appendix B), there are several reasons why it may be interesting to study PostScript.

1. PostScript does not require compilation, and, at the time of this writing, it is one of the simplest methods for drawing diagrams.

2. PostScript is device-independent; it is available on all operating systems. A hardcopy of the execution of a PostScript program can also be obtained by sending a file to almost any printer.

3. The PostScript imaging model defines the notion of a stack. It is easy to practice stack-based visualization on PostScript before moving to OpenGL's stacks.

It is possible to study PostScript as simply an output device. One writes a collection of statements controlling a virtual pen and these statements are interpreted by a machine, a PostScript interpreter. But PostScript is also a full-fledged programming language. Because PostScript was aimed at interpretation by programs embedded in printers, the structure of the language, by today's standards, will appear somewhat arcane. Yet it is fascinating to observe that one can write a geometric or a graphical system entirely in PostScript. Because the language is interpreted, programs written in PostScript will not be efficient and one will, in general, want to minimize the amount of code written in PostScript and instead write a system in another programming language that generates the PostScript output.

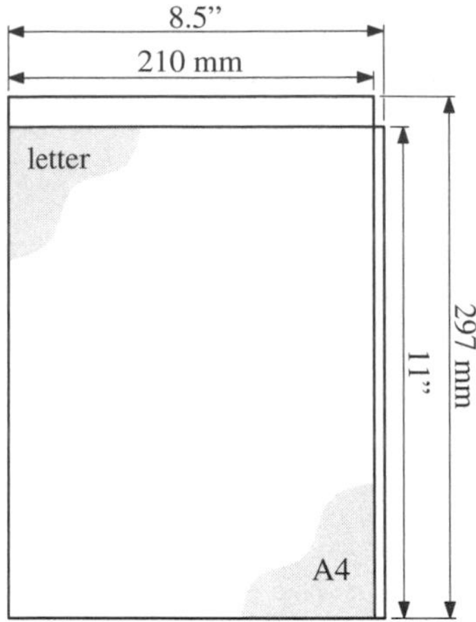

Thus, our sole aim from running a PostScript program is to obtain one or more pages of output. Because the printed page will in effect be considered a fragment of the Euclidean plane and because PostScript was specifically designed for the printed page, the unit of measure is one that predates the computing age: An inch consists of 72 "*points*." A PostScript program running on the interpreter of a printer fed with letter-sized, or 8.5×11-inch, paper, will print the program on a virtual drafting board with the dimensions 612×792. If the same program is run on the interpreter of a printer fed with A4, or 210×297-millimeter, paper, the area that can be printed will have the PostScript dimensions 595.3×841.9. PostScript printers will in practice not be able to print on a small margin around the page and some small margin should therefore remain empty.

When the command **showpage** is encountered by the PostScript program, one page is passed to the output device and processing of the next page, if any, begins.

To identify that a file does indeed contain a PostScript program, the first few characters in the file should be the four letters **%!PS**. The % sign is a comment marker and marks the rest of this, or any other, line as a comment that should not be interpreted. The following three letters **!PS** declare the type of the document. In general, a PostScript program will have the following structure:

```
%!PS
<first–page>
showpage
<second–page>
showpage
etc...
```

where `<first-page>` is the sequence of commands that result in the first page, and so on.

The PostScript language went through two main revisions and the definition intended of the possible three can be specified by writing one of **PS-Adobe-1.0**, **PS-Adobe-2.0**, or **PS-Adobe-3.0** instead of the more generic **PS** label.

Because PostScript is a *vector* drawing language, no information related to rasterization or to the raster resolution of the printing device appears in the PostScript program. The idea is that the same PostScript program can be interpreted on an arbitrary printer with a PostScript engine. A printer capable of rasterizing at 600 dpi, or dots per inch, will produce a higher-quality output than one printing at 300 dpi.

A.2 Encapsulated PostScript

What if one wishes to use a PostScript program not as an end in itself and directly send it to a printing device, but as an intermediate step toward producing a printed document? A file consisting of a PostScript program with that intention is termed an *encapsulated* PostScript file. The convention of the initial marker still holds, but since encapsulation was defined in version 2.0 of

PostScript, the identifying string should signal version (**%!PS-Adobe-2.0**) or a later one. A second line must also appear in the file indicating the bounding box of the printed page to which one intends to output. Of course, the bounding box could be easily deduced by a PostScript interpreter, but declaring such a box helps intermediate programs know the extent of the diagram and thus be able to apply an arbitrary transformation to it *without* forcing such intermediate programs to interpret PostScript themselves, which is not a trivial task. The general form of an encapsulated PostScript file—to be saved with the file extension **.eps**—is the following:

```
%!PS-Adobe-2.0
%%BoundingBox: xmin ymin xmax ymax
<commands>
showpage
```

Yet an encapsulated PostScript file (and implicit program) can itself be sent directly to a printer. In that case a portion of the drawing will appear on the printable bounds of the physical page. Even if the encapsulated PostScript file is not intended to be printed by itself, it is wise to anticipate the need to debug the output by printing the file and setting the four **BoundingBox** values to ones that make the drawing fit comfortably inside a region of the smaller of the two dimensions of A4 and letter-sized paper, or a box of 595.3×792 points.

If a bounding box close to these limits is chosen, one obtains both advantages. The file can be embedded and when directly printed will use the available space on the printed page.

A.3 Drawing Lines and Filling Polygonal Regions

To draw a chain of line segments, or a so-called polyline, one issues the command **newpath** to start the path and the command **stroke** to signal the end of the polyline. The commands **moveto**, **lineto**, **rmoveto**, and **rlineto** move the virtual pen to the location specified by the two preceding coordinates. The pen is moved up, or without leaving a trace, if a move variation is issued and it is moved down, while drawing a line segment, if a line variation is issued. Commands that start with the letter r move the pen relative to its last position.

The dimension provided to the **setlinewidth** command is in the same scale as the drawing and the line width is scaled along with the figure if the output is magnified or minified. Writing **0 setlinewidth** is possible, but it is a poor choice to do so since a printer would use the thinnest line possible *on that printer*, which varies from one to another. The line generated by a 2400-dpi printer, for instance, would be one fourth the width of the same line on a 600-dpi printer. Lines too thin also risk being imperceptible. The line width can be set to a fraction of one point.

The cap and the join styles determined by the **setlinecap** and the **setlinejoin** commands refer to how the line looks at the extreme and at the intermediate vertices, respectively.

The following table illustrates the style of line caps and line joins possible. Line caps and joins can only be distinguished for lines of sufficiently large widths.

```
%!PS-Adobe-2.0
%%BoundingBox: 150 150  350 350

0 setlinecap
2 setlinejoin

10 setlinewidth
```

```
newpath
200 200 moveto
100 100 rlineto
0 −100 rlineto
stroke
```

	Line Join 0	Line Join 1	Line Join 2
Line Cap 0			
Line Cap 1			
Line Cap 2			

Figure A.1
Line caps and line joins in PostScript

The following program illustrates that the last position of the pen may be joined to the first by issuing **closepath**:

```
%!PS-Adobe-2.0
%%BoundingBox: 150 185  300 265

1 setlinejoin

10 setlinewidth
```

```
newpath
200 200 moveto
250 250 lineto
200 250 lineto
closepath
stroke
```

The closed area may be filled by issuing **fill** instead of closepath. In that case, and as the program and accompanying figure below show, the interior of the polygon is rendered, but its boundary is not (thus, the current line width and line cap/join are irrelevant).

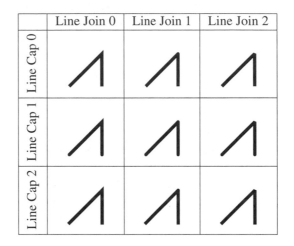

```
%!PS-Adobe-2.0
%%BoundingBox: 150 150  300 300

1 setlinejoin

20 setlinewidth
```

```
newpath
200 200 moveto
250 250 lineto
200 250 lineto
closepath
fill
```

Parameters appear before the corresponding commands because PostScript uses an execution *stack*, discussed in § A.6.

PostScript uses Reverse Polish notation

A.4 Region Overlap

The shade of the pen used may be set using the **setgray** command. A floating point number in the range 0..1 is expected, where 0.0 is the darkest shade (black) and 1.0 is the lightest (white). A related command for color is available. Preceding drawing commands with **0 0.8 0.8 setrgbcolor**, for instance, sets the pen color to cyan.

```
%!PS-Adobe-2.0
%%BoundingBox: 150 150 450 450

% 0 is black; 1 is white

.6 setgray

200 275 moveto
200 0 rlineto
0 50 rlineto
-200 0 rlineto
closepath
fill
```

Consider this PostScript program consisting of two closed paths and two **fill** commands. Figures overwrite what has preceded them. Because the PostScript interpreter is a rasterizing engine, its (grayscale) output is a large array of Boolean flags describing the individual pixels to be rendered. Embedding the interpreter in the printer considerably reduces the bandwidth needed to the printer. This feature of the PostScript language means that a 3D drawing can be properly rendered if depth order is computed and the planar projection of the polygons appears in back-to-front order in the PostScript file (see BSP trees in Chapters 29 and 28).

A.5 Arcs

Circles or circular sections are drawn using the **arc** command, which takes the coordinates of the center, the radius, and the angle of the beginning and the end of the section in degrees. **arcn** generates the complement of the circular section by drawing the arc in counterclockwise order between the two angles.

```
%!PS-Adobe-2.0
%%BoundingBox: 10 50 290 350

% cntr.x cntr.y rad start_angle end_angle arc
% cntr.x cntr.y rad start_angle end_angle arcn

4 setlinewidth
0.5 setgray
    100 300 20   0 360  arc fill

0.0 setgray
    200 300 20   0 360  arc stroke
1 setlinecap
    100 200 20  30 150  arc stroke
    200 200 20  30 150  arc fill
0 setlinecap
    100 100 20  30 150  arcn stroke
```

A.6 The Execution Stack and Function Definitions

PostScript's convention of parameters preceding operators simplifies the task of writing an interpreter. A PostScript interpreter maintains an execution stack consisting of heterogeneous data elements. When an operator is interpreted, the parameters it needs are popped from the stack. **moveto** pops two items from the stack and **arc** pops five items.

In addition to the execution stack, PostScript also maintains the *current-path*, the sequence of path-drawing instructions issued so far by the program. When either **stroke** or **fill** is encountered, the current path is rasterized onto the output device and the current path becomes empty.

String versions of arithmetic operators are used: **add**, **mul**tiply, **sub**tract, and **div**ide. Each of these binary operators pops its two operands from the stack, computes the result, and pushes back the result on the stack. There is also the unary operator **neg**ate.

PostScript maintains a *dictionary*, which associates literals with either values or procedures. As in more traditional programming, one may, for instance, declare constants in one place to ensure that changing their value does not require more than localized editing. If one needs to repeatedly multiply by the value 2.54, for instance, one can use the **def**ine command:

```
\inch 2.54 def
```

The **def**ine command causes two items to be popped from the stack and causes the literal inch to be associated with the number 2.54 in the dictionary. A backslash (\) precedes the literal to signal that the following string acts as a literal and should not be interpreted. If inch (without a backslash) is encountered thereafter, the interpreter will replace the variable by its value in the dictionary.

Each PostScript procedure consists of a sequence of commands, which, when executed, modifies the state of the interpreter—by modifying the stack, by sending elements to the output device, by modifying the dictionary, or by modifying the current path. Since procedures can modify the stack, they can also behave as functions.

If one wishes to draw in metric dimensions, for instance, it would be convenient to define a procedure to convert from centimeters to PostScript's space (in which drawing a 72×72 square results in a square with one inch at each side). The following procedure associates the commands `72 mul 2.54 div` with the literal `cm`. When the procedure is invoked, which is done whenever `cm` appears in the program, the number at the top of the stack is multiplied by 72 and then divided by 2.54.

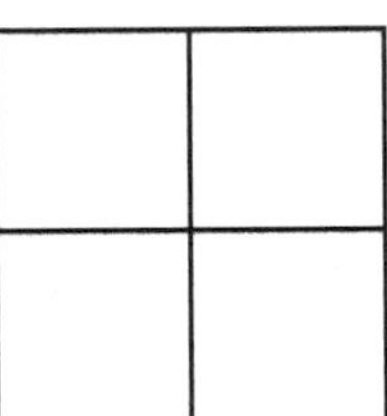

```
/cm { 72 mul 2.54 div } def
```

A 2×2 grid can be drawn using the following program. The grid cells have 4-cm sides.

```
%!PS−Adobe−2.0
%%BoundingBox: 90 120 306 276

/cm { 72 mul 2.54 div } def

0.05 cm setlinewidth

5 cm 5 cm moveto
4 cm 0 cm rlineto
0 cm 4 cm rlineto
−4 cm 0 cm rlineto
closepath stroke

7 cm 5 cm moveto
0 cm 4 cm rlineto
stroke

5 cm 7 cm moveto
4 cm 0 cm rlineto
stroke

showpage
```

A.7 Loops

Looping Using Variables

Looping with the **repeat** command is achieved in PostScript using syntax similar to procedure definitions. One writes `n { commands }` **repeat** to repeat the sequence of commands between the two curly braces n times. The following program illustrates repeat loops by drawing a set of vertical and horizontal lines forming a grid.

```
%!PS-Adobe-2.0
%%BoundingBox: 68 68 220 148

2 setlinecap
0.5 setlinewidth

/zeroCounter { /counter 72 def } def
/incrementCounter {/counter counter 12 add def} def

% ----horizontal Lines----
/horizLine { moveto 144 0 rlineto stroke } def

zeroCounter
7 {
  72 counter horizLine
  incrementCounter
} repeat

% ----Vertical Lines----
/vertLine { moveto 0 72 rlineto stroke } def

zeroCounter
13 {
  counter 72 vertLine
  incrementCounter
} repeat

showpage
```

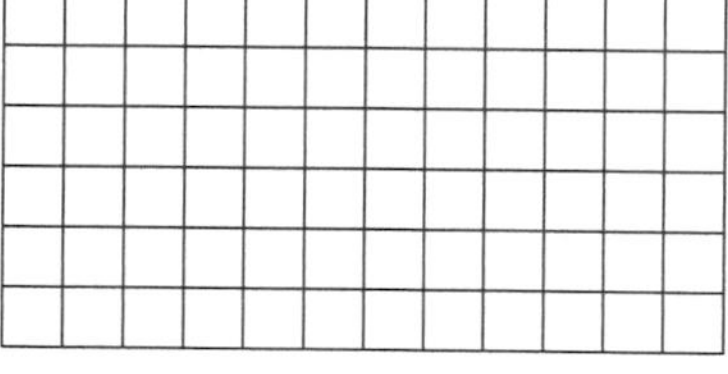

Notice that the definition inside the body of incrementCounter itself redefines the value of counter—after incrementing it.

Looping without Variables—Stack Operators

Using variables in PostScript programs is possible and legal, but it defeats the spirit of the language. Could the stack not be used to store temporary variables? It can, but several operators become necessary and three are illustrated in the revised program on the left below, which draws the same grid. One **exch**anges, or swaps, the two top items on the stack. Another **dup**licates the top item.

There are also the **pop** command, which discards the top item, and **clear**, which clears the stack.

A more elegant, but also more wordy, version of the program on the left is shown on the right. That version extracts the first iteration out of the loop. Clearing the stack at the end is no longer necessary.

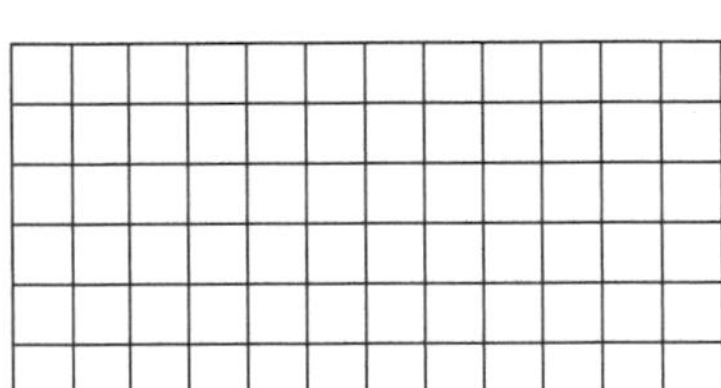

```
%!PS–Adobe–2.0
%%BoundingBox: 68 68 220 148

2 setlinecap
0.5 setlinewidth

/LLx 72 def
/LLy 72 def

newpath LLy
7 {
  dup LLx exch moveto
  144 0 rlineto stroke
  newpath 12 add
} repeat
closepath
pop
```

```
newpath LLx
13 {
  dup LLy moveto
  0 72 rlineto stroke
  newpath 12 add
} repeat
closepath
pop

showpage
```

For Loops

The **for** loop is also available to iterate between two numbers with the following structure:

initial–value increment end–value { commands } **for**

At each iteration of the for loop, a value in the range is pushed on the stack. All values, including the end value, are popped from the stack by the for loop. The same grid can once again be generated using the following concise program:

```
%!PS–Adobe–2.0
%%BoundingBox: 68 68 220 148

2 setlinecap
0.5 setlinewidth
72 12 144 { newpath 72 exch moveto 144 0 rlineto stroke } for
72 12 216 { newpath 72 moveto 0 72 rlineto stroke } for
showpage
```

A.8 Rendering Text

Three commands are needed to render text on the display device. **findfont** searches for the font named on top of the stack and replaces the name by its description; **scalefont** scales the font description; and **setfont** sets the current font.

Command	Pop	Push	Action
findfont	font name literal	font description	
scalefont	font description; size	font description (scaled)	
setfont	font description	—	set font

```
%!PS-Adobe-2.0
%%BoundingBox: 29 19 190 190

0.1 setlinewidth
newpath
  50 50 moveto
  70 50 lineto stroke
newpath
  60 40 moveto
  60 60 lineto stroke

/mytext {
   (Cogito ergo sum) show
} def

/Helvetica-Bold findfont
10 scalefont
setfont
60 50 moveto
mytext

/Times-Bold findfont
8 scalefont
setfont
60 80 moveto
mytext

showpage
```

Cogito ergo sum

Cogito ergo sum

The size of the font indicates the minimum distance between consecutive lines to ensure that the *risers* of characters such as t and l do not interfere with the *descenders* of characters such as g and p. Thus, at most six lines at font size 12 or eight lines at font size 9 can fit in one inch. The origin of the axes system remains at the lower left corner and the initial location of the pen is the lower left corner of the printed text.

A.9 Writing a PostScript Class

In both this and the next appendix, we will be concerned with writing *wrapper classes*—classes that encapsulate all Postscript or OpenGL output statements. There are several advantages for writing wrapper classes:

1. If the decision to switch to a different output device is eventually made, the burden of migrating would be significantly lower if all communication with the initial output device is localized.
2. It is easier to detect that two or more sections of a system need to produce output of similar kind if the output is localized.
3. The initialization and much of the output code is repetitive from one application requiring a certain output device to the next and the possibility of reusing code is increased if initialization and output code is encapsulated.

Just as with programming the OpenGL machine (see Appendix B), PostScript maintains a *state*. The state can be both written to and read from. Dual to the commands **setlinejoin** and **setlinewidth**, for example, are the commands **currentlinejoin** and **currentlinewidth**, which push on the respective value on the stack. The operator == can be subsequently used to pop one item from the stack and print the value on the terminal. Unlike programming the OpenGL machine, however, the session with a PostScript interpreter will normally not be interactive, but one would rather save the PostScript program in a file to be later printed or viewed. Thus, if we wish to have the ability to retrieve the PostScript state, we must duplicate or cache some PostScript variables in a wrapper class.

The mapping between commands in the class API and PostScript commands is straightforward, but a complication arises if we decide to output an Encapsulated PostScript file. Outputting an EPS file means that no output can be made until the output file is complete and is ready to be closed. Because EPS files require the appearance of a bounding box at the top of the file, and because any output command can cause a change to the bounding box, the bounding box needs to be adjusted incrementally as drawing commands are sent to the PostScript class, which in turn means that the drawing instructions must be cached in the PostScript class.

Many programs that output to PostScript write a file header. The objective of the file header is to reduce the size of the output file, and since output commands repeat frequently, it is an inexpensive way of compressing the output file to write a brief header, even if it only has the objective of shortening the subsequent commands.

```
/N {newpath} def
/M {moveto} def
...
```

B OpenGL

This appendix introduces the main functions of the interface defined by OpenGL, a widely implemented and used layer of software separating the application programmer from the raster graphics hardware. To remain operating system-neutral, OpenGL does not define an event model. One user interface library, GLOW, makes it possible to write portable programs following the model-view-controller design pattern and is introduced in Appendix C. If interactivity is not a design objective and only static images, especially ones intended for inclusion in print, are needed, the PostScript language (Appendix A) should be considered as a more suitable interface.

B.1 OpenGL Types and Functions

OpenGL renames many C types via **typedef**s. It is not necessary to use the new typenames, but because all have the prefix GL, using them puts the reader in context and makes the intention of the type evident. GLenum is one such type redefinition. This ubiquitous type is used throughout OpenGL for various heterogeneous tasks. Using the official identifiers also shields programs from potential future changes, although it is quite common to see **float** used rather than GLfloat, presumably on the assumption that the two will remain synonymous.

```
typedef unsigned int GLenum;
typedef unsigned char GLboolean;
typedef unsigned int GLbitfield;
typedef void GLvoid;
typedef signed char GLbyte; /* 1-byte signed */
typedef short GLshort; /* 2-byte signed */
typedef int GLint; /* 4-byte signed */
typedef unsigned char GLubyte; /* 1-byte unsigned */
typedef unsigned short GLushort; /* 2-byte unsigned */
typedef unsigned int GLuint; /* 4-byte unsigned */
typedef int GLsizei; /* 4-byte signed */
typedef float GLfloat; /* single precision float */
typedef float GLclampf; /* single precision float in [0,1] */
typedef double GLdouble; /* double precision float */
typedef double GLclampd; /* double precision float in [0,1] */
```

In addition to types, OpenGL also consists of a (large) set of functions; together the two constitute the *OpenGL API*. Because OpenGL functions do not lie in their own namespace, they have the prefix *gl*. Also, the API does *not* use overloading. The client programmer needs to choose the variant of a function suited to the particular type passed as parameter. For instance, **glVertex2f** is one variant of many functions. A brief excerpt is shown below.

```
void glVertex2d( GLdouble x, GLdouble y );
void glVertex2f( GLfloat x, GLfloat y );
void glVertex2i( GLint x, GLint y );
void glVertex2s( GLshort x, GLshort y );
```

B.2 glBegin–glEnd Blocks

A central part of the OpenGL API is the **glBegin–glEnd** tuple. The OpenGL state may be modified (using gl commands) and the state of the program itself may also be modified using statements between these two statements, but the central command is **glVertex**, which passes vertex coordinates for processing. The correspondence of a visual effect to OpenGL statements is illustrated using the diagram in Figure B.1 and the accompanying code excerpt.

```
void mydraw2(
        float x, float y, float gray,
        Point2f* points, int numPoints, GLenum type)
{
    bool colorFlag = false;
    glPushMatrix(); {
        glTranslatef(x,y,0.0f);
        glBegin(type); {
            int i=-1;
            while(++i != numPoints) {
                if((type==GL_TRIANGLES && i%3==0) ||
                   ((type==GL_TRIANGLE_STRIP ||
                     type==GL_TRIANGLE_FAN) && i>2))
                    colorFlag = !colorFlag;
                float f = colorFlag ? gray : gray + 0.2;
                glColor3f(f,f,f);
                glVertex2f( points[i].x, points[i].y );
            }
        } glEnd();
    } glPopMatrix();
}

virtual void OnEndPaint()
{
    glClearColor(1.0, 1.0, 1.0, 0.0);
    glClear(GL_COLOR_BUFFER_BIT);
    glLoadIdentity();

    drawBoard();

    float i[] = { 0.0f, 160.0f, 320.0f, 480.0f };

    mydraw1( i[0], i[2], 0.0f, points1, 6, GL_POINTS);
    mydraw1( i[1], i[2], 0.0f, points1, 6, GL_LINES);
    mydraw1( i[2], i[2], 0.0f, points1, 6, GL_LINE_STRIP);
    mydraw1( i[3], i[2], 0.0f, points1, 6, GL_LINE_LOOP);

    mydraw2( i[0], i[1], 0.6f, points1, 6, GL_TRIANGLES);
    mydraw2( i[1], i[1], 0.6f, points1, 6, GL_TRIANGLE_STRIP);
    mydraw2( i[2], i[1], 0.6f, points2, 6, GL_TRIANGLE_FAN);

    mydraw3( i[0], i[0], 0.6f, points3, 8, GL_QUADS);
    mydraw3( i[1], i[0], 0.6f, points4, 8, GL_QUAD_STRIP);
    mydraw3( i[2], i[0], 0.6f, points2+1, 5, GL_POLYGON);
}
```

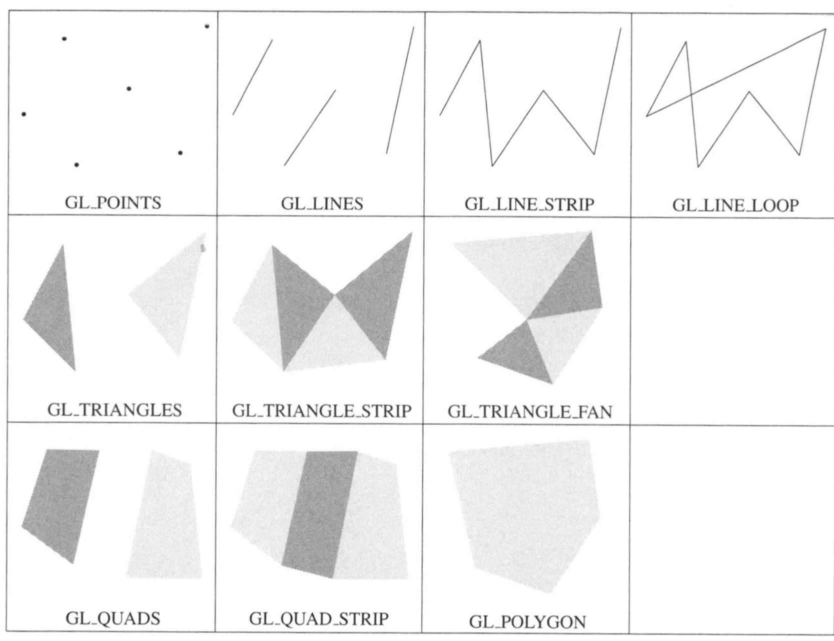

Figure B.1
Types of geometric primitives (options to **glBegin**) [99]

Because **glBegin** and **glEnd** should always appear in tandem, one can improve the readability (and reduce the chance of forgetting to close the block of code) by adding an extra block of curly braces.

B.3 Error Checking

Including error-checking code can often pinpoint errors or potential errors during program development. The function GLenum **glGetError**() returns one at a time the codes of the errors made so far. One would call **glGetError** in a loop until the number returned is 0, signaling that there are no further errors to report. If an error is found,

```
const GLubyte * gluErrorString(GLenum i)
```

can be used to generate a literal byte array describing the error among the following possible ones:

```
#define GL_NO_ERROR 0x0
#define GL_INVALID_ENUM 0x0500
#define GL_INVALID_VALUE 0x0501
#define GL_INVALID_OPERATION 0x0502
#define GL_STACK_OVERFLOW 0x0503
#define GL_STACK_UNDERFLOW 0x0504
#define GL_OUT_OF_MEMORY 0x0505
```

The following code tests whether an error has been made after drawing each frame and terminates the program if one has:

```
static void checkErrors()
{
    GLenum errorCode = 0;
    bool errorFound = false;
    while((errorCode = glGetError()) != GL_NO_ERROR) {
        errorFound = true;
        cerr << "ERROR: "
            << errorCode << endl;
        cerr << "       "
            << gluErrorString(errorCode) << endl;
    }
    if(errorFound)
        exit(1);
}
```

The nonsensical GLenum passed to **glBegin**(..), for instance, causes the following output.

```
OpenGL ERROR: 1280
        invalid enumerant
```

B.4 Display Lists

Whenever a **glBegin**(...)–**glEnd**(...) pair is encountered, the central processing unit (CPU) processes the intervening statements and sends the corresponding data to the graphics processing unit (GPU). To avoid the penalty of repeated processing on the CPU and also, particularly, the penalty of repeated CPU–GPU communication, it is possible to define a *display list*. The signature of the relevant commands is as follows:

```
GLuint glGenLists( GLsizei range );
void glNewList( GLuint list, GLenum mode );
void glEndList( void );
void glCallList( GLuint list );
```

Before creating a display list, a request needs to be made for one or more *display list numbers* by invoking **glGenLists** and passing as parameter the number of display lists requested. The integer returned is the identifier of the first display list reserved.

The parameter passed to **glNewList** determines the display list being defined and the **glNewList**–**glEndList** pair is used to wrap the gl commands constituting the display list. The list can then be invoked in the rendering loop using **glCallList**.

```
{
    GLuint display_list_number = glGenLists(1);

    initialize_root_level();

    while(number_of_subdivisions--)
        recursive_subdivide();
```

```
glNewList(display_list_number, GL_COMPILE); {
  typedef vector<Triangle_S2d>::iterator VIT;
  for(VIT vit = V.begin(); vit != V.end(); ++vit) {
    glColor3f((drand48()>0.5) ? 0.0 : 1.0,
              (drand48()>0.5) ? 0.0 : 1.0,
              (drand48()>0.5) ? 0.0 : 1.0);
    glBegin(GL_TRIANGLES); {
      glVertex3f(vit−>P0().x(), vit−>P0().y(), vit−>P0().z());
      glVertex3f(vit−>P2().x(), vit−>P2().y(), vit−>P2().z());
      glVertex3f(vit−>P1().x(), vit−>P1().y(), vit−>P1().z());
    } glEnd();
  }
```

Other features of display lists not illustrated here are nesting of display lists by calling a display list while defining another and the use of **glCallLists** to trigger multiple display lists using one invocation.

B.5 Colors

The Color Buffer

To display an image on a raster output device, a color buffer is stored in dedicated memory. Each raster sample, or pixel, is saved in a quadruple of red, green, blue, and alpha *channels* as illustrated in Figure B.2. The *depth* of the color buffer is the number of bits needed to represent the four channels.

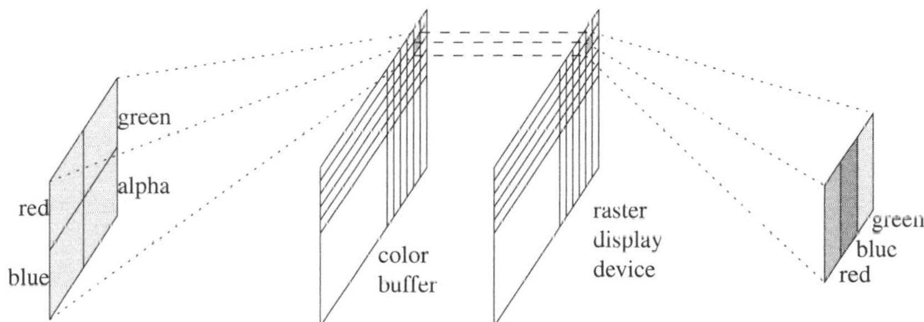

Figure B.2
The color buffer and the raster display device

The command

```
void glClearColor(
        GLclampf red,
        GLclampf green,
        GLclampf blue,
        GLclampf alpha )
```

is used to set the color the buffer is set to upon clearing. If **glClearColor** is not issued, the default of (0,0,0,1), or opaque black, is used. The clear color only needs to be set once for each rendering context. Thereafter, the command

```
glClear(GL_COLOR_BUFFER_BIT);
```

is used prior to rendering each frame to clear the color buffer to the preset color.

Using Colors Without Shading

Most often one wishes to *shade*, or generate colors as functions of the lights and the materials in a scene (§ B.8), but in many cases simply rendering an object using a solid color suffices. Prominently, but not exclusively, this is the case for user interface widgets.

The current color, including an alpha (or transparency) channel, can be set using the function **glColor4f**:

void glColor4f(GLfloat red, GLfloat green, GLfloat blue, GLfloat alpha)

The default alpha value of 1.0 is used if we instead call **glColor3f**, which takes three variables of type **float**, hence the 3f suffix:

void glColor3f(0.0f, 0.0f, 0.0f);

If a pointer to an array of three **float**s is needed, **glColor3fv** would be called instead:

void glColor3fv(**const** GLfloat *v)

B.6 Double Buffering

A projector in a cinema achieves the effect of seamless motion by projecting 24 images per second, but since these images are consecutive on a film, there must be a lag during which no image is projected. When rendering an animation on a raster display device, one may wonder whether a similar principle might be appropriate. One would display one image, clear the color buffer, display the next image, and so on. Clearing the color buffer has the effect of displaying a blank image at the current *clear color*. But however brief, the time it takes to clear the display would be perceptible.

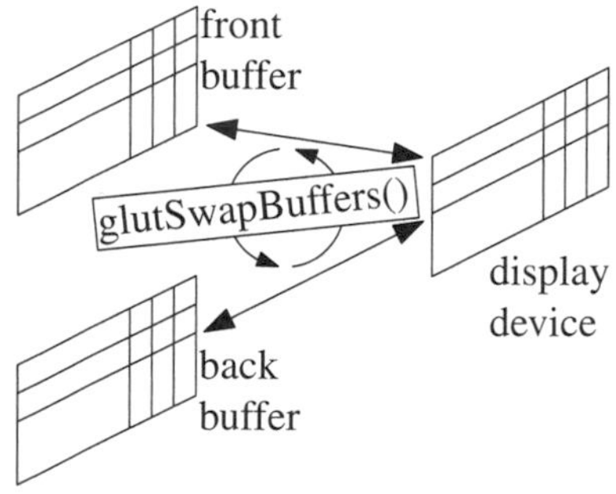

Worse yet, if the color buffer is visible while being written, the primitives drawn first will be displayed longer and will thus be brighter. The answer to both problems lies in double buffering.

The solution is to use *two* color buffers and to use a toggle that decides which of the two buffers currently feeds the raster display device. Reading and writing to the color buffer would be performed on the buffer not currently connected to the display. When drawing of the current frame is complete, the client issues the command **glSwapBuffers**() to request toggling the color buffer connected to the display. The following instructions then resume reading to and writing from the disconnected color buffer.

The two buffers are referred to as the *front buffer* and the *back buffer*. The front buffer is the one currently connected to the display and the back buffer is the one accessed. The distinction is logical, not physical; after the buffers are swapped, their names are also swapped.

One needs to request the allocation of storage for either one or two buffers during the creation of the rendering context. Rendering is then said to be in either *single-buffer* or *double-buffer* mode. The creation of windows and related functions are delayed to Appendix C, which discusses GLOW, one library that allows OpenGL to interact with the operating system and the window system.

As will be seen in § C.4, it is not necessary to issue **glSwapBuffers**() when using GLOW since that command is issued automatically at the conclusion of rendering each frame.

B.7 The Pipeline and Transformations

The Graphics Pipeline

At its most abstract, OpenGL is a transformation and rasterization engine. The engine is fed point coordinates describing a set of primitives (lines, polygons, etc.) and a geometric transformation is applied to the coordinates. The transformed primitives are then passed to the rasterization engine, which produces a raster image. The hardware view, shown in Figure B.3, is never fully isolated from the programmer while interfacing with OpenGL.

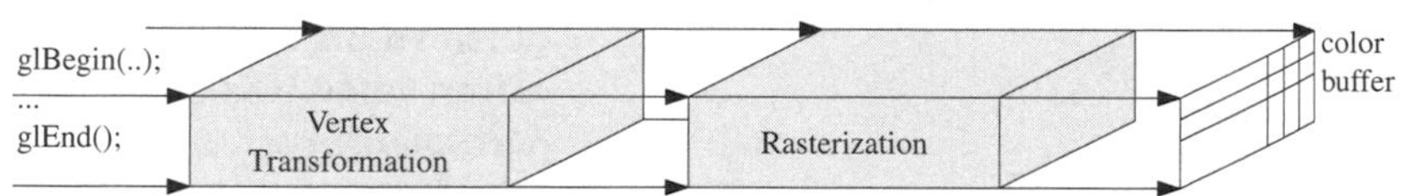

Figure B.3
A simplified view of the graphics pipeline

To pass vertex coordinates to the pipeline, the command

void glBegin(GLenum mode)

is used. The mode can be one of the constants GL_POINTS, GL_LINES, etc. Even if one selects the single-buffer rendering mode, it is likely that some primitives will remain invisible. Because OpenGL is only an application programmer's interface, it allows implementations to cache data at various stages for efficiency. The command **void glFlush**(**void**) is used to flush the intermediate buffers. In double-buffer mode, **glSwapBuffers**() issues an implicit **glFlush**().

Transformations and the Matrix Stack

Two transformation matrices, the modelview matrix and the projection matrix, are used (in that order) to map vertex coordinates. Rasterization follows. The two transformation matrices are stored at the top of two *matrix stacks*. Only the two matrices at the top of their respective stack can be accessed or modified and also only the two top matrices affect vertices. Using a stack makes it easy to use a new matrix temporarily and then pop the stack and resume with the matrix now on top.

The next duo of instructions we encounter, and ones that will almost always appear in tandem (and so it is convenient to create a redundant block to match them), is the **glPushMatrix**()–**glPopMatrix**() pair. Both commands affect only the *current matrix stack*, which can be set by calling either

glMatrixMode(GL_MODELVIEW)

or

glMatrixMode(GL_PROJECTION)

Calling **glLoadIdentity**() initializes the top of the current matrix stack to the identity transformation.

```
virtual void OnEndPaint()
{
  // Set up projection matrix
  glMatrixMode(GL_PROJECTION);
  glLoadIdentity();
  gluPerspective(
          fovAngle,
          1.0 /*aspect*/,
          1.0 /*Znear*/,
          1000.0 /*Zfar*/);

  // set up modelview matrix
  glMatrixMode(GL_MODELVIEW);
  glLoadIdentity();
  gluLookAt(
        viewerxyz/1.5,
        viewerxyz,
        viewerxyz*1.4,
        0,0,0, 0,1,0);

  // draw
  glClearColor(1.0, 1.0, 1.0, 0.0);
  glClear(
        GL_COLOR_BUFFER_BIT |
        GL_DEPTH_BUFFER_BIT);

  glDisable(GL_LIGHTING);
  glColor3f(0.7,0.7,0.7);
  glCallList(meshList);

  glEnable(GL_LIGHTING);

  glPushMatrix(); {
    glRotated(angle1, 0,1,0);

    glPushMatrix(); {
      glScaled(5,0.5,0.5);
      glTranslated(0.5,0,0);
      glutSolidCube(1.0);
    } glPopMatrix();

    glTranslated(5,0,0);
    glRotated(angle2, 0,0,1);

    glPushMatrix(); {
      glScaled(5,0.5,0.5);
      glTranslated(0.5,0,0);
      glutSolidCube(1.0);
    } glPopMatrix();

    glTranslated(5,0,0);
    glRotated(angle3, 0,0,1);

    glPushMatrix(); {
      glScaled(5,0.5,0.5);
      glTranslated(0.5,0,0);
      glutSolidCube(1.0);
    } glPopMatrix();

  } glPopMatrix();
}
```

Two details of matrix stack manipulation are crucial. The first is that matrix pushing results in the matrix currently at the top of the stack being *duplicated*. Since hierarchical structures such as articulated figures are common, the duplication of the matrix on top of the stack is particularly useful because it makes it easy to mirror the transformations needed to process a *scene graph*—the directed acyclic graph containing a mix of geometry and transformation nodes that is commonly used to represent 3D scenes. The second important detail is that matrix multiplication is applied on the *right* of the matrix currently on top of the stack. Even though the rightmost matrix appears at the end of the list of matrices, it is the first to be multiplied by the vertex coordinates. Figure B.4 shows the succession of the lists of matrices on the stack. Note that, in fact, only the product of such matrices appears on the stack.

Two additional commands are frequently useful. An explicit matrix may be loaded using **glLoadMatrix**[df] or multiplied by the current matrix using **glMultMatrix**[df].

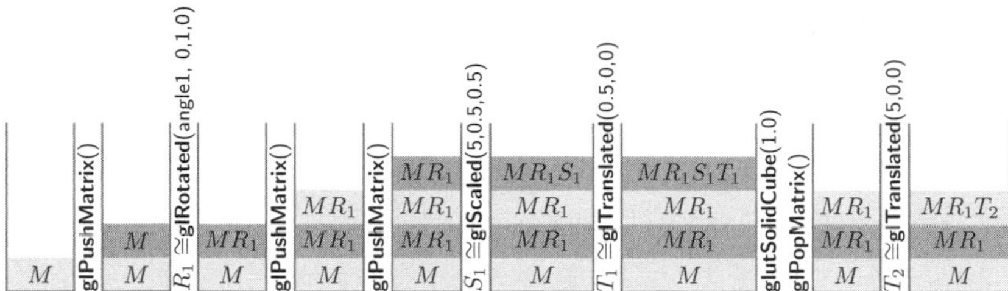

Figure B.4
The effect of **glPushMatrix**() and **glPopMatrix**() on the matrix stack

B.8 Lights and Shading

The functions **glLight**[fi][v] and **glMaterial**[fi][v] set the light parameters (geometry and color) and the material (ambient, diffuse, specular, and emission) parameters, respectively.

Designing and implementing a shading function is discussed in Chapter 21 and is, in this context, rather simple. The important detail to study concerning lights is the transformations affecting it. One way to remember that the coordinates of the light position are multiplied by the modelview transformation and stored without multiplication with the projection transformation is that the latter destroys angles. If angles are not kept intact, the resulting (diffuse) shading will depend on the relative position of the light with the surface to be shaded *and* the viewer, rather than with the surface only.

If the light source follows an object in the scene, then
glLightfv(GL_POSITION, ...),
which sets the location of the light, is used just as ordinary geometry would. The current modelview transformation will act on both.

```
if (movingLightIsOn)
{
  glPushMatrix (); {
    glRotated(lightAngle, 0.0, 1.0, 0.0);
    GLfloat position[] = { 30.0, 10.0, 0.0, 1.0 };
    glLightfv(GL_LIGHT0, GL_POSITION, position);

    glDisable(GL_LIGHTING);
    glColor3f(1.0, 0.0, 0.0);

    glTranslated(movingLP[0],movingLP[1],movingLP[2]);
    glutSolidSphere(1.0,8,8);
  } glPopMatrix ();
}
```

If the light source is a "headlight" that follows the viewer, the modelview matrix is reset to the identity matrix before **glLightfv**(GL_POSITION, ...) is invoked with zero coordinates. The light position then coincides with the eye position.

```
glMatrixMode(GL_MODELVIEW);
glLoadIdentity();

GLfloat position[] = { 0.0, 0.0, 0.0, 1.0 };
glLightfv(GL_LIGHT2, GL_POSITION, position);
```

The light may also be static relative to the viewer, but at a location other than the eye.

B.9 Encapsulating OpenGL Calls

Even though interspersing OpenGL API calls inside a system as done in the examples above may be adequate if the system is small, such interspersing of functionality can lead to

Difficult migration—The decision to adopt a graphics API other than OpenGL would be costly.

Code duplication—The same or similar graphical requests may be made by several parts of a system. If the requests are scattered, it will be difficult to observe that they can be grouped to shrink the code base and resulting executable, and to simplify maintenance.

Test difficulty—Testing of a graphics feature is significantly harder if the graphics code is scattered since no well-defined set of entry and exit points exists for writing test code.

For these reasons it is wise to design an intermediate layer of software that isolates OpenGL from the rest of a system. In the simplest setting, a class GLdraw_E2 would consist of a set of **static** functions that interface with OpenGL.

```
template<typename NT>
struct GLdraw_E2 {
  static void drawPoint(
                const Point_E2<NT>& p,
          const GlowColor& pointColor = GlowColor::yellow,
          float pointSize = 7.0)
  {
  glPointSize(pointSize);
  pointColor.Apply();
  glBegin(GL_POINTS); {
    glVertex2f(p.x(), p.y());
  } glEnd();
  }

  ...
};
```

This elementary design can be augmented in two ways. If a set of points will be repeatedly drawn, we would add functionality for setting and calling drawing lists. If the interface is also expected to maintain a state, the member functions could be made nonstatic.

C The GLOW Toolkit

OpenGL does not, by design, define an event model or a widget set. Yet these are essential features in most interactive applications. This appendix introduces GLOW, a portable, free, concise, and well-designed user interface library.

C.1 Introduction—Why GLOW?

The user interface library most often used in OpenGL programs appears to be the glut library, which defines a simple event model and menu widgets. Due to its simplicity and because its interface has been implemented on multiple operating systems, glut is the semi-official windowing toolkit linking the operating system to the OpenGL rendering context. The portability and simplicity of glut and its lack of widgets can be tolerated in many applications. In larger applications the model-view-controller design pattern becomes essential. MVC separates the object or objects that describe the application itself (the model) from those performing the visualization (the view) and also from those handling user events (the controller). Such a strict separation leads to far more enjoyable programming when the size of the application ceases to be trivial.

GLOW is not the only user interface library available for C++ and OpenGL. Because the C++ standard, unlike Java, lacks a definition for an event model or a widget set, several libraries have been independently developed to fill the gap. At the time of this writing, the list includes fltk, glGUI, plib, wxWidgets, GLUI, AGAR, gtk, SDL, ClanLib, Qt, and GLOW.

This appendix discusses the main ideas behind designing an application using GLOW. Despite its simplicity, there are several reasons why GLOW may be preferred over more vast, yet still well-crafted, C++ graphical user interface libraries such as Qt:

- GLOW is open-sourced. One may fix problems arising during installation or development on one's own.
- GLOW is licensed under the Lesser Gnu Public License (LGPL); it is possible to use it for closed-source commercial applications. LGPL stipulates making public only the modifications one implements in GLOW itself.
- Even though its set of widgets is fairly complete, GLOW takes little time to install and takes a small footprint on disk and in resulting executables.

- GLOW's simplicity derives in part because it is built only on glut. GLOW thus benefits from the latter's implementation on many operating systems.

Shortcomings in one library and advantages in another aside, it is in any case preferable to adopt one C++ GUI library and use it as a, possibly short-term, standard.

C.2 Window Creation and Drawing

Our first example consists of concise code—a brief main function and one class—for creating a window and drawing a segment. We initialize GLOW in main by calling Glow::Init(argc, argv), create an instance of a descendant of GlowWindow (HelloSegmentWindow), and then ask GLOW to handle events by invoking the infinite loop Glow::MainLoop(), which GLOW manages. The program only returns from MainLoop when it is terminated.

```
int main(int argc, char **argv)
{
  glow::Glow::Init(argc, argv);
  new HelloSegmentWindow();
  glow::Glow::MainLoop();
}
```

In this first example each member of the GLOW library is preceded by the namespace qualifier **glow**::. In subsequent examples the statement

```
using namespace glow;
```

is used. Doing so is not a severe pollution of the global namespace since classes in the **glow** namespace are in any case identified by a prefix with the same name.

Notice that we do not bother to store the pointer to the instance of

```
HelloSegmentWindow
```

created; we have no need for it. Notice also that it is easy to create two windows; it suffices to duplicate the line

```
new HelloSegmentWindow();
```

A HelloSegmentWindow object captures, evidently, a window. This is done by deriving from the class GlowWindow. The constructor of the latter takes a few parameters: the title of the window, coordinates to tell the windowing system where the window should be drawn, the desired size of the window, the type of buffer needed, and whether GLOW should "listen" to any events. In this case we request no event handling but ask for a Glow::rgbBuffer—a color buffer. The constructor then initializes the projection and modelview matrices.

The function GlowComponent::OnEndPaint(), where GlowComponent is an ancestor of HelloSegmentWindow, is empty. This function is called for each GlowComponent whenever a Refresh() request is received. Segment drawing is performed by HelloSegmentWindow simply by overriding the behavior of

OnEndPaint(). Even though we have asked for no event handling, a Refresh() event is issued whenever a portion of the window ceases to be hidden by another window.

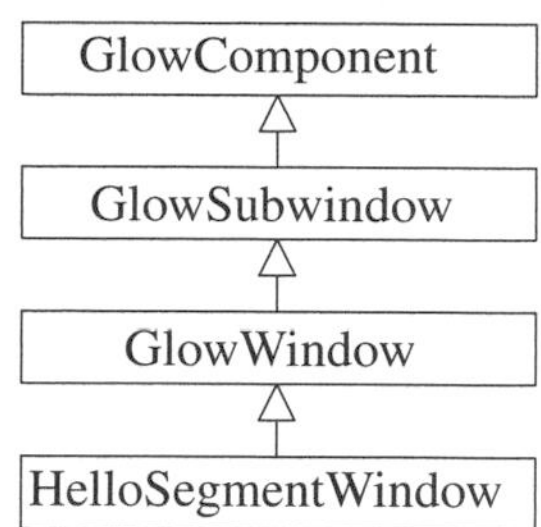


```
const int initViewportWidth = 640;
const int initViewportHeight = 480;

class HelloSegmentWindow : public glow::GlowWindow
{
public:
  HelloSegmentWindow()
    : glow::GlowWindow(
              "HelloSegmentWindow",
              glow::GlowWindow::autoPosition,
              glow::GlowWindow::autoPosition,
              initViewportWidth, initViewportHeight,
              glow::Glow::rgbBuffer,
              glow::Glow::noEvents)
  {
    glMatrixMode(GL_PROJECTION);
    glPushMatrix();
    glLoadIdentity();
    gluOrtho2D(0, initViewportWidth, 0, initViewportHeight);

    glMatrixMode(GL_MODELVIEW);
    glPushMatrix();
    glLoadIdentity();
  }

protected:
  virtual void OnEndPaint()
  {
    glClear(GL_COLOR_BUFFER_BIT);
    glBegin(GL_LINES); {
      glVertex2f(160.0f, 120.0f);
      glVertex2f(480.0f, 360.0f);
    } glEnd();
  }
};
```

C.3 Keyboard Events

A few minor modifications to the simple example above will serve as an incremental introduction to GLOW. We express interest in handling events sent from the keyboard by listing the constant Glow::keyboardEvents in the parameter initialization list of the parent class GlowWindow. The function OnKeyboard will now be invoked whenever a keyboard event is issued (signaling a keyboard key is pressed). In the **switch** statement the program simply exits when either "q" or the escape key is pressed. An empty **default** case is used to avoid compiler warnings about underhandling an enumeration.

```
class HelloSegmentWindow : public GlowWindow
{
public:
  HelloSegmentWindow()
    : GlowWindow(
          "HelloSegmentWindow",
          GlowWindow::autoPosition, GlowWindow::autoPosition,
          initViewportWidth, initViewportHeight,
          Glow::rgbBuffer,
          Glow::keyboardEvents)
  { ... }

  virtual void OnKeyboard(
              Glow::KeyCode key, int x, int y,
              Glow::Modifiers modifiers)
  {
    switch (key) {
    case 'q': case 'Q': case Glow::escapeKey:
      Close();
    default:
      ;
    }
  }
};
```

It would be possible to just write exit(0) instead of Close(), but then the program would exit even if some windows remain open, which may be the desired behavior. The behavior used here is to close any window that receives the "q" key-press event, and then quit the program when the last such window is closed. This behavior is achieved by setting "auto-quitting" to **true**. We confirm that the exit-on-last-close behavior using the following code:

```
int main(int argc, char **argv)
{
  Glow::Init(argc, argv);
  Glow::SetAutoQuitting(true);
  new HelloSegmentWindow();
  new HelloSegmentWindow();
  Glow::MainLoop();
}
```

C.4 Idle Events

Idle events are events that run in the background without being triggered by the user—the motion of a cloud or of a flock of birds may be candidates. In this example we use an idle event to rotate the line segment. The line could be incrementally rotated by a fixed angle at each frame, or whenever a new image is drawn, but that would lead to the rather undesirable effect that the rotation speed would depend on the speed of the machine.

An identical performance can be obtained by rotating the line at a rate relative to a wall clock, modeled by the class Clock. Because Clock uses the POSIX function gettimeofday, it would need to be modified if the system is intended to compile under multiple operating systems.

```
#include <sys/time.h>
class Clock
{
protected:
  timeval startTime;

public:
  Clock()
  {
    // gettimeofday assumes we're on POSIX.
    gettimeofday(&startTime, NULL);
  }

  float
  getSecondsSinceStart()
  {
    timeval currentTime;
    gettimeofday(&currentTime, NULL);
    float elapsedTime =
      currentTime.tv_sec − startTime.tv_sec +
      ((currentTime.tv_usec − startTime.tv_usec) / 1.0E6);
    return elapsedTime;
  }
};
```

Three modifications need to be made to HelloSegmentWindow to activate an idle rotation of the segment. HelloSegmentWindow is derived from

GlowIdleReceiver

(in addition to GlowWindow); the object is set to receive idle events by invoking Glow::RegisterIdle(**this**); and the function OnMessage triggers a redraw if some time (0.01 seconds) has passed since the last redraw (with a parameter of type GlowIdleMessage). Sending a Refresh request inside an idle loop risks consuming all the cycles available from the CPU and/or from the GPU. The instance variable lastSeconds stores the time when the last Refresh request was issued and the two consecutive times are compared before another request is sent.

A new rotation matrix could be appended to the current modelview matrix by issuing a **glRotatef** with a minuscule angle, but doing so would slowly get the matrix out of orthogonality. The identity matrix is loaded and a fresh rotation matrix is generated instead.

This is a suitable point to add an important improvement. Using a single color buffer means that the top part of the line may appear brighter than the bottom (if rasterization is performed from top to bottom). To set OpenGL for double buffering, it suffices to write Glow::rgbBuffer | Glow::doubleBuffer while constructing GlowWindow. There is no need to explicitly swap the front and back buffers; GLOW performs buffer swapping following the conclusion of OnEndPaint.

```
class HelloSegmentWindow : public GlowWindow, public GlowIdleReceiver
{
  Clock myClock;
  float lastSeconds;
public:
  HelloSegmentWindow()
      ...
      Glow::rgbBuffer | Glow::doubleBuffer,
      ... ,
      lastSeconds(0)
  {
   ...
    Glow::RegisterIdle(this);
  }

protected:
  virtual void OnEndPaint()
  {
    glClear(GL_COLOR_BUFFER_BIT);
    glLoadIdentity();
    glRotatef(myClock.getSecondsSinceStart() * 100.0, 0,0,1);
    glBegin(GL_LINES); {
      glVertex2f(-160.0f, -120.0f);
      glVertex2f( 160.0f, 120.0f);
    } glEnd();
  }
  ...
  virtual void OnMessage(const GlowIdleMessage& message)
  {
    float seconds = myClock.getSecondsSinceStart();
    if(seconds - lastSeconds > 0.01) {
      Refresh();
      lastSeconds = seconds;
    }
  }
};
```

C.5 Modifying the Projection upon Window Resizing

GLOW's default behavior of reacting to a user's request for modifying the size of the window, or *reshaping* it, is to simply forward the request to OpenGL via **glViewport**(0, 0, width, height). Modifying only viewport mapping changes the scale and the aspect ratio. If the desired effect is to modify neither, the function OnReshape needs to be overriden to update the projection matrices accordingly. Modifying (as well as initializing) the matrices is done using the function setupProjectionMatrices. For clarity the functions Width(), Height(), and OnReshape() are qualified with their defining superclass, GlowSubWindow.

```
void setupProjectionMatrices()
{
    glMatrixMode(GL_PROJECTION);
    glPushMatrix();
    glLoadIdentity();

    int W = GlowSubwindow::Width()/2;
    int H = GlowSubwindow::Height()/2;
    gluOrtho2D(−W, W, −H, H);

    glMatrixMode(GL_MODELVIEW);
    glPushMatrix();
    glLoadIdentity();

    glEnable(GL_LINE_SMOOTH);
    glEnable(GL_BLEND);
    glBlendFunc(GL_SRC_ALPHA, GL_ONE_MINUS_SRC_ALPHA);
}
virtual void OnReshape(int width, int height)
{
    GlowSubwindow::OnReshape(width, height);

    setupProjectionMatrices();
}
```

C.6 Widgets—Checkboxes and Buttons

Those who believe that object orientation is mere flourish over fundamental (algorithmic) computing may still dismiss GLOW as an object-oriented ornament over the glut library, but the ease of creating widgets under GLOW should make the case more convincing. A *widget* is an object capturing a user interface unit such as a checkbox, a radio button, or a push button. To add an instance of the first and the last, we add derivations to HelloSegmentWindow from GlowCheckBoxReceiver and from GlowPushButtonReceiver.

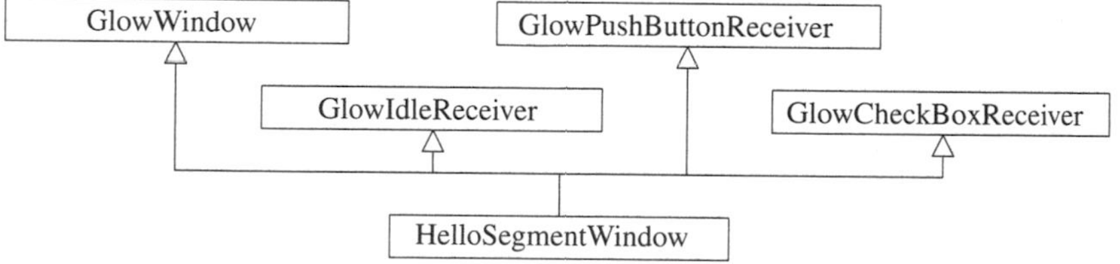

Figure C.1
Parent classes of HelloSegmentWindow

We then store pointers to the two widgets as well as one to a control window holding them.

```
GlowQuickPaletteWindow* controlWindow;
GlowCheckBoxWidget*     circleCheckbox;
GlowPushButtonWidget*   quitButton;
```

The control window is separate from the main drawing window. The window manager will likely decorate it as an individual window.

```
controlWindow = new GlowQuickPaletteWindow(
    "Controls",
    GlowWindow::autoPosition,
    GlowWindow::autoPosition,
    GlowQuickPalette::vertical);

GlowQuickPanelWidget* panel = controlWindow->AddPanel(
    GlowQuickPanelWidget::etchedStyle, "Panel");

circleCheckbox = panel->AddCheckBox(
    "Circle", GlowCheckBoxWidget::on, this);
quitButton     = controlWindow->AddPushButton("Quit", this );

controlWindow->Pack();
```

The OnEndPaint function now tests whether the box is checked before it draws a circle.

```
if(circleCheckbox->GetState() == GlowCheckBoxWidget::on)
    gluPartialDisk(qobj, 200.0, 201.0, 60, 1, 0.0, 360.0);
```

Finally, we listen to incoming events via the following two functions.

```
void OnMessage(const GlowCheckBoxMessage& message)
{
    Refresh();
}

void OnMessage(const GlowPushButtonMessage& message)
{
    if(message.widget == quitButton)
        exit(0);
}
```

C.7 Arcball Manipulation

Just as multiplication by a unit-length complex number effects a rotation about the origin in the plane, a suitably-defined multiplication by a quaternion effects a rotation about an axis passing by the origin in space. The generalization from complex numbers to quaternions was understood in the 19th century. A more recent development [98] defines a user interface for manipulating 3D solids using a pointing device. By mapping a point inside a circle on the screen with a quaternion, it is possible to rotate a 3D scene with the important constraint that the rotation is reversible: The rotation corresponding to moving the pointer from a point A to a point B is the reverse of the one defined by moving from B to A.

Since the arcball interface is widely useful, it is a good candidate for encapsulation in a library and GLOW provides an implementation. As before, the following code creates a window and initiates GLOW's main loop:

```
int main(int argc, char **argv)
{
  Glow::Init(argc, argv);
  Glow::SetAutoQuitting(true);
  /* ArcballWindow* myArcballWindow = */ new ArcballWindow();
  Glow::MainLoop();
}
```

The window ArcballWindow remains a child class of GlowWindow, but it now holds a pointer to an instance of GlowViewManipulator, which in turn holds a pointer to an instance of Scene. As illustrated in Figure C.2, Scene is derived from GlowComponent.

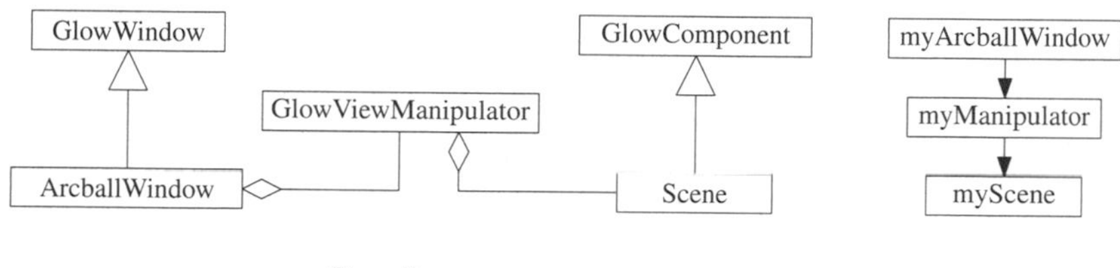

Figure C.2
GlowViewManipulator class and object diagrams

The links between the three objects (appearing in the object diagram) are created in the constructor of ArcballWindow. The two links in the object diagrams are established during the creation of the manipulator and the scene. GlowViewManipulator is passed **this** pointer to identify the arcball window and the scene is passed a pointer to the manipulator. The events that ArcballWindow expresses interest in are now mouse key presses and mouse drag (motion while a button is pressed) in addition to keyboard events.

```
ArcballWindow::ArcballWindow()
  : GlowWindow(
        "Arcball Window",
        GlowWindow::autoPosition, GlowWindow::autoPosition,
        initViewportWidth, initViewportHeight,
        Glow::rgbBuffer | Glow::doubleBuffer | Glow::depthBuffer,
        Glow::mouseEvents | Glow::dragEvents | Glow::keyboardEvents)
{
  setupProjectionMatrices();
  //----Manipulator----
  GlowViewManipulatorParams manipparams =
    GlowViewManipulatorParams::defaults;
  manipparams.draw = true;
  myManipulator = new GlowViewManipulator(this, manipparams);
  myManipulator->SetSpinnable(true);
  /* Scene* myScene = */ new Scene(myManipulator);
}
```

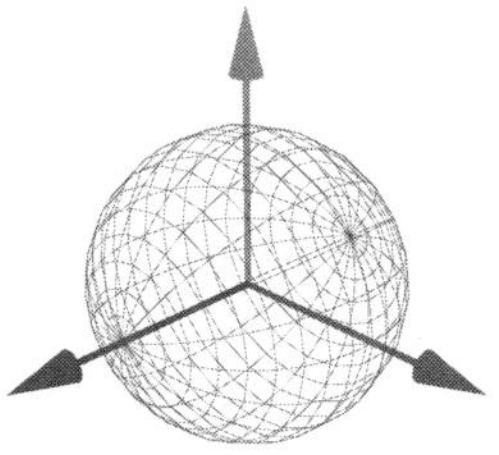

A GlowWindow draws its children hierarchy by invoking OnBeginPaint() followed by OnEndPaint(). OnEndPaint() in a manipulator only modifies the modelview matrix, thus affecting what is drawn by an instance of Scene. Since an ArcballWindow object is the one receiving mouse events, it needs to delegate the events to the manipulator. The function for handling mouse drag events is shown below:

```
virtual void ArcballWindow::OnMouseDrag(int x, int y)
{
  if (myManipulator->IsDragging())
  {
    float xn, yn;
    GlowSubwindow::NormalizeCoordinates(x, y, xn, yn);
    myManipulator->InDrag(xn, yn);
  }
}
```

C.8 Displaying the Frame Rate

We make an improvement to the arcball manipulation program. Displaying the number of frames per second has little, if anything, to do with GLOW, but it is frequently needed in interactive graphical systems.

The first idea for determining the rate of rendering frames is likely to find the inverse of the time needed for each frame. That the clock is accurate to the nearest microsecond would suggest this approach is suitable. It fails because frame rendering times will likely fluctuate significantly, resulting in a fast-changing display of numbers. A more suitable approach is encapsulated in the class FPScounter. After creating an instance of FPScounter, the function showFrameRate will display the last computed frame rate until one second has elapsed, during which the number of frames would be counted.

```
class FPScounter
{
protected:
  timeval start_time;
  std::string str;

public:
  FPScounter() : str("")
  {
    gettimeofday(&start_time, NULL);
  }

  std::string
  getFrameRate()
  {
    static int number_of_frames = 0;
    static timeval current_time;
    float elapsedTime;

    gettimeofday(&current_time, NULL);

    elapsedTime =
      current_time.tv_sec −
      start_time.tv_sec +
      ( (current_time.tv_usec −
        start_time.tv_usec) / 1.0E6 );

    number_of_frames++;

    if(elapsedTime >= 1.0) {
      start_time = current_time;

      std::ostringstream os;
      os << number_of_frames << "_fps";
      str = os.str();

      number_of_frames = 0;
    }
    return str;
  }
};
```

The usual benefits of object orientation should be evident from this quick introduction to GLOW. Encapsulating various functionalities makes it easy to debug a system and promotes reuse through either inheritance or through simple file and class copying.

C.9 Displaying the Frame Rate in a Widget

An alternative method for showing the frame rate will illustrate the use of GlowLabelWidget. After constructing a GlowLabelWidget object and invoking AddLabel on either a GlowQuickPaletteWindow or a GlowQuickPanelWidget, the text passed as parameter is displayed in a label. The label can subsequently be modified by invoking SetText on the GlowLabelWidget object. Only as much space as was initially provided through AddLabel is available in the GlowQuickPaletteWindow, and so the initial string should exceed in length any that is expected to appear.

C.10 GLUT Primitives

Some of the functionality defined in GLUT has not been duplicated in GLOW since it carries over with no difficulty. One such set of functions are those for rendering solid or wireframe objects:

- **glutSolidSphere**, **glutWireSphere**
- **glutSolidTorus**, **glutWireTorus**
- **glutSolidTetrahedron**, **glutWireTetrahedron**
- **glutSolidCube**, **glutWireCube**
- **glutSolidOctahedron**, **glutWireOctahedron**
- **glutSolidDodecahedron**, **glutWireDodecahedron**
- **glutSolidIcosahedron**, **glutWireIcosahedron**
- **glutSolidTeapot**, **glutWireTeapot**

Bibliography

[1] M.H. Austern. *Generic Programming and the STL: Using and Extending the C++ Standard Template Library*. Addison-Wesley, 1998. [265]

[2] A. Baer, C. Eastman, and M. Henrion. Geometric modelling: A survey. *Computer-Aided Design*, 11(5):253–272, September 1979. [238]

[3] C. Bajaj. The algebraic degree of geometric optimization problems. *Discrete Comput. Geom.*, 3:177–191, 1988. [65]

[4] C. Baldazzi and A. Paoluzzi. Dimension-independent BSP (1): Section and interior to boundary mapping. Technical Report RT-DIA-26-97, Roma Tre, July 1997. [267]

[5] T.F. Banchoff. *Beyond the Third Dimension*. W. H. Freeman, 1996. [240]

[6] J.J. Barton and L.R. Nackman. *Scientific and Engineering C++: An Introduction with Advanced Techniques and Examples*. Addison-Wesley, 1994. [186]

[7] B.G. Baumgart. A polyhedron representation for computer vision. In *Proc. AFIPS Natl. Comput. Conf.*, volume 44, pages 589–596, 1975. [247, 249]

[8] E.T. Bell. *Men of Mathematics*. Simon & Schuster, 1937. [109, 175]

[9] M. Berger. *Géométrie*, volume 2. Nathan, 2nd edition, 1977. [87]

[10] M. Berger. *Géométrie*, volume 5. Nathan, 2nd edition, 1977. [98, 107]

[11] H. Bieri and W. Nef. Elementary set operations with d-dimensional polyhedra. In *Computational Geometry and Its Applications*, volume 333 of *LNCS*, pages 97–112. Springer-Verlag, 1988. [276]

[12] J.F. Blinn. Models of light reflection for computing synthesized pictures. *Computer Graphics*, pages 192–198, July 1977. [206]

[13] J.F. Blinn. Me and my (fake) shadow. *IEEE Comput. Graph. & Appl.*, 8(1):82–86, 1988. [35]

[14] J.F. Blinn. A trip down the graphics pipeline: Line clipping. *IEEE Comput. Graph. & Appl.*, pages 98–105, January 1991. [57, 161]

[15] J.F. Blinn. A trip down the graphics pipeline: The homogeneous perspective transform. *IEEE Comput. Graph. & Appl.*, pages 75–80, May 1993. [114, 156]

[16] J.F. Blinn. Vectors and geometry and objects, oh my! *IEEE Comput. Graph. & Appl.*, 25(3):84–93, 2005. [39, 189]

[17] J.F. Blinn and M.E. Newell. Clipping using homogeneous coordinates. *Comput. Graph.*, pages 245–251, August 1978. [140, 156]

[18] J.-D. Boissonnat and F.P. Preparata. Robust plane sweep for intersecting segments. Report TR 3270, INRIA, Sophia Antipolis, September 1997. [65]

[19] C.B. Boyer. *History of Analytic Geometry*. Dover, 2004. [80, 119]

[20] C.B. Boyer and U.C. Merzbach. *A History of Mathematics*. Wiley, 2nd edition, 1991. [127]

[21] J.E. Bresenham. Algorithm for computer control of a digital plotter. *IBM Systems Journal*, 4(1):25–30, 1965. [77, 193, 196]

[22] C. Buchheim, M. Jünger, and S. Leipert. Drawing rooted trees in linear time. *Software—Practice and Experience*, 36:651–665, 2006. [224]

[23] *CGAL Reference Manual*, CGAL 3.3.1 edition, September 2007. http://www.-cgal.org/. [v, 19, 41, 189, 265]

[24] H. Chiyokura. *Solid Modelling with Designbase: Theory and Implementation*. Addison-Wesley, 1988. [v]

[25] H.S.M. Coxeter. *Regular Polytopes*. Dover, 1973. [240, 241]

[26] H.S.M. Coxeter. *Projective Geometry*. Springer, 2nd edition, 1987. [136, 189]

[27] H.S.M. Coxeter. *Introduction to Geometry*. Wiley, 2nd edition, 1989. [34, 38, 82, 118, 142, 146, 241, 243]

[28] H.S.M. Coxeter. *Non-Euclidean Geometry*. MAA, 6th edition, 1998. [114, 129, 190]

[29] M.J. Crowe. *A History of Vector Analysis*. Dover, 1994. [119]

[30] M. de Berg, M. van Kreveld, M. Overmars, and O. Schwarzkopf. *Computational Geometry: Algorithms and Applications*. Springer-Verlag, 2nd edition, 2000. [276]

[31] P. de Casteljau. *Les quaternions*. Hermes, 1987. [104]

[32] G. di Battista, P. Eades, R. Tamassia, and I.G. Tollis. *Graph Drawing: Algorithms for the Visualization of Graphs*. Prentice-Hall, 1999. [v, 224]

[33] L. Dorst, D. Fontijne, and S. Mann. *Geometric Algebra for Computer Science*. Morgan Kaufmann, 2007. [ix, 18, 155]

[34] C.M. Eastman and K. Weiler. Geometric modeling using the Euler operators. In *Proceedings of 1st Ann. Conf. on Comput. Graph. in CAD/CAM Systems*, pages 248–254, April 1979. [249]

[35] H. Edelsbrunner, M.H. Overmars, and D. Wood. Graphics in Flatland: A case study. In F.P. Preparata, editor, *Computational Geometry*, volume 1 of *Adv. Comput. Res.*, pages 35–59. JAI Press, 1983. [290]

[36] A. Fabri, G.-J. Giezeman, L. Kettner, S. Schirra, and S. Schönherr. On the design of CGAL: A computational geometry algorithms library. *Software— Practice and Experience*, 30(11):1167–1202, September 2000. [183, 265]

[37] S. Fortune and C.J. Van Wyk. Efficient exact arithmetic for computational geometry. In *Proc. 9th Ann. ACM Symp. Comput. Geom.*, pages 163–172, May 1993. [80]

[38] H. Fuchs, Z.M. Kedem, and B. Naylor. On visible surface generation by a priori tree structures. *Comput. Graph.*, 14(3):124–133, July 1980. [255]

[39] E. Gamma, R. Helm, R. Johnson, and J. Vlissides. *Design Patterns: Elements of Reusable Object–Oriented Software*. Addison-Wesley, 1995. [278]

[40] R. Goldman. Illicit expressions in vector algebra. *ACM Trans. on Graphics*, 4(3):223–243, 1985. [10, 181]

[41] R. Goldman. On the algebraic and geometric foundations of computer graphics. *ACM Trans. on Graph.*, 21(1):52–86, January 2002. [155]

[42] H. Gouraud. Continuous shading of curved surfaces. *IEEE Trans. on Computers*, C-20(6):623–629, June 1971. [207]

[43] I. Grattan-Guinness, editor. *Landmark Writings in Western Mathematics 1640–1940*. Elsevier, 2005. [143]

[44] L. Guibas, M. Karavelas, and D. Russel. A computational framework for handling motion. In *ALENEX*, 2004. [ix]

[45] P. Hachenberger, L. Kettner, and K. Mehlhorn. Boolean operations on 3d selective Nef complexes: Data structure, algorithms, optimized implementation and experiments. *Computational Geometry: Theory and Applications*, 38(1-2):64–99, September 2007. [276]

[46] A.J. Hanson. *Visualizing Quaternions*. Morgan Kaufmann, 2006. [46, 105, 107]

[47] I. Herman. *The Use of Projective Geometry in Computer Graphics*. Springer-Verlag, 1992. [140]

[48] S. Hert, M. Hoffmann, L. Kettner, S. Pion, and M. Seel. An adaptable and extensible geometry kernel. *Computational Geometry: Theory and Applications*, 38:16–36, 2007. [183]

[49] D. Hilbert and S. Cohn-Vóssen. *Geometry and the Imagination*. Chelsea Publishing, 1952. [114]

[50] F.S. Hill. *Computer Graphics*. Prentice-Hall, 2nd edition, 2001. [138]

[51] C. Hoffmann. *Geometric and Solid Modeling*. Morgan Kaufmann, 1989. [v]

[52] C. Hoffmann. Robustness in geometric computation. *Journal of Comp. and Inf. Sci. in Eng.*, 1:143–155, June 2001. [80, 275]

[53] A. Holden. *Shapes, Space, and Symmetry*. Columbia University Press, 1971. [253]

[54] *IEEE Standard for binary floating-point arithmetic ANSI/IEEE Std 754—1985*. Reprinted in SIGPLAN notices, 22(2):9–25, 1987. [11, 74]

[55] M.J. Katz, M.H. Overmars, and M. Sharir. Efficient hidden surface removal for objects with small union size. *Comput. Geom. Theory Appl.*, 2:223–234, 1992. [289]

[56] T. Keeler, J. Fedorkiw, and S. Ghali. The spherical visibility map. *Computer-Aided Design*, 39(1):17–26, January 2007. [275, 293]

[57] L. Kettner, K. Mehlhorn, S. Pion, S. Schirra, and C. Yap. Classroom examples of robustness problems in geometric computations. In *Proc. 12th Ann. Euro. Symp. Algorithms (ESA'04)*, pages 702–713. Springer-Verlag, September 2004. LNCS 3221. [76, 82]

[58] F. Klein. *Le programme d'Erlangen*. Gauthier-Villars, 1974. [166]

[59] J.R. Kline. Double elliptic geometry in terms of point and order. *Annals of Math.*, 18(1):31–44, 1916. [151]

[60] D.E. Knuth. Optimum binary search trees. *Acta Informatica*, 1:14–25, 1971. [221]

[61] D.E. Knuth. *The Stanford GraphBase: A Platform for Combinatorial Computing*. Addison-Wesley, 2003. [227]

[62] S. Laveau and O. Faugeras. Oriented projective geometry for computer vision. In *European Conference on Computer Vision*, 1996. [150]

[63] D.T. Lee and B.J. Schachter. Two algorithms for constructing Delaunay triangulations. *Intl. J. Comput. Inform. Sci.*, 9(3):219–242, 1980. [82, 276]

[64] S. Mann, N. Litke, and T. DeRose. A coordinate free geometry ADT. Technical Report CS-97-15, University of Waterloo, July 1997. [62, 181]

[65] M. Mäntylä. *An Introduction to Solid Modeling*. Computer Science Press, 1988. [v, 246]

[66] M. Mäntylä and R. Sulonen. GWB: A solid modeler with Euler operators. *IEEE Comput. Graph. & Appl.*, 2(5):17–31, 1982. [249]

[67] K. Mehlhorn and S. Näher. *LEDA: A Platform for Combinatorial and Geometric Computing*. Cambridge University Press, 1999. [v, 24, 72, 93, 183, 227]

[68] K. Mehlhorn and M. Seel. Infimaximal frames: A technique for making lines look like segments. *Intl. J. Comput. Geometry Appl.*, 13(3), 2003. [271]

[69] J.R. Miller. Vector geometry for computer graphics. *IEEE Comput. Graph. & Appl.*, 19(3):66–73, 1999. [181]

[70] A.F. Möbius. *Der barycentrische Calcul*. Johann Ambrosius Barth, 1827. [143]

[71] N.C. Myers. Traits: A new and useful template technique. *C++ Report*, June 1995. [184, 228, 266]

[72] B. Naylor, J.A. Amanatides, and W. Thibault. Merging BSP trees yields polyhedral set operations. *Comput. Graph.*, 24(4):115–124, August 1990. [268]

[73] W.M. Newman and R.F. Sproull. *Principles of Interactive Computer Graphics*. McGraw-Hill, 2nd edition, 1979. [57]

[74] J. O'Rourke. *Computational Geometry in C*. Cambridge University Press, 1998. [82, 276]

[75] M.H. Overmars. Designing the computational geometry algorithms library CGAL. In *Proc. 1st ACM Workshop on Appl. Comput. Geom.*, pages 113–119, May 1996. [183]

[76] A. Paoluzzi. *Geometric Programming for Computer Aided Design*. Wiley, 2003. [267]

[77] R. Parent. *Computer Animation: Algorithms and Techniques*. Morgan Kaufmann, 2001. [ix]

[78] M.S. Paterson and F.F. Yao. Efficient binary space partitions for hidden-surface removal and solid modeling. *Discrete Comput. Geom.*, 5:485–503, 1990. [288]

[79] M.A. Penna and R.R. Patterson. *Projective Geometry and Its Applications to Computer Graphics*. Prentice-Hall, 1986. [110, 112, 136]

[80] B.-T. Phong. Illumination for computer generated pictures. *Commun. of the ACM*, 18(6):311–317, June 1975. [205, 207]

[81] J. Plücker. Fundamental views regarding mechanics. *Proc. of the Royal Soc. of London*, 15:204–208, 1866-1867. [8]

[82] H. Preece, Y. Rogers, and J. Sharp. *Interaction Design*. Wiley, 2002. [253]

[83] F.P. Preparata and M.I. Shamos. *Computational Geometry: An Introduction*. Springer-Verlag, 1985. [81, 210, 276]

[84] E. Reingold and J. Tilford. Tidier drawing of trees. *IEEE Trans. Software Eng.*, SE-7(2):223–228, 1981. [223]

[85] A.A.G. Requicha. Representations of rigid solids: theory, methods, and systems. *ACM Computing Surveys*, 12:437–464, 1980. [vi]

[86] A.A.G. Requicha. Geometric modeling: A first course, 1999. Web notes. [249, 281]

[87] R.F. Riesenfeld. Homogeneous coordinates and projective planes in computer graphics. *IEEE Comput. Graph. & Appl.*, pages 50–55, January 1981. [114, 123]

[88] J.R. Rossignac and M.A. O'Connor. SGC: A dimension-independent model for pointsets with internal structures and incomplete boundaries. In M.J. Wozny, J.U. Turner, and K. Preiss, editors, *Geometric Modeling for Product Engineering*, pages 145–180. Elsevier, 1990. [248]

[89] G.-C. Rota. *Indiscrete Thoughts*. Birkhäuser, 1997. [156]

[90] S. Schirra. Robustness and precision issues in geometric computation. In J.-R. Sack and J. Urrutia, editors, *Handbook of Computational Geometry*, chapter 14. Elsevier, 1999. [80]

[91] S. Schirra. Real numbers and robustness in computational geometry. *Real Numbers and Computers*, pages 7–21, November 2004. Dagstuhl. [80]

[92] J. Schwerdt and M. Smid. Computational geometry on the unit sphere. Technical report, Carleton University, 2002. LEDA Extension Package. [93]

[93] R. Sedgewick. *Algorithms in C++ Part 5: Graph Algorithms*. Addison-Wesley, 2002. [227]

[94] J.G. Semple and G.T. Kneebone. *Algebraic Projective Geometry*. Oxford University Press, 1952. [129, 133]

[95] M. Serfati. René Descartes, *Geometria*. In I. Grattan-Guinness, editor, *Landmark Writings in Western Mathematics 1640–1940*, pages 1–22. Elsevier, 2005. [80]

[96] P. Shirley. *Fundamentals of Computer Graphics*. AK Peters, 2nd edition, 2005. [146]

[97] K. Shoemake. Animating rotations with quaternion curves. *Comput. Graph.*, 19(3):245–254, July 1985. [107]

[98] K. Shoemake. ARCBALL: A user interface for specifying three–dimensional orientation using a mouse. In *Graphics Interface*, pages 151–156, May 1992. [326]

[99] D. Shreiner, M. Woo, J. Neider, and T. Davis. *OpenGL Programming Guide*. Addison-Wesley, 4th edition, 2004. [311]

[100] J.G. Siek, L.-Q. Lee, and A. Lumsdaine. *The Boost Graph Library User Guide and Reference Manual*. Addison-Wesley, 2001. [227, 265]

[101] A.R. Smith. A pixel is not a little square. Technical report, Microsoft, July 1995. Technical Memo 6. [193]

[102] R.F. Sproull. Using program transformations to derive line-drawing algorithms. *ACM Trans. on Graphics*, 1(4):259–273, 1982. [77, 193, 196]

[103] J. Stillwell. *The Four Pillars of Geometry*. Springer-Verlag, 2005. [107]

[104] J. Stolfi. *Oriented Projective Geometry: A Framework for Geometric Computations*. Academic Press, 1991. [99, 141, 149, 151, 166, 215]

[105] B. Stroustrup. *The C++ Programming Language*. Addison-Wesley, special 3rd edition, 2000. [4]

[106] D.J. Struik. *A Concise History of Mathematics*. Dover, 4th revised edition, 1987. [118]

[107] I.E. Sutherland and G.W. Hodgman. Reentrant polygon clipping. *Commun. of the ACM*, 17:32–42, 1974. [17, 59, 140, 156, 261]

[108] I.E. Sutherland, R.F. Sproull, and R.A. Schumacker. A characterization of ten hidden-surface algorithms. *ACM Computing Surveys*, 6(1):1–55, March 1974. [293]

[109] W.C. Thibault and B.F. Naylor. Set operations on polyhedra using binary space partitioning trees. *Comput. Graph.*, 21(4):153–162, 1987. [256, 268]

[110] F. Toth. *Regular Figures*. Macmillan, 1964. [244]

[111] W.T. Tutte. How to draw a graph. *Proc. London Math. Soc.*, 3(13):743–767, 1963. [231]

[112] D. Vandevoorde and N.M. Josuttis. *C++ Templates*. Addison-Wesley, 2002. [184, 228, 266]

[113] G.S. Watkins. *A Real Time Visible Surface Algorithm*. PhD thesis, University of Utah, June 1970. UTECH-CSc-70-101. [210]

[114] K. Weiler. Edge-based data structures for solid modeling in a curved surface environment. *IEEE Comput. Graph. & Appl.*, 5(1):21–40, 1985. [238]

[115] K. Weiler. The radial edge structure: A topological representation for nonmanifold geometric boundary modeling. In *Geometric Modeling for CAD Applications*, pages 3–36. IFIP, May 1988. [248]

[116] M. Wenninger. *Spherical Models*. Cambridge University Press, 1979. [99]

[117] C. Wetherell and A. Shannon. Tidy drawing of trees. *IEEE Trans. Software Eng.*, SE-5(5):514–520, 1979. [223]

[118] T. Whitted. An improved illumination model for shaded display. *Commun. of the ACM*, 23(6):343–349, June 1980. [213]

[119] J. Woodwark. *How to Run a Paper Mill*. Information Geometers, 1992. [ix]

[120] C.K. Yap. Towards exact geometric computation. *Comput. Geom. Theory Appl.*, 7:3–23, 1997. [80]

Index

Springer
Secaucus, NJ
(on-line mail order)
Tue 29 July 2008
$ 69.95 + 4.90 tax
= $ 74.85
+ $ 5.00 delivery fee

Printed in the United States